Pollution and Physiology of Marine Organisms

ACADEMIC PRESS RAPID MANUSCRIPT REPRODUCTION

Pollution and Physiology of Marine Organisms

EDITED BY

F. John Vernberg
Winona B. Vernberg

Belle W. Baruch Institute for Marine Biology
and Coastal Research
University of South Carolina
Columbia, South Carolina

ACADEMIC PRESS   New York   San Francisco   London   1974

A Subsidiary of Harcourt Brace Jovanovich, Publishers

ACADEMIC PRESS, INC.
111 Fifth Avenue, New York, New York 10003

United Kingdom Edition published by
ACADEMIC PRESS, INC. (LONDON) LTD.
24/28 Oval Road, London NW1

LIBRARY OF CONGRESS CATALOG CARD NUMBER: 74-27490

ISBN 0–12–718250–0

PRINTED IN THE UNITED STATES OF AMERICA

Contents

v

CONTENTS

PART IV. GENERAL

PART V. FACTOR INTERACTION—SYNERGISTIC EFFECTS OF POLLUTANTS AND VARIOUS ENVIRONMENTAL PARAMETERS

Contributors

D. G. Ahearn, Department of Biology, Georgia State University, Atlanta, Georgia 30303

Jonathan Allen, Mount Desert Island Biological Laboratory, Salsbury Cove, Maine 04672

J. W. Anderson, Biology Department, Texas A & M University, College Station, Texas 77843

L. H. Bahner, U. S. Environmental Protection Agency, Gulf Breeze Environmental Research Laboratory, Gulf Breeze, Florida 32561

Dale Benos, Mount Desert Island Biological Laboratory, Salsbury Cove, Maine 04672

G. W. Bryan, Marine Biological Association Laboratory, Citadel Hill, Plymouth, England

B. A. Cox, Biology Department, Texas A & M University, College Station, Texas 77843

Anthony Calabrese, National Marine Fisheries Service, Middle Atlantic Coastal Fisheries Center, Milford Laboratory, Milford, Connecticut 06460

Richard S. Caldwell, Department of Fisheries and Wildlife, Marine Science Center, Oregon State University, Newport, Oregon 97365

Janice Chambers, Department of Biochemistry and Department of Zoology, Mississippi State University, Mississippi State, Mississippi 39762

Anthony D'Agostino, New York Ocean Science Laboratory, Montauk, New York 11954

Margaret A. Dawson, National Marine Fisheries Service, Middle Atlantic Coastal Fisheries Center, Milford Laboratory, Milford, Connecticut 06460

Patricia J. DeCoursey, Belle W. Baruch Institute for Marine Biology and Coastal Research, University of South Carolina, Columbia, South Carolina 29208

Thomas W. Duke, U. S. Environmental Protection Agency, Gulf Breeze Environmental Research Laboratory, Gulf Breeze, Florida 32561

David P. Dumas, U. S. Environmental Protection Agency, Gulf Breeze Environmental Research Laboratory, Gulf Breeze, Florida 32561

A. Dunning, Department of Zoology, University of Maine, Orono, Maine 04473

Tina Echeverria, National Marine Fisheries Service, Tiburon Fisheries Laboratory, Tiburon, California 94920

Maxwell B. Eldridge, National Marine Fisheries Service, Tiburon Fisheries Laboratory, Tiburon, California 94920

Colin Finney, New York Ocean Science Laboratory, Montauk, New York 11954

J. S. Gray, Wellcome Marine Laboratory, Robin Hood's Bay, Yorkshire, England

James R. Heitz, Department of Biochemistry, Mississippi State University, Mississippi State, Mississippi 39762

G. M. Hightower, Department of Biology, Texas A & M University, College Station, Texas 77843

Eugene Jackim, Environmental Protection Agency, National Marine Water Quality Laboratory, Narragansett, Rhode Island 02882

H. Kleerekoper, Department of Biology, Texas A & M University, College Station, Texas 77843

Marian L. Klein, Mote Marine Laboratory, Sarasota, Florida 33581

Lancelot Lewis, Department of Biochemistry and Department of Zoology, Mississippi State University, Mississippi State, Mississippi 39762

Jeffrey L. Lincer, Mote Marine Laboratory, Sarasota, Florida 33581

C. W. Major, Department of Zoology, University of Maine, Orono, Maine 04473

David Miller, Mount Desert Island Biological Laboratory, Salsbury Cove, Maine 04672

J. M. Neff, Biology Department, Texas A & M University, College Station, Texas 77843

D. R. Nimmo, U. S. Environmental Protection Agency, Gulf Breeze Environmental Research Laboratory, Gulf Breeze, Florida 32561

James O'Hara, Applied Biology, Decatur, Georgia 30030

John B. Pritchard, Department of Physiology, Medical University of South Carolina, Charleston, South Carolina 29401

J. Larry Renfro, Mount Desert Island Biological Laboratory, Salsbury Cove, Maine 04672

William T. Roubal, Northwest Fisheries Center, National Marine Fisheries Service, National Oceanic and Atmospheric Administration, Seattle, Washington 98112

Bodil Schmidt-Nielsen, Mount Desert Island Biological Laboratory, Salsbury Cove, Maine 04672

John J. Stegeman, Department of Biology, Woods Hole Oceanographic Institution, Woods Hole, Massachusetts 02543

Jeannette W. Struhsaker, National Marine Fisheries Service, Tiburon Fisheries Laboratory, Tiburon, California 94920

H. E. Tatem, Department of Biology, Texas A & M University, College Station, Texas 77843

Frederick P. Thurberg, National Marine Fisheries Service, Middle Atlantic Coastal Fisheries Center, Milford Laboratory, Milford, Connecticut 06460

M. R. Tripp, Department of Biological Sciences, University of Delaware, Newark, Delaware 19711

Winona B. Vernberg, Belle W. Baruch Institute for Marine Biology and Coastal Research, University of South Carolina, Columbia, South Carolina 29208

Michael Waldichuk, Pacific Environment Institute, West Vancouver, B. C. Canada

Douglas A. Wolfe, National Oceanic and Atmospheric Administration, National Marine Fisheries Service, Atlantic Estuarine Fisheries Center, Beaufort, North Carolina 28516

James D. Yarbrough, Department of Biochemistry and Department of Zoology, Mississippi State University, Mississippi State, Mississippi 39762

Preface

Coastal and estuarine areas tend to be increasingly heavily populated and industrialized, and the great majority of estuarine systems today are polluted to some extent with a wide variety of toxicants. In fact, it must be recognized that man-induced environmental changes are now a part of what classically has been called a "normal" environment. While biologists have long recognized that capacity and resistance adaptations of organisms are a function of many environmental stimuli, it is only within recent years that much attention has been directed toward the effect of man-induced environmental stimuli. To pinpoint and highlight some recent advances in studies on the physiological effects of pollutants on marine organisms, a symposium was organized on the effects of pollution on the physiological ecology of estuarine and coastal water organisms, sponsored by the Environmental Protection Agency and the Belle W. Baruch Institute for Marine Biology and Coastal Research, University of South Carolina. The papers in this volume are the result of that symposium. The studies are subdivided into five sections: Heavy Metals, Pesticides and PCBs, Oil and Dispersants, General, and Factor Interaction—Synergistic Effects of Pollutants and Various Environmental Parameters.

The editors are indebted to the authors for their promptness in turning in their papers and to the staff of the Belle W. Baruch Institute for Marine Biology and Coastal Research for assistance in many ways. We are particularly indebted to Ms. Hilda Merritt for typing the manuscript and to Ms. Bettye Dudley for the indexing.

F. John Vernberg
Winona B. Vernberg

Pollution and Physiology
of Marine Organisms

SOME BIOLOGICAL CONCERNS
IN HEAVY METALS POLLUTION

MICHAEL WALDICHUK

Pacific Environment Institute
West Vancouver, British Columbia, Canada

Next to wood, clay and stone, metals were one of
the first meterials available in nature which pre-
historic man could extract from the earth's crust and
work into implements and trinkets. First, he may
have found exposed in pure form, precious metals,
such as gold and silver, which he fashioned into
items of primitive jewelry and possibly coins as a
means of exchange. He learned how to extract copper
from ores, and then by adding tin, was able to produce a durable alloy, bronze. Thus came the Bronze
Age. Sometime later he learned how to smelt iron ore
to produce iron, a much harder metal which could be
ground and honed to a sharp edge for weapons. The
Iron Age appeared. It must have been obvious even to
early man that some metals, such as iron, were subject
to corrosion when left exposed to the elements, while
other metals and alloys, such as gold, copper and
bronze were more resistant.

With the early use of metals, there little
concern about environmental contamination. The oxides
of the metals from corrosion products were hardly
sufficient to be a cause for alarm. However, salts
of the metals began to find their way into commercial

1

and industrial applications. Then it became evident
that metallic salts possess certain biocidal properties. For example, it has been known for a long time
that copper sulphate is an effective algicide.
Portuguese fishermen, by tradition, wear copper
bracelets to ward off ill luck, particularly from
injuries. The leached copper probably provides a
certain degree of antisepticity against infection in
the event of a cut or bruise. Formulations of copper
compounds, such as Paris Green, were used against
insects long before the more exotic organic compounds
became available for combating insects.

A first property of metals of which we must be
aware, when dealing with them as environmental contaminants, is that they are immutable. They can
neither be created nor destroyed, nor can one metal
be transformed into another, as the medieval al-
chemists soon learned. This means that once a metal
is mobilized in the environment, its total amount
there remains the same, regardless of form, until it
is immobilized again. Its form may be altered by
biogeochemical processes, so that the particular salt
in which it originally entered the environment no
longer exists, but the total amount of the metal present as other compounds or ions remains unchanged.

The importance of metals in the marine environment emerged from studies of radionuclides resulting
from fallout in the oceans during the 50's and 60's
(NAS, 1971a). It became apparent that certain
nuclides were being accumulated by organisms in large
concentrations, particularly in certain organs, *e.g.*,
cobalt-60 in the kidney of the giant clam *Tridacna*
(Lowman, 1960). Since in uptake by organisms the non-
radioactive species of the metals behave in much the
same way as the radioactive species, it was suspected
that the stable metals were accumulating in marine
organisms to high concentrations wherever the metals
were present in seawater in higher than ambient concentrations. This proved to be the case, and high
concentrations of zinc and copper were found in
oysters dwelling in waters receiving effluents

2

containing these metals. The high concentrations of heavy metals in shellfish are now used to trace movements of metal-containing pollutants (Environment Canada, 1973), and to generally identify the presence of heavy metal pollution (Huggett et al., 1973). Table 1 summarizes concentrations of metals in seawater and marine organisms.

Although there has been a great deal of concern about heavy metals pollution in the last few years, it takes only a little probing in the literature to discover that most of this concern has revolved about the effects of metals on humans (Anon., 1967; Council on Environmental Quality, 1971; U. S. Department of Interior, 1970; Warren, 1972; Fulkerson and Goeller, 1973). Starting with the effects of Minimata Disease, caused by the consumption of mercury-contaminated shellfish and finfish taken from Minimata Bay in Japan, and continuing with the affliction known as Itai itai, caused by the consumption of foods contaminated by cadmium, again in Japan (Ui, 1972), and continuing with concerns about such metals as lead in the atmosphere from smelters and from the use of tetraethyl lead in gasoline, it has been the awareness of the effects of these metals on human beings that has caused action to be taken for their control. In the United Kingdom, a Working Party on the Monitoring of Foodstuffs for Heavy Metals has issued a series of White Papers (U. K., 1971, 1972a, 1973a, 1973b) on the problems of heavy metals concentrations in seafoods, among other foodstuffs.

The current alarm of metal pollution in the sea, however, started with the tragedy of Minimata and later, Niigata, in Japan. In these two areas there were 50 mortalities and over 100 permanently disabled victims of mercury poisoning. This tragedy resulted in an awareness of the problem of bioaccumulation of mercury for the fabrication of felt hats.

In these two areas there were 50 mortalities and over 100 permanently disabled victims of mercury poisoning. The current alarm of metal pollution in the sea, however, started with the tragedy of Minimata

TABLE 1

Metals in Seawater and Their Effects on Fish Life and/or the Food Chain

Metal	Natural Concentration in Seawater[a] μg/l (ppb)	Bioaccumulation, Concentration Factor[b] Plankton Pp	Zp	Macro-invertebrates	Fish	Stimulatory[c] Concentrations for Algae μg/l	Inhibitory[c] Concentrations for Algae μg/l
Aluminum	1	745	1,150				1,500-2,000
Antimony	0.33				77-4,100[d]		
Arsenic	2.6			⎰ 300 ⎱ 423-810[d] ⎰ 3,300			
Barium	20	62(31)	110				34,000
Beryllium	0.0006						
Bismuth	0.02			1,000			
Cadmium	0.11	1,694	9,440	82,000-182,000[d]	180-730[d]	0	100-140,000
Chromium	0.2	<34	<65			0.04	
Cobalt	0.05	<190	<365		50-250[d]	6-200	500
Copper	2	38	437	24,000-35,000[d]	130-660[e]	0.006-10	6.4-1,000
Iron	3.4	2,400	5,430				
Lead	0.03	2,087	15,500	7,000-100,000[d]	<6,000-10,000[d]		
Manganese	1.9	158	290		16-26[d]	0.05-50	
Mercury	0.15	180	172		530-12,300	0.01-10	30-50
Molybdenum	10	<10	<20				54,000
Nickel	2	41	149				
Rubidium	120						
Silver	0.28	98	117				50
Thallium	<0.01						
Thorium	<0.0005						400-800

TABLE 1—Continued
Metals in Seawater and Their Effects on Fish Life and/or the Food Chain

Metal	Natural Concentration in Seawater[a] μg/l (ppb)	Bioaccumulation, Concentration Factor[b] Plankton Pp	Zp	Macro-invertebrates	Fish	Stimulatory[c] Concentrations for Algae μg/l	Inhibitory[c] Concentrations for Algae μg/l
Titanium	1	<290		<550			2,000
Uranium	3.3						
Vanadium	1.9	<100		<200		100	10,000–20,000
Zinc	2	113		1,800	{148,000[f] 172,000–290,000[d]	6.5	500–5,000

[a]From Skinner and Turekian (1973).

[b]Concentration factor (CF) is the ratio of the concentration of the element in the organism to that in ambient water. Data are mainly from Lowman et al. (1971) and are based in some cases on the bioaccumulation of the radioactive species. Pp = phytoplankton; Zp = zooplankton.

[c]Wide ranges in concentrations arise from use of different algal species in the tests. For species tested, see North et al. (1972).

[d]CF values for macroinvertebrates (mainly Pacific oysters, *Crassostrea gigas*) and fish (mainly Pacific halibut, *Hippoglossus stenolepis*) were taken from tissue analyses of the Fish Inspection Laboratory, Fisheries and Marine Service, Vancouver, British Columbia, assuming average ambient seawater concentrations.

[e]Analyses for copper in plaice *Pleuronectes platessa* (Vink, 1972) were used for CF values, assuming average concentrations of copper (2 μg/l) in ambient seawater, and a factor of five for conversion of dry weight values in fish to wet weight.

[f]For oysters from Pringle et al. (1968).

5

and later, Niigata, in Japan. This tragedy resulted in an awareness of the problem of bioaccumulation of mercury by aquatic organisms, and spurred research in examination of the levels of metals in aquatic organisms and other foods for humans. Such studies became more of a matter for chemists to satisfy the needs of food and drug agencies than for physiologists examining the effect of the metals on the aquatic organisms, or for ecologists investigating the effects on the food chain in the sea and on the ecosystem in general.

A survey of the literature on effects of metals on aquatic organisms in revising the book *Water Quality Criteria* in 1972 showed there was little information available on the effects of these substances in seawater, although there was considerable body literature on the effects of metals in fresh water, both soft and hard. However, extrapolation of this information to the marine environment is questionable under the best of circumstances. Hence, LC50 data for metals in the marine environment are based on rather fragmentary seawater information in the literature, and sometimes were based on the best data of freshwater bioassays available (NAS, 1974). A literature retrieval exercise run through the National Science Library in Ottawa, Canada, covering some 2,500 relevant journals dealing with problems of water pollution, was expected to provide some recent references on the effects of metals in the marine environment. Rather surprisingly, even with the current interest in metals pollution and in the environmental problem generally, there was little information on either acute or sublethal effects of metals on marine organisms.

It is clear that a great deal of research still has to be done on the effects of metals on aquatic organisms. National and international legislation is being prepared on the control of all substances entering our waters from an advanced technological society. Many of these substances involve metals either as inorganic salts or as metallo-organic compounds. Mercury and cadmium compounds have been

banned from dumping into the sea, except in trace
concentrations, in the Convention for the Prevention
of Marine Pollution by the Dumping of Wastes at Sea
1972 (United Kingdom, 1972b), and in the very recent
International Convention for the Prevention of Pollution of the Sea from Ships, signed in London on
November 2, 1973 (IMCO, 1973). Other metals, such as
zinc, lead and copper, will require strict control.
Regulations for the control of pollution by these
substances must be based on a certain degree of
scientific knowledge concerning their effects to
marine organisms. Preliminary attempts to develop
regulations based on known effects of metals, along
with many of the other potential pollutants of the
sea, shows that our knowledge even on acute toxicity
is sadly lacking. The man-induced mobilization of
metals is shown in Table 2 for the sea, and in Table 3
for the atmosphere.

PHYSICAL AND CHEMICAL CHARACTERISTICS OF HEAVY METALS

It is difficult enough to define a metal, even
more so to define "heavy metals." A metal generally
is recognized as a substance having
"certain characteristic and almost unique
physical properties. Among these are: high
electrical and thermal conductivity, attributed to free electrons, great opacity
and high reflectivity for light, due to the
same cause and responsible for the lustre
'commonly associated with metals'; malleability—a sort of plasticity by virtue by
which a metal may be cold-worked and rolled
into thin sheets; ductility—a combination
of malleability and toughness which permits
a metal to be drawn into wire. Metals in
their normal, pure state are crystalline"
(*Van Nostrand's Scientific Encyclopedia*,
1947, p. 933).

Some elements like selenium, not chemically a metal, exhibit some of the physical properties of metals. The heavy metals are normally regarded as the ones having an atomic number of 22 to 92 in all groups from period 3 to 7 in the Periodic Table, but there is no really satisfactory grouping by which they can be identified in the Periodic Table. For the foregoing reason, I wondered whether my title of this talk should not be changed to "Some Biological Concerns in Metals Pollution." For this reason, too, we avoided the designation of "heavy metals" or even "metals" in the Panel on Marine Aquatic Life and Wildlife during the revision of the document, *Water Quality Criteria*

TABLE 2
*Man-Induced Mobilization of Metals Exceeding Natural Rates by Geological Processes, As Estimated from Annual River Discharges to the Oceans**

Element	Geological Rate (in Rivers)	Man-Induced Rate (Mining)
	10^3 Metric Tons Per Year	
Iron	25,000	319,000
Manganese	440	1,600
Copper	375	4,460
Zinc	370	3,930
Nickel	300	358
Lead	180	2,330
Molybdenum	13	57
Silver	5	7
Mercury	3	7
Tin	1.5	166
Antimony	1.3	40

*From *Study of Critical Environmental Problems*, MIT, 1970.

(1974), and instead considered the broad group of pollutants, including metals and other inorganic chemicals, under the heading "Inorganics."

Metals in their pure state present little hazard, except those having a high vapor pressure, such as mercury, and those which may be present in the particulate form in the atmosphere, such as vanadium. It is the soluble compounds of the metals which create the problems in the aquatic environments. Some of the metallo-organic compounds are the most toxic compounds known, e.g., methyl mercury and tetraethyl lead. The danger of discharging some of the metals into the environment in inorganic form lies in their conversion into the highly poisonous metallo-organic compounds through biological action, as was discovered not too long ago with mercury (Jensen and Jernelöv, 1969).

Because certain metals are required in life processes, most organisms have a capability of concentrating them. This capability is enhanced by certain feeding and metabolic processes which can lead to enormously high concentration factors. The invertebrates appear to have a particularly high capability for concentrating metals, along with other foreign materials found in their environment, when they filter plankton during feeding. Because of the ability of many metals to form complexes with organic substances, there is a tendency for them to be fixed in the tissue and not to be excreted. In other words, they have a large biological half-life. This is perhaps one of the major problems that metals pose with respect to their effects on aquatic organisms.

Metals also have certain properties which render them amenable to treatment and/or removal from waste waters. Metals can be removed with cation exchangers. Generally, the metals form highly insoluble sulphides and can be precipitated from solution. Because of their propensity towards the formation of complexes, metals can be chelated by chelating agents such as Ethylenediaminetetraacetic acid (EDTA) and the sodium salt of Nitrilotriacetic acid (NTA). These agents

TABLE 3

Mobilization of Metallic Elements into the Atmosphere by Burning of Fossil Fuels[a] and Concentrations in Natural Waters[b] (N.D. = No data or reasonable estimates available)

Element	Release to Atmosphere (10³ Metric Tons Per Year)			Concentration in Natural Waters	
	Coal	Oil	Total	Oceans (µg/l)	Rivers (µg/l)
Aluminum	1400	0.08	1400	0-2900	6,100
Arsenic	0.7	0.0002	0.7	2.6	2
Barium	70	0.02	70	20	20
Beryllium	0.41	0.00006	0.41	0.0006	N.D.
Bismuth	0.75	N.D.	N.D.	0.02	N.D.
Cadmium		0.002		0.11	N.D.
Calcium	1400	0.82	1400	411,000	15,000
Cerium	1.6	0.002	1.6	0.0012	(0.06)
Chromium	1.4	0.05	1.5	0.2	1
Cobalt	0.7	0.03	0.7	0.05	0.1
Copper	2.1	0.023	2.1	2	7
Erbium	0.085	0.0002	0.085	0.00087	0.05
Gallium	1	0.002	1	0.03	0.09
Germanium	0.7	0.0002	0.7	0.06	N.D.
Gold		0.0002	0.7	0.011	0.002

Element					
Holmium	0.042	0.41	---	0.00022	0.01
Iron	1400	0.0008	1400	3.4	670
Lanthanium	3.5	0.05	1.4	0.0034	0.3
Lead	9	0.02	3.6	0.03	3
Lithium		0.02		170	3
Magnesium	280	1.6	280	1,290,000	4,100
Manganese	7	1.6	7	1.9	7
Mercury	0.0017	1.6	1.6	0.15	0.07
Molybdenum	0.7	N.D.	2.3	10	0.6
Neodymium	0.65	N.D.		0.0028	0.2
Nickel	2.1	N.D.	3.7	2	0.3
Potassium	140	N.D.		392,000	2,300
Praeseodymium	0.31	N.D.	---	0.00064	0.03
Rhenium	0.007	N.D.	---	0.0084	N.D.
Rubidium	14	N.D.	---	120	1
Ruthenium	---		---	0.0007	N.D.
Samarium	0.22	N.D.		0.00043	0.03
Scandium	0.7	0.0002	0.7	0.0004	0.004
Silver	0.07	0.00002	0.07	0.28	0.3
Sodium	280	0.33	280	10,800,000	6,300
Strontium	70	0.02	70	8,100	70
Terbium	0.042			0.00014	0.008
Thulium	0.014	0.002		0.00017	0.009
Tin	0.28	0.02	0.28	0.81	N.D.
Titanium	70	0.02	70	1	3
Tungsten	N.D.	N.D.	---	<0.001	0.03

TABLE 3—Continued
*Mobilization of Metallic Elements into the Atmosphere by Burning of Fossil Fuels[a]
and Concentrations in Natural Waters[b] (N.D. = No data or reasonable estimates
available)*

Element	Release to Atmosphere (10³ Metric Tons Per Year)			Concentration in Natural Waters	
	Coal	Oil	Total	Oceans (µg/l)	Rivers (µg/l)
Uranium	0.14	0.001	0.14	3.3	0.3
Vanadium	3.5	8.2	12	1.9	0.9
Ytterbium	0.07			0.00082	0.05
Yttrium	1.4	0.0002	1.4	0.013	0.07
Zinc	7	0.04	7	2	20
Zirconium	N.D.	N.D.	---	0.026	N.D.

[a]From Bertine and Goldberg, 1971.
[b]From Skinner and Turekian, 1973.

12

appear to render the heavy metals less toxic (Sprague, 1968). In nature, the organic substances found in waters of high humic acid content appear to perform to some extent the same function. Even organic substances in sewage may render heavy metals less toxic than they would be in pure, uncontaminated water (Lewis *et al.*, 1972).

The different oxidation states of metals determine, in a certain degree, their toxicity to aquatic organisms. Again, this is related to the position of these elements in the Periodic Table. The power of the elements to attract electrons, the so-called electronegativity, no doubt has some bearing on its ecological effects, with respect to toxicity to aquatic organisms. The electronic orbitals being filled in the atom of the element, which determines its position in Mendeleev's periodic arrangement of the elements, is a factor which may determine its toxicity. Unfortunately, this factor has not been fully explored with respect to aquatic organisms, but the lethal dose to mice where the substance is administered intraperitoneally suggests this to be the case, at least when added to the bloodstream of mammals (Bienvenu *et al.*, 1963). The 48-hr LC50 values for a number of metals to which aquatic animals were exposed have been superimposed on a graph giving the lethal doses for mice in Figure 1. Although not clearly related, there is a trend in LC50 values for fish and other aquatic organisms somewhat similar to that of toxicity to mice.

It is quite clear at the present time at least that the form of the heavy metal to which aquatic organisms are exposed is extremely important in its overall toxicity. Whether it is present in ionic form, in an oxidized or reduced state (valence), complexed by organic substances such as chelating agents in added or natural form, adsorbed on inorganic or organic particulate material, or whether it is acting singly or in combination with other cations are factors determining its uptake by aquatic organisms and its toxicity to them.

13

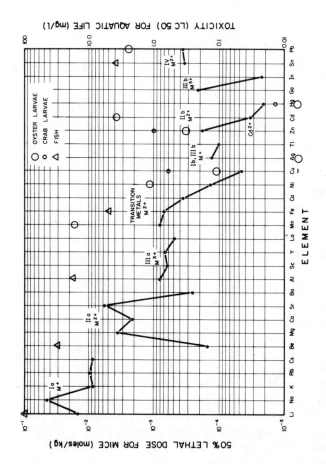

Fig. 1. Lethal dose of various metal ions to mice, administered intraperitoneally (∂), and 48-hr LC50 for aquatic life (⊙, ⊘, △), in relation to the position of the metal in the periodic table and its charge. [Adapted from Baes (1973). Bioassay data from Calabrese et al. (1973) for oyster larvae; Connor (1971) for crab larvae; and McKee and Wolf (1963) for fish.]

14

ACUTE TOXICITY

Standard bioassays on metals in seawater, where the organism is exposed to a concentration of the metal salt in a static setup or in a continuous-flow system, have not been reported widely in the literature. Perhaps this is not surprising, inasmuch as seawater facilities are not always readily available to biological laboratories and the use of seawater presents certain problems which are not present with fresh water. It is always simpler to start with distilled water and add the particular substance being tested so that toxicity can be strictly attributed to the material added. Once there are other ions in the water, the bioassay results obtained are subject to some interpretation.

There are a number of problems that must be resolved in any bioassays of metals, much as required with bioassays of any other substances considered as pollutants. It has been clearly shown in studies with some of the metals in freshwater pollution that the toxicity depends in large part on whether the water is soft or hard. This is related to the antagonistic effect of calcium and magnesium salts to the heavy metals, such as zinc (Lloyd, 1960) or lead, which might be added. How the salts of magnesium and calcium and other major constituents in seawater affect the toxicity of metals is uncertain. However, one would expect some interference from seawater salts with the toxicity of at least certain heavy metals.

Another factor that must be considered in terms of toxicity of metals to aquatic organisms is synergism. Studies done on the effects of zinc, cadmium and copper in combination have clearly shown that there is more than just an additive effect of toxicity (La Roche, 1972; Eisler and Gardner, 1973). Again, synergism may be related to certain chemical characteristics of the elements, and the effects of other metals in combination should be explored.

15

Bioassays have often been conducted in static tests involving a hardy species over a period of 48 or 96 hrs. Although the solutions are usually changed once a day, not too much is known about the alteration in concentration as the tests continue between solution changes. It is assumed that the concentration of the substance being tested remains the same from the beginning of the test to the end, and chemical tests are seldom carried out to confirm or to provide analytical data to go along with the bioassay results. Techniques for acute bioassays in sea are not substantially different from those in fresh water. Edwards and Brown (1967) described the methods used extensively and effectively at the Water Pollution Research Laboratory, Stevenage, U. K. Sprague (1969, 1970, 1973) has reviewed in some detail the principles and techniques of acute bioassay determinations, as well as interpretation of results.

Despite efforts to control temperature, salinity and dissolved oxygen in the water at a fairly uniform level throughout the tests, the effects of varying these variables is poorly understood. Often tests conducted at a particular salinity, temperature and dissolved oxygen concentration are applied in situations which are environmentally quite different. The tendency has been to overload tanks with fish in bioassay tests, so that frequently there is a crowding effect that introduces a spurious factor into the bioassay results. Recent studies of factors affecting bioassay results, including concentration of fish in tests with underyearling Pacific salmon, suggest that a loading density of 2.5 1/g fish should not be exceeded (Davis and Mason, 1973).

As far as protection of aquatic organisms in a body of water is concerned, there is a need to know something about the effect of the toxic substance on the most sensitive species to be protected, as well as on the more sensitive stages in its life history. For example, very often the juveniles of salmonids are tested in bioassay experiments, partly because salmonids are quite sensitive to pollutants, but also

16

because there is a great deal of experience in holding and acclimating these fish. What may be more
important for the survival of the species is the
effect of a pollutant on the egg and larval stages of
the fish. Certainly with salmonids, freshwater studies on metal toxicity should include the egg stages
inasmuch as these fish spawn in fresh waters. Studies
conducted at the Sweltzer Creek Laboratory of the
International Pacific Salmon Fisheries Commission,
Cultus Lake, British Columbia, indicate that mercury
concentrations as low as 10 ppb have a teratogenic
effect when preeyed sockeye salmon eggs are exposed
(Servizi and Gordon, 1974). These fish will either
die before they reach a juvenile stage, or if they
mature, they will certainly not reproduce. Hence,
the toxic effect, while not acute in that it does not
kill the salmon eggs, certainly has the same net
effect as direct mortality.

Bioassay studies are also required on the food
organisms of commercially valuable species. If concentrations of metals are not high enough to kill the
fish, but high enough to destroy the organisms on
which the fish feeds, then there is still substantial
damage to the fishery. Some studies on the effects
of metals on phytoplankton have been reported (North
et al., 1972) (see Table 1). However, little has
been done on the effects of metals on zooplankton,
except in special experiments (Lewis et al., 1972).
Investigations have been conducted on the effects of
metals on freshwater aquatic insects (Warnick and
Bell, 1969), which are an important component of food
of freshwater fish, as well as estuarine and anadro-
mous species.

Standard bioassays are normally conducted for a
48-hr or 96-hr period to obtain a measure of acute
toxicity. However, the mode of toxicity of substances
such as the metals, which bioaccumulate, is not such
that an LC50 can be clearly obtained over a 96-hr
period. Because of the uptake of metal by the organisms being tested, the toxicity continues over a
prolonged period and there is no so-called toxic

threshold. Thus, for practical purposes it may be necessary to conduct tests for at least a month or longer to obtain a longer-term toxicity than that usually acquired from 96-hr tests. Studies at the Ministry of Agriculture, Fisheries and Food Laboratory in Burnham-on-Crouch, U. K., indicate from some of the tests with metals that 1500-hr LC50s of mercury (as mercuric chloride) and cadmium (as cadmium chloride) to brown shrimp *(Crangon crangon)* are about 1/1000 and 1/100, respectively, of the 48-hr LC50s (Wilson and Connor, 1971). Their studies showed, moreover, that larvae of the invertebrates *Ostrea edulis*, *C. crangon*, *Carcinus maenus* and *Homarus gammarus* were from 14 to 1000 times more susceptible to toxicities of mercury, copper and zinc (Connor, 1971).

It is clear that bioassay tests even for acute toxicity of heavy metals require special precautions. In addition to the usual care in maintaining concentrations of the metals in solution, which can be best achieved through continuous-flow type bioassays (Wilson and Connor, 1971), one must be certain of maintaining uniform concentrations of dissolved oxygen, salinity and temperature throughout the tests. Bioassays should be conducted for at least 1 month, or at least until such time as mortality ceases at "sublethal concentrations." For fish spawning in the marine environment, the tests should be extended to include the egg and larval stages, which are generally considered to be the stages most sensitive to pollutants.

Because of all the foregoing problems, it has been extremely difficult to obtain consistent data for the marine environment on some of the metals examined. Nevertheless, it has been concluded, on the basis of acute toxicity using more reliable 48-hr LC50 values for fish, synergism of some metals in combination, sublethal effects, bioaccumulation and the danger to man in consumption of seafood, that the pollution hazard of some of the more prominent metals in seawater can be ranked as follows: $Hg^{2+} > Cd^{2+} > Ag^+ > Ni^{2+} > Pb^{2+} > As^{3+} > Cr^{3+} > Sn^{2+} > Zn^{2+}$ (NAS, 1974).

The toxicity to aquatic organisms alone provides a somewhat different order of hazard. Choosing LC50 values for the most sensitive stages (larvae) of some of the marine species, obtained in one laboratory if possible, we can find a considerable displacement of some metals, *e.g.*, cadmium in the ranking of toxicities, although mercury still heads the list and silver is virtually as toxic. A somewhat expanded list, based on some recent toxicity data and on some much older, gives the toxicity ranking as follows: $Hg^{2+} > Ag^+ > Cu^{2+} > Zn^{2+} > Ni^{2+} > Pb^{2+} > Cd^{2+} > As^{3+} > Cr^{3+} > Sn^{2+} > Fe^{3+} > Mn^{2+} > Al^{3+} > Be^{2+} > Li^+$.

The accepted LC50 values for these metals in the marine environment have been superimposed on the diagram of Baes (1973) (Fig. 1).

The accepted 48-hr LC50 values for these metals in the marine environment have been superimposed on the diagram of Baes (1973) (Fig. 1). In order to provide some degree of compatibility, as many of the bioassay data as possible were taken from Calabrese *et al.* (1973), who conducted their toxicity tests with embryos of the American oyster *Crassostrea virginica*. A comparison is given for three metals (Cu, Zn and Hg) with larvae of the crab *Carcinus maenus* (Connor, 1972), which in each case appears to be less sensitive than the oyster larvae. If all the bioassay data had been obtained in one laboratory, where consistency in experimental technique and conditions could have been maintained, perhaps the relationship with position of the metal in the Periodic Table would have been better. However, as shown in Table 4, LC50 data taken even in one laboratory can show a wide range.

It is becoming increasingly evident that the embryonic stages of the organism are usually the most sensitive to metals, among other pollutants. Techniques for rearing larvae are perhaps more specialized than for fish. However, there is a growing fund of experience in bioassays with invertebrate larvae (Okubo and Okubo, 1962; Woelke, 1967, 1972). Even with oyster larvae bioassays, one must be cautious

TABLE 4
Median Lethal Concentrations (LC50) of a Number of Selected Substances for Marine and Some Freshwater Organisms

Substance	Description	48-Hr[a] LC50 in ppm[b]		Fish[c]	Reference
		Brown Shrimp (Crangon crangon) except as noted	European Cockle (Cardium edule) except as noted		
Inorganic Ions					
Aluminum	3+d			0.27[1]	McKee and Wolf, 1973
Arsenic	4+			8.4[2]	McKee and Wolf, 1973
Beryllium	2+			31.0[3]	Jackim et al., 1970
Cadmium	2+			27.0[3]	Jackim et al., 1970
Chromium	6+	100	100-300	33-100[4]; 17.85	McKee and Wolf, 1973; Portmann, 1972
Copper	2+	10-33	1	1-3.36; 3.2[3]	Jackim et al., 1970; Portmann, 1972
Iron	3+	33-100	100-300	5[7]	McKee and Wolf, 1973; Portmann, 1972
Lead	2+		2.45[e]	0.34[f]; 188.0[3]	Jackim et al., 1970; McKee and Wolf, 1973
Mercury	2+	3.3-10	3.3-10	3.36; 0.23[3]	Jackim et al., 1970; Portmann, 1972
Nickel	2+	100-330	> 330; 1.54[e]	0.8[g]	Portmann, 1972
Silver	1+		0.006[e]	0.019[h]; 0.04[3]	Jackim et al., 1970; McKee and Wolf, 1973; Portmann, 1972
Zinc	2+	100-330	100-330	3.3[10i]	McKee and Wolf, 1973; Portmann, 1972

Footnotes on next page.

20

TABLE 4—Continued

a Concentrations underlined are very close to the actual LC50 values. LC50 data taken from Jackim et al. (1970) are for 96-hr exposure.

b In most cases, concentrations are in mg/l (weight/volume), but where the raw material was available in liquid form only, as in the case of oil dispersants, concentrations are in ml/l (volume/volume).

c Species: 1 young eels, *Anguilla* sp.; 2 chum salmon fry, *Oncorhynchus keta*; 3 killifish, *Fundulus heteroclitus*; 4 sea-poacher, *Agonus cataphractus*; 5 coho salmon, *Oncorhynchus kisutch*; 6 flat fish, *Pleuronectes flesus*; 7 dogfish, *Squalus* sp.; 8 stickleback, *Gasterosteus aculeatus*; 9 goldfish, *Carassius auratus*; 10 rainbow trout, *Salmo gairdneri*.

d Valence state of elements tested.

e Studies were conducted on eggs of the oyster *Crassostrea virginica* (Calabrese et al., 1973).

f Tests were conducted on sticklebacks, *Eucatia inconstans*, and coho salmon, *Oncorhynchus kisutch*, in fresh water with 1000–3000 mg/l of dissolved solids.

g Concentration given as the lethal limit of nickel nitrate as nickel for sticklebacks, *Gasterosteus aculeatus*.

h Given as the concentration for which there was a 96-hr average survival time for sticklebacks, *Gasterosteus aculeatus*.

i Tests were conducted in hard fresh water with about 320 mg/l CaCO₃ (Edwards and Brown, 1966).

with results obtained, since the toxicity of a substance to which they are exposed can vary according
to whether the larvae are metabolizing or not (Woelke,
personal communication).

BIOACCUMULATION

Their tendency to be bioaccumulated is perhaps
one of the most important biological properties of
metals. Some of the earliest studies on the bioaccumulation of elements in the sea were compiled by
Vinogradov (1953), where many marine organisms were
noted to be capable of highly concentrating the metals
present in seawater. More recently (Lowman, 1960),
the concentration by marine organisms, of neutron-
induced radionuclides of the metals from fallout in
nuclear weapons tests of the South Pacific has emphasized the bioaccumulative ability of certain species.
From the point of view of commercial fisheries,
metals have had a devastating economic impact, particularly on the freshwater fisheries in Canada and
the United States, arising out of the uncovered
mercury contamination episode of the 1970-71 period.
When the Food and Drug Directorate of Canada and the
U. S. Food and Drug Administration placed a maximum
acceptable level for mercury in food at 0.5 ppm, a
large part of the freshwater fisheries catch had to
be confiscated. Later, there was also great concern
about mercury in marine fishes, tuna and swordfish
bearing the brunt of this ban, although mercury in
dogfish and Pacific halibut did not go unnoticed.
Mercury, of course, was a special problem. It
has a neurological effect on its victims that may be
fatal, but worse still, it leaves those surviving
with permanent nervous disorders. The alarm about
mercury poisoning in Japan has been more responsible
perhaps than anything else for the subsequent energy
in many institutions devoted to combating the metals
pollution problem.

At the other end of the scale of effects of metals to humans, we have zinc in oysters at concentrations as high as 17,000 ppm (dry weight) in some zinc-contaminated waters (Fig. 2). So far, this has apparently not produced overt symptoms of poisoning in normal consumers of oysters. The high concentration of zinc-65 in oysters exposed to radionuclides released from nuclear reactors (Seymour and Lewis, 1964) indicated that this neutron-induced radionuclide was accumulated by molluscs.

Zinc, copper and cadmium concentrations (dry weight) in Pacific oysters, *Crassostrea gigas*, are shown in Figure 2, in the vicinity of a pulp and paper mill in British Columbia. Zinc has been released for some time in effluents from newsprint mills, owing to the use of zinc hydrosulphite (more correctly zinc dithionite, ZnS_2O_4) as a brightening agent for the groundwood pulp used in producing this type of paper (Environment Canada, 1973). As a result, oysters which concentrate zinc by a factor of 150,000-300,000 (see Table 1) over that present in ambient water, were found to far exceed the level permitted by the Food and Drug Directorate. For this reason, the industry decided to switch in 1973 to a less ecologically hazardous brightening agent, sodium hydrosulphite (sodium dithionite, NaS_2O_4) (Joyette and Nelson, 1973).

Obviously, zinc affects humans in a different way than mercury does, and the same may be true for its effects on aquatic organisms. This may be partly related to the fact that zinc is required in animal nutrition, being normally present in many tissues and enzymes. At least in mammals, it is considered indispensable to growth and function of the vital organs such as liver, kidneys and spleen, as well as to the normal function of the blood. Zinc may also be a helpful pharmacological agent in treatment of antherosclerotic patients and has been long recognized as being beneficial in the treatment of wounds.

The need for mercury in human metabolism has never been identified. It is strictly a poison,

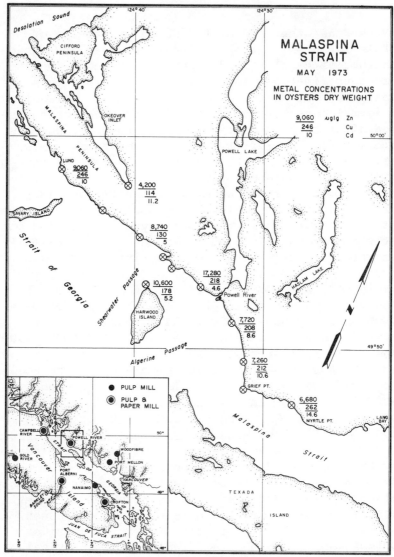

Fig. 2. *Chart of the Malaspina Strait, in the
vicinity of a major newsprint mill in Powell
River, British Columbia, showing concentration
of zinc, copper and cadmium in Pacific oysters
Crassostrea gigas. Zinc originated from zinc
dithionite (ZnS_2O_4) used for brightening
groundwood. Inset: Location map showing other
major pulp and paper mills in the region.*

24

having a serious neurophysiological effect when present in the methyl mercury form, attacking the central nervous system. However, the mode of action of mercury even in humans is not clearly understood. There are situations of natives consuming large quantities of fish from mercury-contaminated waters of Clay Lake in Northern Ontario, where neurophysiological disorders might be expected, but none has been found so far. The form in which zinc is present in the body may be extremely important in its effects on organisms when compared to that of mercury. Whereas some of the mercury is converted in fish to the highly dangerous methyl mercury (Jensen and Jernelöv, 1969) that affects the brain cells, most of the zinc consumed in a normal diet is excreted. That which remains in the body may be in an organic combination and is comparatively nontoxic in the animal's physiological processes. Recent studies show that mercury can be bound by protein in marine organisms (Olafson and Thompson, 1974), and in this form, it too can be less biologically damaging.

The bioaccumulation of metals varies from metal to metal and differs among the various organisms. Some of the metal is bioaccumulated through the food chain, so that predators have the highest concentrations; but the highest concentrations of metals appear to occur in the invertebrates. Cadmium concentration in fish exposed to 10 mg/l Cd^{2+} was found to be 113 mg/kg in the whole body ash (Eisler, 1971). Marine molluscs accumulated Cd^{2+} up to 352% of controls from water containing 10 mg/l Cd^{2+}, and the lobster *Homarus americanus* increased their cadmium content about 41%, with most of the accumulation occurring in the gills (Eisler *et al.*, 1972). Fiddler crabs, *Uca pugilator*, accumulated cadmium in the gills up to a level of 110 mg/g of Cd^{2+} wet weight (O'Hara, 1973a). The crabs died shortly after this level was reached.

Studies of uptake by the mussel, *Mytilus edulis*, of [65]Zn, [54]Mn, [58]Co and ^{59}Fe showed that the principal accumulation of these metal radionuclides occurs

25

through the food and that the transfer directly from the water is relatively minor (Pentreath, 1973a). A further study on the accumulation and retention of ^{65}Zn and ^{54}Mn by the plaice *Pleuronectes platessa*, led to the same conclusion (Pentreath, 1973b).

Metals in seawater, suspended sediments and biota in coastal waters of the United Kingdom have been examined during the last 15 years (Preston *et al.*, 1972). Although high concentrations were found in both waters and biota in certain areas, no significant changes were recognized in either level or distribution or in characteristics of organisms during the 1960-70 decade (Preston, 1973a). Concerning cadmium in the marine environment of the United Kingdom, Preston (1973b) summed up that concern about human intake of this metal, is restricted to a few "hot spots" in coastal waters and that the main items that exhibit high concentrations of cadmium are molluscan shellfish and brown meat of crabs. But the United Kingdom Working Party on the Monitoring of Foodstuffs in their Fourth Report (U. K., 1973b) concluded that the average human diet would contain an insignificant amount of cadmium from these sources.

Other human health problems that have arisen out of bioaccumulation of metals in Japan include *Itai itai* disease, resulting from cadmium released from mining activities and introduced into irrigation waters (Yamagata and Shigematsu, 1970). Curiously, this particular disease involves a brittleness of the bones, accompanied by pain, tenderness, muscular weakness, and loss of appetite in those victims who consumed large quantities of food contaminated by cadmium, but it also has a serious effect on the circulatory system of humans. Hypertension has been attributed to high cadmium concentrations in the diet (Hise and Fulkerson, 1973), although effects of other substances such as salt have not been fully evaluated to separate the various causal factors.

Bioaccumulation of copper by oysters and other invertebrates has been known for a long time. Copper tends to give the oysters not only a green color, but

26

a rather unpleasant metallic taste. While the concentration of copper in seafood has never been considered a serious threat to human health, because the
color and flavor imparted to such items as oysters
usually deter consumers before the concentration of
copper is high enough to cause harm, the marketing of
such seafood is seriously affected. In regulations
for controlling noxious substances other than oil
transported by ships developed in the new Convention
on Prevention of Pollution from Ships (IMCO, 1973),
close control must be maintained over those substances
which bioaccumulate and produce a taint or discolora-
tion in aquatic organisms.

So far in this discussion, the effects of bioaccumulation have centered largely on the aspect of
human consumption of seafood. But what of the effects
of bioaccumulation on the organism itself and on the
ecosystem? Surely there must be at least subtle
physiological changes in an organism with a high
burden of heavy metals. There is much current interest in the problem, but little research has been conducted on it in the past. Some of the significant
results of this research will be reviewed in the next
subsection.

It is clear that the bioaccumulation aspect of
metals in the marine environment has to be at least
partly taken out of the realm of human epidemiology
and etiology and placed into the realm of the marine
physiologist. Sublethal effects of bioaccumulation
by organisms and their consequences in the marine
ecosystem merit intensive investigation.

SUBLETHAL EFFECTS

Some of the thoughts explored in the previous
section with respect to effects of bioaccumulation by
aquatic organisms will be expanded here. Although
the acute toxicity of metallic salts in the marine
environment is not yet fully established, it is time
perhaps that subacute effects of these salts were

examined in detail for some of the more marginal
toxicity considerations. These include the neurophysiological effect, the influence on enzyme activity, the effect on vigor of organisms, endocrinology,
parasitology and disease, the teratogenic, carcinogenic and mutagenic effects, as well as the other
impacts of metals at the cellular level.

It has been said by Professor Robert B. Clark,
of the University of Newcastle-upon-Tyne (personal
communication), that the only observed effect of some
of the metals present in high concentrations in estuaries of the United Kingdom is a tendency for the
organisms to be more sluggish in response. This may
be a clue to some rather interesting sublethal effects
on the organisms. In connection with metals pollution
in estuaries of the United Kingdom, there is some evidence of adaptation by certain organisms over several
centuries of exposure (Bryan, this volume).

It is not the intent in this paper to go into any
detail on the sublethal effects of the heavy metals
on aquatic organisms, inasmuch as there is rather
little known of these effects. Moreover, in the
papers that follow, recent findings will be discussed
by scientists working in this field. Sprague (1971)
reviewed the status of understanding and methods of
determination of sublethal effects of pollutants to
1970. There is now much research in progress in the
sublethal range. Clearly, there are physiological
effects from all the metals, some of the effects being
even favorable at low concentrations. The problem
that often exists, however, is to quantify the physiological changes that occur in an organism.

ENVIRONMENTAL FACTORS

One of the simpler ways of examining sublethal
effects of a metal salt on aquatic organisms is to
use a concentration just below the lethal threshold.
Then vary the physical and chemical properties, such
as salinity, temperature and dissolved oxygen, of the
water.

Recent work (MacLeod and Pessah, 1973) has demonstrated temperature effects on mercury accumulation, toxicity and metabolic rate in rainbow trout, *Salmo gairdneri*, suggesting a certain interaction of environmental factors with the toxicity of mercury in low concentrations.

Vernberg and O'Hara (1972), in a study of mercury uptake by *U. pugilator* under a temperature-salinity stress, found the translocation of mercury from gills to hepatopancreas highest at higher temperatures. The cause of higher mortality at lower temperatures was believed to be the high mercury residue left in the gills of these crabs.

O'Hara (1973b) exposed *U. pugilator* to 10 mg/l of Cd^{2+} at various salinity and temperature regimes, and found the maximum cadmium accumulation in a combination of high temperature and low salinity. Under these conditions, the amount of cadmium present in the hepatopancreas was also greatest, the high temperature considered to be contributing to the high rate of cadmium translocation from gills to hepatopancreas. The 24-hr cadmium concentration ratio in gill/hepatopancreas was found to be 305.9% at 10°C, 10 o/oo, compared to 87.8% at 33°C and 10 o/oo.

NEUROPHYSIOLOGY

One of the underlying problems with mercury pollution is the effect on the nervous system through attack of vital brain centers by methyl mercury in humans and other mammals (Suzuki, 1969). Subtle behavioral effects can arise in offspring following exposure of pregnant mothers to low methyl mercury concentrations (Spyker *et al.*, 1972). Whether this also occurs in aquatic organisms has not been clearly identified, but no doubt similar effects prevail. Certainly, some of the functions that are controlled by the nervous system, such as maintenance of equilibrium, have been already demonstrated with low concentrations of mercury in fish (Lindahl and Schwanbom, 1971a, 1971b).

The effects of metals, particularly mercury, cadmium and lead, on the behavior of organisms merits examination. Neurophysiological examination in the laboratory may also reveal some profound effects on certain important brain centers which control learnability, response to temperature changes, and general reaction to adverse conditions. There may be a neurophysiological effect of methyl mercury and lead on fish, in the same way as there appears to be in humans, but this has not yet been widely reported.

Work done in Sweden (Lindahl and Schwanbom, 1971a, 1971b) has shown by a rotary-flow technique the effect of accumulation of sublethal concentrations of mercury on fish. Fish with concentrations of mercury as low as 0.1 ppm were not able to adjust to a torque applied on them in a rotating cylinder.

Kleerekoper *et al.* (1973) have examined the interaction of temperature and copper ions as orientation stimuli in the locomotor behavior of the goldfish, *Carassius auratus*. The fish chose water containing 0.010 mg/l $CuCl_2$ less frequently and spent less time there per entry than in copper-free water at 21.1°C, but the situation was reversed, with an increase of temperature to 21.5°C, the fish entering the water containing 0.010 mg/l $CuCl_2$ more frequently than the copper-free water. Dr. Kleerekoper's approach to the behavioral manifestations of sublethal concentrations of pollutants will be presented later (Kleerekoper, this volume).

ENZYME ACTIVITY

There are indications of depressed or accelerated enzyme activity in aquatic organisms exposed to low concentrations of metals (Hewitt and Nicholas, 1963). Studies have shown that cadmium ion has an effect on respiration and ATPase activity of the pulmonary alveolar microphage (Cross *et al.*, 1970). Brown and coworkers at the College of Fisheries, University of Washington, Seattle, have demonstrated the activation and inhibition by heavy metals of allantoinase,

extracted from the polychaete *Eudistylia vancouveri*
(May and Brown, 1973). They found this enzyme to be
most sensitive to metal ions in the order: Cd^{2+} >
Pb^{2+} > Zn^{2+} > Hg^{2+} > Cu^{2+}. Activity of the enzyme
alpha-glycerophosphate dehydrogenase found in fish
(trout) muscle tissue was found to be inhibited by a
number of metals in the following descending order:
Hg^{2+} > Cd^{2+} > Zn^{2+} > Pb^{2+} > Ni^{2+} > Co^{2+}. As little
as 0.001 millimoles (0.2 ppm) of Hg, compared to 2 Mm
(120 ppm) of Ni, caused a 50% reduction of the enzyme
activity (Bargmann and Brown, 1974).

Some of the classical work in this regard has
been conducted by Jackim *et al*. (1970) with respect
to metal poisoning of a number of liver enzymes in
killifish *(Fundulus heteroclitus)*. More recently the
influence of lead and other metals on δ-aminolevulin-
ate dehydrase activity in fish has been studied
(Jackim, 1973). I shall say little more here,
inasmuch as Dr. Jackim will be discussing the effect
of metals on selected enzymes in fish. It may be an
important effect in determining how well fish can
perform certain physiological functions.

Related to this type of action of metals, recent
studies at the Vancouver Laboratory of the Fisheries
and Marine Service of the Department of Environment
(Bilinski and Jonas, 1973) have shown that the oxidation of lactate by gills in rainbow trout *(Salmo
gairdneri)* is inhibited by over 50% when the fish
are exposed to 1.12 mg Cd/l for 24 hrs or 0.064 mg
Cu/l for 48 hrs. Of course, these concentrations
were high enough to kill the fish during longer periods of exposure. In fact, exposure of the trout to
even lower metal concentrations resulted in mortalities, but there was no detectable effect on the oxidative activity in gills. Therefore, this can hardly
be considered as a useful diagnostic tool for measuring sublethal effects of a subtle nature, but it does
point to some of the biochemical alterations which
may very well lead to mortality.

ENDOCRINOLOGY

There is little doubt that the effects of certain metals may be quite profound on the hormone activity in marine organisms, as well as in terrestrial animals. In one of our associate laboratories in Vancouver, Dr. E. M. Donaldson, an endocrinologist, has conducted experiments on the effects of copper on corticosteroid levels in yearling sockeye salmon, *Oncorhynchus nerka*, of 20-50 g weight, held in soft fresh water (hardness of 12 mg/1 of calcium ion) over a period of a day (Donaldson and Dye, 1974). The salmon were exposed to copper concentrations of 0.0064, 0.064, and 0.64 mg Cu/1 (10^{-7}M, 10^{-6}M, 10^{-5}M) as cupric sulfate. Cortisol and cortisone levels were measured in the plasma of fish at 1, 2, 4, 8 and 24 hrs after exposure to the different copper concentrations, as well as in controls kept in clean fresh water. The cortisol is the active steroid while the cortisone is the metabolite, and both give a measure of stress in the fish.

It was found that there was a rise in both cortisol and cortisone levels during the first hour of exposure, particularly for the cortisol at 0.064 and 0.64 mg/1 copper. In the highest copper concentration, the fish were dead by 24 hrs. It is interesting that at 0.064 mg/1, the cortisol level reached a maximum at 4 hrs, declined to a level almost as low as that at 2 hrs, and then increased again at 24 hrs. The cortisone level, on the other hand, peaked at 2 hrs, declined to about two-thirds of the first maximum at 8 hrs and then reached its greatest level, 22.5 μg cortisone/100 ml plasma, at 24 hrs.

The total corticosteroids in these salmon stressed by copper at 0.64 mg/1 of Cu were comparatively uniform, ranging from 64 to 71 μg/100 ml during the first 4 hrs of exposure, and then increasing to 93 μg/100 ml at 8 hrs. Death occurred between 8 and 24 hrs exposure. At 0.064 mg/1 of Cu, the corticosteroids reached a first maximum of 66 μg/100 ml in 2 hrs, dropping to 31 μg/100 ml at 8 hrs and then

32

increasing again to 59 µg/100 ml at 24 hrs. It is
noteworthy that the fish may be able to adjust to
the stress imposed by the low concentrations of
copper. This is indicated by the lowered corticosteroid level reached in 8 hrs at the lowest concentration of copper used in the experiment (i.e.,
0.0064 mg/l). However, the total corticosteroids
increased again (12.5 µg/100 ml) at 24 hrs even at
this low copper concentration, although the level was
not as high as it was at 2 hrs (19.5 µg/100 ml).

Earlier work in Michigan on exposure of trout to
chromium VI (Hill and Fromm, 1968) indicated an effect
on corticosteroid levels, although the controls also
showed unusually high levels which suggested some
other stressing factor in the experiment. Other
metals may be capable of causing similar biochemical
effects on fish which affect its vitality, capability
to carry out normal life functions, and ultimately
to survive and reproduce.

CELLULAR EFFECTS

There can be an effect of metals on marine
organisms at the cellular level. Again, I am not too
familiar with the literature on cell and molecular
biology to know how much has been studied on the
effects of metals at the cell level of biological
material. However, recent work has shown that there
is certainly an effect on the embryonic stages of
some aquatic organisms which can lead to teratogenesis,
mutagenesis and/or carcinogenesis. Servizi and Gordon
(1974), in their studies on the effects of selected
heavy metals on early life of sockeye and pink salmon,
have shown that mercury at a concentration of 10 ppb
gives a teratogenic effect to pink and sockeye salmon
when their preeyed eggs are exposed at these concentrations. Cadmium, on the other hand, was not as
toxic to the eggs, but was very toxic to the fry,
especially once they emerged from the gravel and were
free-swimming.

33

Exposure of fish to cadmium has induced lesions (Gardner and La Roche, 1973), and it may be possible to obtain similar effects with other metals. Lesions have been reported in fish taken from polluted coastal waters (Halstead, 1972). Studies being conducted at the Cancer Research Institute, University of British Columbia (Dr. H. F. Stich, personal communication), suggest that certain pollutants cause tumors in juvenile flatfish, and that these flatfish probably die the first year of their life. It is hoped to compare flatfish species in different life stages from polluted and unpolluted areas. At this stage in the investigation, however, it is uncertain what the pollutant might be, although it is more likely petroleum or petroleum products than heavy metals.

The effects of metals on the genetic material of cells is still unclear, but no doubt, certain metals have an effect on DNA and RNA, which could lead to mutations. There is a suggestion of genetic adaptation to heavy metals in plants (Creed, 1973) and this may be true to some extent also in the marine fauna.

There is some evidence of histological and hematological response of certain fish after long-term exposure to certain metals (McKim et al., 1970; Gardner and Yevich, 1970), and this may be a good tool for quantitatively assessing sublethal effects without having to resort to sophisticated techniques of cell culture and electron microscopy.

Karbe (1972) found marine hydroids, colonies of Eirene viridula, to be excellent test organisms for assessing toxicity of metals. Morphological changes and tissue reorganization of hydranths occurred at low doses of heavy metals (3 to 10 mg/l Zn, 1 to 10 mg/l Pb, 0.3 to 10 mg/l Cd, 0.06 to 1 mg/l Cu, 0.003 to 0.1 mg/l Hg). The degree of damage could be assessed from six discrete steps of tissue reorganization.

RESPIRATION, OSMOREGULATION, BLOOD CHARACTERISTICS, AND CIRCULATION

Techniques are available now with a respirometer (Brett, 1964), and with other clinical devices for measuring blood pressure and circulation (Davis, 1973), and for recording the effects of pollutants on the respiration and circulation of aquatic organisms. Some work has already been reported on the respiration and osmoregulation in rainbow trout (Salmo gairdneri), where the gills were damaged by zinc sulphate (Skidmore, 1970). Sparks et al. (1972) noted the presence of zinc at 8.7, 5.22, 4.16 and 2.55 mg/l, in dechlorinated tap water, caused an increase in breathing rate or a change in breathing rate variance of bluegills, Lepomis macrochirus. Oxygen consumption rate by the mud snail, Nassarius obsoletus (Say), has been shown to be lowered when exposed to arsenic, copper, silver and zinc individually and elevated when exposed to cadmium alone and to a combination of copper and cadmium (MacInnes and Thurberg, 1973).

Blood studies have been conducted in a number of laboratories to investigate changes following short- and long-term exposure to metals. McKim et al. (1970) examined the blood of brook trout, Salvelinus fontinalis, after exposure to copper, and noted changes in several blood characteristics at concentrations of 38 to 70 µg/l over a 6-day period. Later work by some of the same authors on the blood of the brown bullhead, Ictalurus nebulosus, showed that the maximum concentration of Cu(II) having no detectable effect was between 11 and 16 µg/l (Christensen et al., 1972).

Much more needs to be done on the effects of various metals which are contaminating our marine environment in their physiological effects on respiration, osmoregulation, and circulation. These are important factors to the survival of the organism in that they determine physical well-being, capability to withstand stresses, and the ability to react in the face of danger, such as predators.

PARASITISM AND DISEASE

There is little doubt that an organism under stress is more susceptible to disease than one free of such effects. It has been found sometimes in holding fish in pens in the sea nearshore at our laboratory that when the conditions are unfavorable, such as during the influx of fresh water from the Fraser River or increasing temperature because of insolation, there is an outbreak of disease. Such diseases probably existed in latent form, but awaited conditions suitable for development in the weakened fish.

It would appear that high concentrations of metals would have essentially the same effect as other stresses, such as unfavorable salinity and temperature. In fact, work by Pippy and Hare (1969) on Canada's east coast suggested greater viral infection in fish exposed to pollution in a river than that in an unpolluted section. It would seem that parasitism would also be more prevalent in fish under the stress of excessive metals concentration. If viruses cause tumors in fish, as some investigators hypothesize, then the tumors may be related indirectly to the presence of virus stimulated by stressful conditions in the fish.

REPRODUCTION

For maintenance of the species, it is as important for fish to be able to reproduce as it is for them to survive following exposure to a high concentration of metal or any other pollutant. All the foregoing sublethal factors ultimately have an effect on reproduction. There are a number of direct effects on the reproductive process which are worthy of mention. The effects of low concentrations of mercury, cadmium and copper on eggs and fry of sockeye and pink salmon, *Oncorhynchus nerka* and *gorbuscha*, have already been noted (Servizi and Gordon, 1974).

The fertilizing capacity of spermatozoa of steelhead trout, *Salmo gairdneri*, was reduced following exposure to 1.0 ppm of methylmercuric chloride for 30 min (McIntyre, 1973). There may be an adverse effect on reproduction at much lower concentrations of mercury with long-term exposure, since Hannerz (1968) found a concentration factor of 1300 for pike gonads in fish exposed to methylmercuric oxide for a month.

Studies of effects of metals on reproduction in the zebra fish, *Brachydanio rerio*, showed that the number of eggs spawned was reduced by 22% following exposure to 1.0 ppb of phenylmercuric acetate, and spawning was almost completely stopped at 20 ppb (Kihlström et al., 1971). Moreover, there was a significantly lower proportion of hatched eggs from the treated fish than from untreated fish when the eggs were placed in clean water. An effect of the mercury compound on the contractile mechanism of the oviduct or of the genital papilla was deduced as the cause of the smaller number of eggs spawned. When eggs from normal zebra fish were exposed to different concentration of phenylmercuric acetate, it was found that the percentage hatched actually increased at 10 ppb, was the same as the controls at 20 ppb, and was zero at 50 ppb. The improved hatching at the low phenylmercuric acetate concentration was attributed to its bacteriocidal and fungicidal activity (Kihlström and Hulth, 1972).

The effect of cadmium on testes in mammals is well known (Gunn et al., 1968). Its interference with spermatogenesis and androgen synthesis in brook trout, *Salvelinus fontinalis*, has been recently revealed (Sanagalang and O'Halloran, 1972). Exposure of brook trout to 25 ppb of cadmium chloride for 24 hrs caused haemorrhagic necrosis and disintegration of the lobule boundary cells. The biosynthesis of 11-ketotestosterone, a major androgen in salmonids, was decreased when trout testes were incubated with 4-^{14}C-pregnenolone in the presence of cadmium chloride.

Although possibly less dramatic, it is anticipated that there are effects on the reproductive
organs and on the viability of eggs and sperm and
fish from other metals.

EFFECTS OF HEAVY METALS ON THE ECOSYSTEM

Whereas a considerable amount of research information has been published on the effects of metals on
individual organisms, little is known on the effects
of metals on aquatic ecosystems. The tendency is to
look only at the commercially important species.
Preston (1973b), in noting high cadmium concentrations
in noncommercial species in areas of high cadmium concentration such as the United Kingdom's Bristol
Channel, asks the question: "What effect, if any,
are these concentrations having?"

Some work has been done, in fact, on the sublethal effects of cadmium in a number of organisms.
Eisler (1971) examined the acute toxicity of cadmium
in a variety of organisms and considered the lethal
concentration at different temperatures.

Phytoplankton are known to be inhibited by
different metals (Mandelli, 1969). They contain
metals at somewhat higher levels (Knauer and Martin,
1973) than present in ambient seawaters (Spencer and
Brewer, 1969; Riley and Taylor, 1972). It is known
that certain metals are concentrated by certain
trophic levels in the sea, and that the metals are
moved up the food pyramid with accumulation in the
top predators. The high concentrations of metals in
filter-feeding invertebrates (Pringle et al., 1968;
Pentreath, 1973a; Bryan, 1973; Topping, 1973) is a
good indication of concentration of metals via
phytoplankton. However, the mechanism of this transfer through the food chain is not well known even for
the more fully investigated metals, such as mercury.
The ultimate fate of the metals in the marine environment is likewise not fully understood.

Such shortcomings in the current scientific knowledge as uptake of pollutants, their movement through the food chain, and ultimate deposition in the sediments, prompted a workshop on Marine Environmental Quality, held in Durham, New Hampshire, in August 1971. To identify those areas requiring research for a better understanding of man's effects on the ocean was a major goal of this workshop (NAS, 1971b). As a result, a base line study was conducted on the concentrations of metals, among other pollutants, in the marine environment during 1971–72, and a workshop was convened in Brookhaven, New York, in May 1972 (Goldberg, 1972) to evaluate the results and draw some conclusions.

Out of the Durham meeting arose a number of monitoring programs, as well as the proposal for the Controlled Ecosystem Pollution Experiment (CEPEX), which started in the summer of 1973 in Saanich Inlet, British Columbia (NSF, 1973). This is an interdisciplinary program of investigation by chemists, zoologists, botanists, microbiologists and mathematical modelers on heavy metals and hydrocarbons in the food chain of the sea under controlled experimental conditions maintained *in situ*, as similar as possible to the natural environment. It involves several United States, Canadian and United Kingdom institutions, and is supported by the National Science Foundation's International Decade of Ocean Exploration Office in Washington, D. C. A primary goal of the experiment is to determine whether deep-sea forms are more sensitive to pollutants than inshore forms, and therefore whether the former require greater protection from pollution.

Preliminary studies, commenced this summer, used quarter-scale models of large inverted plastic silos or underwater "greenhouses," as they have been sometimes referred to. The full-scale experiment will consist of nine of these silos with a diameter at the surface of 30 feet, extending to 90 feet below the surface. Each silo will contain some 2350 tons of

water for the duration of the experiment, and controlled amounts of pollutants, such as heavy metals
and/or hydrocarbons, will be added to the trapped
column of water, and the effects on life in the
contained water column will be studied. Water will
slowly upwell through these columns at a rate of
displacement of about 1% per day. It is planned to
carry out the experiments for a duration of 100 days,
during which time the concentrations of pollutants
will be measured in the water and in the phytoplankton
and zooplankton.

It is hoped with this experiment to gain a
fuller understanding of the effects of metals and
other pollutants on different trophic levels and of
their movement through the food chain in the sea,
under experimental conditions fashioned to resemble
as closely as possible those occurring in nature.
A better appreciation of the weak links in the food
chain, which are affected more than others by
specific pollutants, will allow better management of
waste disposal and of renewable resources in the sea.

AQUACULTURE

Aside from the aquatic and ecosystem effects of
metals in the natural marine environment, there are
also practical considerations for aquaculture.
Perhaps even of more importance than in the natural
environment, water quality is vital in aquaculture.
A major threat that looms over the enterpreneur in
aquaculture is disease. As noted earlier, this can
be aggravated by adverse water conditions which
impose a stress on the cultured animal. Metals can
contribute to this stress if they are present in
excessive concentrations.

Culturing of shellfish and finfish requires
particular care in the egg, larval and immediate
postlarval stages. The effects of metals on eggs of
salmonids has been noted earlier. The sensitivity to
pollutants of larvae of the Pacific oyster *Crassostrea*

gigas has been known for some time (Woelke, 1967).
The effect of zinc on their development and growth
has been recently reported by Brereton *et al.* (1973)
and to embryos of the American oyster, *Crassostrea
virginica*, by Calabrese and coworkers (Calabrese,
1972; Calabrese *et al.*, 1973).

Metals in fish reared by aquacultural techniques
can pose a serious problem in marketing the product,
because of Food and Drug regulations. If mercury,
cadmium, zinc or lead are present above the accepted
maximum concentration in the fish tissue, the aquacultural products may be banned from the market. An
example of mercury concentrations exceeding the maximum accepted level of 0.5 ppm was found in rearing
sablefish *Anoplopoma fimbria* by feeding them with
dogfish *Squalus suckleyi* in culturing experiments at
the Pacific Biological Station in Nanaimo, British
Columbia (Kennedy and Smith, 1972, 1973).

Preliminary tests conducted in 1971, where
sablefish were reared for a 28-month period on a diet
of 50:50 herring plus dogfish, gave a mercury content
of 1.29 ppm for three specimens weighing an average
of 3.7 kg. The dogfish and herring, *Clupea pallasii*,
used in the diet of the sablefish had mercury concentrations of 0.39 and 0.05 ppm, respectively (Kennedy
and Smith, 1972).

In a later experiment (Kennedy and Smith, 1973),
ten sablefish averaging 2.82 kg in weight and kept
for 31.5 months in captivity on a whole dogfish diet
gave a concentration of 1.77 ppm mercury in their
tissue. In another experiment of the same series,
six sablefish, averaging 3.36 kg and kept for 37
months on a 50:50 dogfish plus herring diet, ended up
with a mercury concentration of 1.14 ppm. It is
obvious that a choice of diet is extremely important
in aquaculture to ensure that bioaccumulation of
metals does not occur to above-acceptable levels.

Portmann (1972b) reported on investigations of
mercury, cadmium, lead, zinc, copper and chromium in
coastal and deep-water fish adjacent to England and
Wales, and in distant-water fishing grounds, in order

to establish areas where concentrations of metals are highest. Cod, whiting, plaice, herring and mackerel were analyzed for metals in the muscle tissue, as well as in the homogenate of the bulked livers from each sample. Although the mean mercury level in fish from the Irish Sea was twice as high (0.20 mg/l wet weight) as that found in fish of the North Sea, copper and zinc appeared to be higher in fish from distant-water regions than from coastal waters.

Elderfield *et al.* (1971), in a study of oyster culture in Wales involving rearing of the larvae of *Ostrea edulis*, concluded that heavy metal contamination is a cause of poor larval performance. The work of Brereton *et al.* (1973) on the effects of zinc on growth and development of larvae of the Pacific oyster, *Crassostrea gigas*, certainly bears out their conclusion at least with this metal. The problem of accumulation of metals by shellfish reared in estuarine areas of the United Kingdom has been reviewed by Boyden (1973), who summed up: "Shellfish can and do accumulate high levels but if we know what, how much and to what degree, and understand the factors influencing the accumulation, then expansion of the oyster fisheries especially of *Crassostrea* will be a safer process for all concerned."

In recirculating systems, the problem of metals enrichment can be particularly acute, especially if there is a source of metals by leaching from the facilities or in the food. Regular monitoring of metals in the product may be advisable, not only from the point of view of its market acceptability, but also for the state of health of the animals.

SUMMARY AND CONCLUSIONS

Metals may be a real problem in pollution of the marine environment. Starting with the tragedy of Minimata in Japan, involving mercury pollution, we have become aware of other problems of metals, such as cadmium, lead, zinc, and copper. The concern of mercury and cadmium pollution has been largely related to its effects on humans, as a result of consumption of these metals in high concentrations, bioaccumulated by seafood organisms. This has been a matter of great human health concern to food and drug agencies in many countries.

However, the metals are acutely and subacutely toxic to aquatic organisms, and unfortunately, less attention has been paid to this aspect of the metals pollution problem. While there is a moderate amount of information on the toxicity of metals to freshwater organisms, there is little in the literature of a consistent nature with respect to effects on marine organisms. Some investigators are now beginning to look at the sublethal effects of metals, which may be more important in the long term than the acute effects. Studies of zinc and lead at low sublethal concentration in streams have shown that there is definitely a repellent effect of these metals on Atlantic salmon in fresh water. This along with other behavioral manifestations may occur in other fishes, under the influence of these and other metals. This may be related to a neurophysiological effect of metals about which we know virtually nothing.

It is clear that enzyme activity of organisms can be altered by metals, judging by research conducted thus far. However, there are other sublethal effects on endocrinology, cell biology, hematology, histopathology and various physiological functions, such as respiration, circulation, and osmoregulation. Genetic changes need to be more fully investigated with a larger series of metals under a wide spectrum of conditions in the estuarine and marine environments.

With the growing interest in the pollution of the marine environment, monitoring studies on metals in both seawater and in marine organisms are being conducted by many countries. Invertebrates are considered excellent indicator organisms because of their capability in concentrating metals, among other pollutants. Besides measuring the concentrations of these metals in aquatic organisms, there is a need to study the effects of the concentrated metals on the organisms and on the ecosystem. This will require careful evaluation of effects of metals on the various trophic levels in the marine ecosystem and movement of metals through the food web. One such experiment being conducted under the auspices of the National Science Foundation's International Decade of Ocean Exploration is CEPEX, being conducted in Saanich Inlet, British Columbia. Other such studies have been underway in coastal lochs of Scotland. By careful mathematical modeling of the system once the experimental data are acquired, it will be possible to set up a predictive model for a better understanding of the effects of metals on the marine ecosystem.

With the growing interest in aquaculture, heavy metals have an important practical aspect. Not only are they important in terms of water quality, but also they are important from the point of view of product acceptability. For example, fish culture studies conducted at the Pacific Biological Station in Nanaimo, British Columbia, have shown that sablefish *Anoplopoma fimbria* when fed dogfish *Squalus suckleyi* acquire a concentration of mercury well above the acceptable levels, over 1 ppm. It is clear that a diet containing low metals concentration must be used if fish cultured by man are to be acceptable under food and drug regulations. Metals, when present in excessive concentrations, may put fish under stress and contribute to an outbreak of disease. Disease control is one of the major problems confounding aquaculture.

In conclusion, one can say that although there has been a great deal of interest in and considerable research on metals pollution during the last few years as a result of human poisoning, there is still much to be learned about the effects of metals on aquatic organisms, particularly at the sublethal level. If we are to be able to manage our renewable resources properly in the face of a growing technology and industrial production, such new learning must be acquired soon.

ACKNOWLEDGMENTS

Assistance has been provided in various ways by many persons in preparation of this paper, for which I am grateful. I should particularly wish to acknowledge Drs. E. M. Donaldson and J. A. Servizi, who permitted the use of unpublished information from their research, and Mr. J. R. Marier, National Research Council, Ottawa, who facilitated in retrieval of literature through the computerized system at the National Science Library.

LITERATURE CITED

Anon. 1967. Trace metals, bane or blessing. *Spectrum 11*:65-66.

Baes, C. F. Jr. 1973. III. The properties of cadmium. In: *Cadmium the Dissipated Element*, pp. 29-59, ed. by W. Fulkerson and H. E. Goeller. Report ORNL NSF-EP-wl. Oak Ridge, Tenn.: Oak Ridge National Laboratory.

Bargmann, G. G. and G. W. Brown, Jr. 1974. Fish muscle alpha-glycerophosphate dehydrogenase and its inhibition by metals. In: *1973 Research in Fisheries*, Annual Report of the College of Fisheries, p. 56. Seattle, Wash.: University of Washington.

Bertine, K. K. and E. D. Goldberg. 1971. Fossil fuel combustion and the major sedimentary cycle. *Science 173*:233-55.

Bienvenu, M. M. P., C. Nofre, and A. Cier. 1963. Toxicité Générale Comparée des Ions Metalligues. Relation avec la Classification Périodique. *Comptes Rendus 256*:1043-44.

Bilinski, E. and R. E. E. Jonas. 1973. Effects of cadmium and copper on the oxidation of lactate by rainbow trout *(Salmo gairdneri)* gills. *J. Fish. Res. Bd. Canada 30*:1553-58.

Boyden, C. R. 1973. Accumulation of heavy metals by shellfish. *Proc. Shellfish Ass. Gt. Brit. (1973)*, Fourth Shellfish Conference, pp. 38-48.

Brereton, A., H. Lord, I. Thornton, and J. S. Webb. 1973. Effect of zinc on growth and development of larvae of the Pacific oyster *Crassostrea gigas. Mar. Biol. 19*:96-101.

Brett, J. R. 1964. The respiratory metabolism and swimming performance of young sockeye salmon. *J. Fish. Res. Bd. Canada 21*:1183-1226.

Bryan, G. W. 1973. The occurrence and seasonal variation of trace metals in the scallop *Pecten maximus* (L.) and *Chlamys opercularis* (L.). *J. Mar. Biol. Assn. U. K. 53*:145-166.

_____. 1974. Adaptation of an estuarine polychaete to sediments containing high concentrations of heavy metals. This volume.

Calabrese, A. 1972. How some pollutants affect embryos and larvae of American oyster and hardshell clam. *Mar. Fish. Rev. 34*:66-77.

_____, R. S. Collier, D. A. Nelson, and J. R. MacInnes. 1973. The toxicity of heavy metals to the embryos of the American oyster *Crassostrea virginica. Mar. Biol. 18*:162-66.

Christensen, G. M., J. M. McKim, W. A. Brungs, and E. P. Hunt. 1972. Changes in the blood of the brown bullhead [*Ictalurus nebulosus* (Lesueur)] following short and long term exposure to copper (III). *Toxicol. Appl. Pharmacol. 23*:417-27.

Connor, P. M. 1971. The acute toxicity of heavy metals to the larvae of some marine animals. International Council for the Exploration of the Sea, C. M. 1971/K116. In manuscript.

_____. 1972. Acute toxicity of heavy metals to some marine larvae. *Mar. Poll. Bull. 3*:190-92.

Council on Environmental Quality. 1971. Toxic substances. U. S. Government Printing Office, Washington, D. C.

Creed, E. R. 1973. Genetic adaptation. In: *Technological Injury*, pp. 131-34, ed. by J. Rose. London: Gordon and Breach Publishers.

Cross, C. E., A. B. Ibrahim, M. Ahmed, and M. G. Mustafa. 1970. Effects of cadmium ion on respiration and ATPase activity on the pulmonary alveolar macrophage: a model for the study of environmental interference with pulmonary cell function. *Environ. Res.* 3:512-20.

Davis, J. C. 1973. Sublethal effects of bleached kraft pulp mill effluent on respiration and circulation in sockeye salmon *(Oncorhynchus nerka)*. *J. Fish. Res. Bd. Canada* 30:369-77.

_____ and B. J. Mason. 1973. Bioassay procedures to evaluate acute toxicity of neutralized bleached kraft pulp mill effluent to Pacific salmon. *J. Fish. Res. Bd. Canada* 30:1565-73.

Donaldson, E. M. and H. M. Dye. 1974. Corticosteroid concentrations in sockeye salmon *(Oncorhynchus nerka)* exposed to low concentrations of copper. *J. Fish. Res. Bd. Canada.* Submitted for publication.

Edwards, R. W. and V. M. Brown. 1967. Pollution and fisheries: a progress report. Water Pollution Control, London. *J. Proc. Inst. Sewage Purif.* 66:63-78.

Eisler, R. 1971. Cadmium poisoning in *Fundulus heteroclitus* (Pisces: Cyprinodontidae) and other marine organisms. *J. Fish. Res. Bd. Canada* 28:1225-34.

_____ and G. R. Gardner. 1973. Acute toxicity to an estuarine teleost of mixtures of cadmium, copper and zinc salts. *J. Fish. Biol.* 5:131-42.

_____, G. E. Zaroogian, and R. J. Hennekey. 1972. Cadmium uptake by marine organisms. *J. Fish. Res. Bd. Canada* 29:1367-69.

Elderfield, H., L. Thornton, and J. S. Webb. 1971. Heavy metals and oyster culture in Wales. *Mar. Poll. Bull.* 2:44-47.

Environment Canada. 1973. Zinc and boron pollution in coastal waters of British Columbia by effluents from the pulp and paper industry. Fisheries and Marine Service and Environmental Protection Service, Pacific Region, Vancouver, B. C.

47

Fulkerson, W. and H. E. Goeller. 1973. *Cadmium the Dissipated Element.* Report ORNL NSF-EP-21. Oak Ridge, Tenn.: Oak Ridge National Laboratory.

Gardner, G. R. and G. La Roche. 1973. Copper induced lesions in estuarine teleosts. *J. Fish. Res. Bd. Canada 30*:363-68.

_____ and P. P. Yevich. 1970. Histological and hematological responses of an estuarine teleost to cadmium. *J. Fish. Res. Bd. Canada 27*:2185-96.

Goldberg, E. D. (Convener). 1972. Baseline studies of heavy metal, halogenated hydrocarbon and petroleum hydrocarbon pollutants in the marine environment and research recommendations. Deliberations of the International Decade of Ocean Exploration (IDOE) Baseline Conference.

Goyette, D. and H. Nelson. 1973. Heavy metal monitoring program with emphasis on zinc contamination of the Pacific oyster *Crassostrea gigas*. In: *Zinc and Boron Pollution in Coastal Waters of British Columbia by Effluents from the Pulp and Paper Industry*. Vancouver, B. C.: Fisheries and Marine Service and Environmental Protection Service, Pacific Region.

Gunn, S. A., T. C. Gould, and W. A. D. Anderson. 1968. Selectivity of organ response to cadmium injury and various protective measures. *J. Pathol. Bacteriol. 96*:89-96.

Halstead, B. W. 1972. Toxicity of marine organisms caused by pollutants. In: *Marine Pollution and Sea Life*, pp. 584-94, ed. by M. Ruivo. London: FAO, Fishing News (Books) Ltd.

Hannerz, L. 1968. Experimental investigations on the accumulation of mercury in water organisms. *Fish. Bd. Sweden, Instit. Fresh Water Res., Drottingholm 48*:120-76.

Hewitt, E. J. and D. J. D. Nicholas. 1963. Cations and anions: inhibitions and interreactions in metabolism and enzyme activity. In: *Metabolic Inhibitors*, Vol. 2, pp. 311-436, ed. by R. M. Hochster and J. H. Quastel. New York: Academic Press.

Hill, C. W. and P. O. Fromm. 1968. Response of the interrenal gland of rainbow trout *Salmo gairdneri* to stress. *Gen. Comp. Endocrin. 11*:69-77.

Hise, E. C. and W. Fulkerson. 1973. Environmental impact of cadmium flow. In: *Cadmium the Dissipated Element*, pp. 203-22, ed. by W. Fulkerson and H. E. Goeller. Report ORNL NSF-EP-wl. Oak Ridge, Tenn.: Oak Ridge National Laboratory.

Huggett, R. J., M. E. Bender, and H. D. Slone. 1973. Utilizing metal concentration relationships in the eastern oyster *(Crassostrea virginica)* to detect heavy metal pollution. *Water Research 7:* 451-60.

IMCO. 1973. International Conference on Marine Pollution, 1973. IMCO Sess. Pap., MP/CONF/1-10 and INF. 1-8. Inter-Governmental Maritime Consultative Organization, London.

Jackim, E. 1973. Influence of lead and other metals on fish δ-aminolevulinate dehydrase activity. *J. Fish. Res. Bd. Canada 30*:560-62.

_____, M. Hamlin, and S. Sonis. 1970. Effects of metal poisoning on five liver enzymes in killifish *(Fundulus heteroclitus). J. Fish. Res. Bd. Canada 27*:383-90.

Jensen, S. and A. Jernelöv. 1969. Biological methylation of mercury in aquatic organisms. *Nature 223*:753-54.

Karbe, L. 1972. Marine Hydroiden als Testorganismen zur Prüfungder Toxizität von Abwasserstoffen. Die Wirkung von Schwermetallen auf Kolonien von *Eirene viridula* [Marine hydroids as test organisms for assessing the toxicity of water pollutants. The effects of heavy metals on colonies of *Eirene viridula*]. *Mar. Biol. 12:* 316-28.

Kennedy, W. A. and M. S. Smith. 1972. Sablefish culture—progress in 1971. *Fish. Res. Bd. Canada*, Tech. Report 309.

_____ and _____. 1973. Sablefish culture—progress in 1972. *Fish. Res. Bd. Canada*, Tech. Report 388.

Kihlström, J. E. and L. Hulth. 1972. The effect of phenylmercuric acetate upon the frequency of hatching of eggs from Zebra fish. *Bull. Environ. Contam. Toxicol.* 7:111-14.

_____, C. Lundberg, and L. Hulth. 1971. Number of eggs and young produced by Zebra fishes *(Brachydanio rerio*, Ham.-Buch.). Spawning in water containing small amounts of phenylmercuric acetate. *Environ. Res.* 4:355-59.

Kleerekoper, H. 1974. Effects of exposure to a subacute concentration of parathion on the interaction between chemoreception and water flow in fish. This volume.

_____, J. B. Waxman, and J. Matis. 1973. Interaction of temperature and copper ions as orienting stimuli in the locomotor behaviour of the goldfish *(Carassius auratus). J. Fish. Res. Bd. Canada 30*:725-28.

Knauer, G. A. and J. H. Martin. 1973. Seasonal variations of cadmium, copper, manganese, lead, and zinc in water and phytoplankton in Monterey Bay, California. *Limnol. Oceanog. 18*:597-604.

La Roche, G. 1972. Biological effects of short-term exposures to hazardous materials. In: *Control of Hazardous Material Spills*, pp. 199-206. Houston, Tex.: Environmental Protection Agency, University of Houston.

Lewis, A. G., P. H. Whitfield, and A. Ramnarine. 1972. Some particulate and soluble agents affecting the relationship between metal toxicity and organism survival in the calanoid copepod *Euchaeta japonica. Mar. Biol. 17*:215-21.

Lindahl, P. E. and E. Schwanbom. 1971a. A method for the detection and quantitative estimation of sublethal poisoning in fish. *Oikos 22*:210.

_____ and _____. 1971b. Rotary-flow technique as a means of sublethal poisoning in fish populations. *Oikos 22*:354-57.

Lloyd, R. 1960. The toxicity of zinc sulphate to rainbow trout. *Ann. Appl. Biol. 48*:84-94.

Lowman, F. G. 1960. Marine biological investigations at the Eniwetok Test Site. In: *Disposal of Radioactive Wastes*, Vol. 2, pp. 105-38. Vienna: International Atomic Energy Agency.

_____, T. R. Rice, and F. A. Richards. 1971. Accumulation and redistribution of radionuclides by marine organisms. In: *Radioactivity in the Marine Environment*, pp. 161-99. Washington, D. C.: National Academy of Sciences.

MacInnes, J. R. and F. P. Thurberg. 1973. Effects of metals on the behaviour and oxygen consumption of the mud snail. *Mar. Poll. Bull. 4*: 185-86.

MacLeod, J. C. and E. Pessah. 1973. Temperature effects on mercury accumulation, toxicity and metabolic rate in rainbow trout *(Salmo gairdneri). J. Fish. Res. Bd. Canada 30*:485-92.

Mandelli, E. F. 1969. The inhibitory effects of copper on marine phytoplankton. *Contrib. Mar. Sci. 14*:47-57.

MIT. 1970. *Man's Impact on the Global Environment— Assessment and Recommendations for Action. Report Study of Critical Environmental Problems (SCEP)*. Cambridge, Mass.: MIT Press.

May, D. R. and G. W. Brown, Jr. 1973. Allantoinase from *Eudistylia vancouveri* (Polychaete): activation and inhibition by heavy metals. In: *1972 Research in Fisheries Annual Report of the College of Fisheries*, pp. 66-67. Seattle, Wash.: University of Washington.

McIntyre, J. D. 1973. Toxicity of methylmercury for steelhead trout sperm. *Bull. Environ. Contam. Toxicol. 9*:98-99.

McKee, J. E. and H. W. Wolf (Eds.). 1973. *Water Quality Criteria*, 2nd ed. State Water Quality Control Board Publication 3-A. Sacramento, Calif.: The Resources Agency of California.

McKim, J. M., G. M. Christensen, and E. P. Hunt.
1970. · Changes in the blood of brook trout
Salvelinus fontinalis after short-term and long-
term exposure to copper. *J. Fish. Res. Bd.
Canada 27*:1883-89.

NAS. 1971a. *Radioactivity in the Marine Environment*.
Washington, D. C.: National Academy of Sciences.

NAS. 1971b. *Marine Environmental Quality. Suggested
Research Programs for Understanding Man's Effect
on the Oceans*. Washington, D. C.: National
Academy of Sciences.

NAS. 1974. *Research Needs in Water Quality Criteria,
1972*. Water Quality Criteria Committee of the
Environmental Studies Board. Washington, D. C.:
National Academy of Sciences-National Academy of
Engineering.

NSF. 1973. International Decade of Ocean Explora-
tion. Program Description. Office for the
International Decade of Ocean Exploration,
Second Report. National Science Foundation,
Washington, D. C.

North, W. J., G. C. Stephens, and B. B. North. 1972.
Marine algae and their relation to pollution
problems. In: *Marine Pollution and Sea Life*,
pp. 330-40, ed. by M. Ruivo. London: FAO,
Fishing News (Books) Ltd.

O'Hara, J. 1973a. The influence of temperature and
salinity on the toxicity of cadmium to fiddler
crab, *Uca pugilator*. *Fishery Bull. 71*:149-53.

_____. 1973b. Cadmium uptake by fiddler crabs
exposed to temperature and salinity stress.
J. Fish. Res. Bd. Canada 30:846-48.

Okubo, K. and T. Okubo. 1962. Study on the bioassay
method for the evaluation of water pollution.
II. Use of fertilized eggs of sea urchins and
bivalves. Bull. Tokai Regional Fish. Res.
Lab. No. 32, pp. 131-40.

Olafson, R. W. and J. A. J. Thompson. 1974. Isolation of heavy metal binding proteins from marine
vertebrates. *J. Fish. Res. Bd. Canada*.
In press.

Pentreath, R. J. 1973a. The accumulation from water of ^{65}Zn, ^{54}Mn, ^{58}Co and ^{59}Fe by the mussel *Mytilus edulis. J. Mar. Biol. Assn. U. K. 53*: 127-43.

_____. 1973b. The accumulation and retention of ^{65}Zn and ^{54}Mn by the plaice, *Pleuronectes platessa* L. *J. Exp. Mar. Biol. Ecol. 12*:1-18.

Pippy, J. H. C. and G. M. Hare. 1969. Relationship of river pollution to bacterial infection in salmon *(Salmo salar)* and suckers *(Catostomus commersoni). Trans. Amer. Fish. Soc. 98*:685-90.

Portmann, J. E. 1972a. Results of acute toxicity tests with marine organisms, using a standard method. In: *Marine Pollution and Sea Life*, pp. 212-17, ed. by M. Ruivo. London: FAO, Fishery News (Books) Ltd.

_____. 1972b. The levels of certain metals in fish from coastal waters around England and Wales. *Aquaculture 1*:91-96.

Preston, A. 1973a. Heavy metals in British waters. *Nature 242*:95-97.

_____. 1973b. Cadmium in the marine environment of the United Kingdom. *Mar. Poll. Bull. 4*: 105-107.

_____, D. F. Jeffries, J. W. R. Dutton, B. R. Harvey, and A. K. Steele. 1972. British Isles coastal waters: the concentrations of selected heavy metals in sea water, suspended matter and biological indicators—a pilot survey. *Environ. Pollut. 3*:69-82.

Pringle, B. H., D. E. Hissong, E. L. Katz, and S. T. Mulawka. 1968. Trace metal accumulation by estuarine molluscs. *J. Sanit. Engng. Div. Am. Soc. Civ. Engnrs. 94*:455-73.

Riley, J. P. and D. Taylor. 1972. The concentrations of cadmium, copper, iron, manganese, molybdenum, nickel, vanadium and zinc in part of the tropical North-East Atlantic Ocean. *Deep-Sea Res. 19*:307-17.

Sangalang, G. B. and M. J. O'Halloran. 1972. Cadmium-induced testicular injury and alterations of androgen synthesis in brook trout. *Nature 240*:470-71.

Servizi, J. A. and R. W. Gordon. 1974. Effects of selected heavy metals on early life of sockeye and pink salmon. Manuscript in preparation.

Seymour, A. H. and G. B. Lewis. 1964. Radionuclides of Columbia River origin in marine organisms, sediments and water collected from the coastal and offshore waters of Washington and Oregon, 1961-1963. USAEC Report UWFL-86. University of Washington, Seattle, Washington.

Skidmore, J. F. 1970. Respiration and osmoregulation in rainbow trout with gills damaged by zinc sulphate. *J. Exp. Biol. 52*:481-94.

Skinner, B. J. and K. K. Turekian. 1973. *Man and the Ocean*. Englewood Cliffs, N. J.: Prentice-Hall, Inc.

Sparks, R. E., J. Cairns, Jr., and A. G. Heath. 1972. The use of bluegill breathing rates to detect zinc. *Water Research 6*:895-911.

Spencer, D. W. and P. G. Brewer. 1969. The distribution of copper, zinc and nickel in sea water of the Gulf of Maine and the Sargasso Sea. *Geochim. Cosmochim. Acta 33*:325-39.

Sprague, J. B. 1968. Promising anti-pollutant: chelating agent NTA protects fish from copper and zinc. *Nature 220*:1345-46.

_____. 1969. Measurement of pollutant toxicity to fish. I. Bioassay methods for acute toxicity. *Water Research 3*:793-821.

_____. 1970. Measurement of pollutant toxicity to fish. II. Utilizing and applying bioassay results. *Water Research 4*:3-32.

_____. 1971. Measurement of pollutant toxicity to fish. III. Sublethal effects and "safe" concentrations. *Water Research 5*:245-66.

_____. 1973. The ABC's of pollutant bioassays using fish. In: *Biological Methods for the Assessment of Water Quality*, ASTM STP 528, pp. 6-30. American Society for Testing and Materials.

Spyker, J. M., S. B. Sparber, and A. M. Goldberg. 1972. Subtle consequences of methylmercury exposure: behavioral deviations in offspring of treated mothers. *Science 177*:621-23.

Suzuki, T. 1969. Neurological symptoms from concentration of mercury in the brain. In: *Chemical Fallout*, pp. 245-56, ed. by M. W. Miller and G. G. Berg. Springfield, Ill.: Charles C. Thomas, Publisher.

Topping, G. 1973. Heavy metals in shellfish from Scottish waters. *Aquaculture 1*:379-84.

Ui, J. 1972. A few coastal pollution problems in Japan. In: *The Changing Chemistry of the Oceans*, pp. 171-76, ed. by D. Dyrssen and D. Jagner. Stockholm: Almkvist and Wiksell.

U. K. 1971. Survey of mercury in food. Working Party on the Monitoring of Foodstuffs for Heavy Metals. First Report, Ministry of Agriculture, Fisheries and Food. H.M.S.O., London.

_____. 1972a. Survey of lead in food. Working Party on the Monitoring of Foodstuffs for Heavy Metals. Ministry of Agriculture, Fisheries and Food. H.M.S.O., London.

_____. 1972b. Final Act Inter-Governmental Conference on the Convention of the Dumping of Wastes at Sea, London, 13 November, 1972. H.M.S.O., London, Cmnd. 5169.

_____. 1973a. Survey of mercury in food: a supplementary report. Working Party on the Monitoring of Foodstuffs for Heavy Metals. Fourth Report. Ministry of Agriculture, Fisheries and Food, H.M.S.O., London.

_____. 1973b. Survey of cadmium in food. Working Party on the Monitoring of Foodstuffs for Heavy Metals. Fourth Report. Ministry of Agriculture, Fisheries and Food. H.M.S.O., London.

U. S. Department of Interior. 1970. Mercury in the Environment. Geological Survey Profess. Paper 713. U. S. Government Printing Office, Washington, D. C.

Van Nostrand Co. 1947. *Van Nostrand's Scientific Encyclopedia*, 2nd ed. New York: D. Van Nostrand Company, Inc.

Vernberg, W. B. and J. O'Hara. 1972. Temperature-salinity stress and mercury uptake in the fiddler crab *Uca pugilator*. *J. Fish. Res. Bd. Canada 29*: 1491-94.

Vink, G. J. 1972. Koper in vis [Copper in fish]. *TNO NIEUWS 27*:493-96.

Vinogradov, A. P. 1953. *The Elementary Chemical Composition of Marine Organisms*. Sears Foundation for Marine Research, Memoir 2. New Haven, Conn.: Yale University Press.

Warnick, S. L. and H. L. Bell, 1969. The acute toxicity of heavy metals to different species of aquatic insects. *J. Water Poll. Control Fed. 41*:280-84.

Warren, H. V. 1972. Geology and medicine. Western Miner.

Wilson, K. W. and P. M. Connor. 1971. The use of a continuous flow apparatus in the study of longer-term toxicity of heavy metals Inter-national Council for the Exploration of the Sea, C. M. 1971/E:8.

Woelke, C. E. 1967. Measurement of water quality with the Pacific oyster embryo bioassay. In: *Water Quality Criteria*, p. 112. ASTM STP 416.

_____. 1972. Development of a receiving water quality bioassay criterion based on the 48-hour Pacific oyster *(Crassostrea gigas)* embryo. Washington Department of Fisheries, Olympia, Washington, Tech. Report 9.

Yamagata, N. and I. Shigematsu. 1970. Cadmium pollution in perspective. *Bull. Inst. Public Health* 19:1-27.

ENZYME RESPONSES TO METALS IN FISH

EUGENE JACKIM

Environmental Protection Agency
National Marine Water Quality Laboratory
Narragansett, Rhode Island   02882

A major obstacle in evaluating the effects of
pollution and establishing criteria for a productive
environment is the inability to measure accurately
stress in fish and other lower aquatic animals. Medical research has provided clinicians with sensitive
tests for diagnosing human illness, and many of these
procedures have quickly spread to veterinary practice.
However, when dealing with fish and other marine life,
the problems become significantly different and the
available techniques cannot be readily adopted without
further experimentation.

Our exploratory work with enzymes is an initial
attempt at using enzymological techniques to evaluate
stress resulting from sublethal metal toxicity in
fish. The rationale for studying enzymes was to see
if they could be used as a biochemical barometer in
diagnosing sublethal metal toxicity or stress. This
has been successfully accomplished with phosphorous
and carbomate insecticides, and nerve gasses where
the toxins are known to bind directly to active sites
on cholinesterase, inhibiting this enzyme. This
occurs when the toxin is added *in vitro* as well as
with whole animal exposure (Cornish, 1971). Metals
can combine with enzymes in many ways, among which

are sulfhydryl binding, chelation and salt formation.
Finding an enzyme that would bind a metal at the
active site would be a chance or occurrence dependant
upon the kinds of functional groups available and
their arrangement within the molecule at this site.
Binding at remote locations on the enzyme molecule
will also influence activity in some way, which could
range from activation to complete inhibition.

There are many ways of looking at the influence
of metals on enzymes. Data on several approaches are
presented.

Generally, *Fundulus heteroclitus* (mummichog) was
the test animal, although codfish was used for *in
vitro* studies with ATPàse and winter flounder were
used in some of the studies on δamino levulinate
dehydrase. Several common enzymes were selected on
the basis of their involvement with metals. In preliminary experiments we observed the effects of
direct addition of toxic metal salts to enzyme preparations from unexposed fish. Then we exposed fish to
a toxic metal and measured enzyme activity in the
organs of the exposed animals. Most of the methods
have been published in detail (Jackim *et al.*, 1970;
Jackim, 1973). Methods for our ATPàse assay were not
published, but were based on those of Bunting *et al.*
(1963) using whole brain homogenates combined with a
phosphate assay procedure of Gibbs *et al.* (1965).
Other than the ATPàse assay, liver was used for all
assays. Kidney as well as liver were used for δamino
levulinate dehydrase. In most instances, enzyme
preparations consisted of 1:25 tissue to water
homogenates which were sonicated and centrifuged
lightly to remove traces of connective tissue. In
some instances, sucrose or buffer were used to prepare the homogenates.

Initially, fish were exposed to 96-hr TL-50
concentrations of toxic metals and the surviving
animals were sacrificed. Tissues from the desired
organs were then removed for assay. After evaluation
of the effects of acute exposure, lower concentrations
were selected for chronic exposure (between one-fourth

and one-tenth of the acute dose). The fish were subjected to these concentrations for periods of up to
5 months. In all cases, we compared enzyme activity
in exposed fish to that of comparable control fish.

Figure 1 shows the influence of direct addition
of 10^{-5}M Cd, Cu, Pb and Ag on enzymes preparations
in vitro, and also the influence of metals on several
enzymes of surviving animals after a 96-hr exposure
of whole fish. From this figure, one would conclude

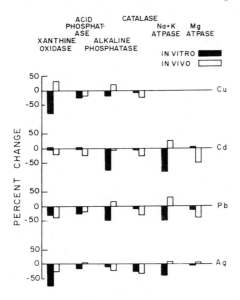

*Fig. 1. Solid bars
represent percent
change in enzyme
activity upon direct
addition of 10^{-5}M
metal. Line bars
show the mean percent
difference in enzyme
activity between
F. heteroclitus
exposed to toxic
metal for 96 hrs
and control fish.*

that there are no consistent relationships between
the direct *in vitro* effects of the metal on enzymes,
which are usually inhibitory, and the effect of
exposing the whole animal to the same metal. Beryl-
lium and mercury were also shown to influence the
activity of various enzymes by direct exposure and by
in vivo animal exposure.

The influence of the time of fish exposure to
metals on ribonuclease at pH 7.8 is seen in Figure 2,
which depicts a biphasic response. Figure 3 shows a
more pronounced initial increase when the ribonuclease
was measured at pH 5. Responses such as these are the
cause of many contradictory reports in the literature.

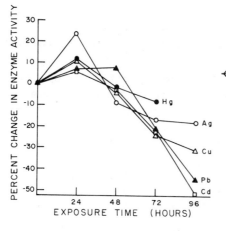

←Fig. 2. Effect of fish exposure time on Fundulus heteroclitus *liver ribonuclease activity measured at pH 7.8.*

Fig. 3. Influence of fish exposure time on ribonuclease activity measured at pH 5.→

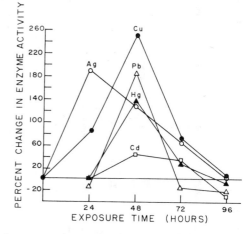

Depending upon its duration, the response for a single exposure time might result in a positive change, no change, or a negative change. Other factors, such as toxicant concentrations, salinity and temperature influence the time course of these responses. Therefore, one must always measure changes in enzyme activity as a function of exposure time as well as exposure concentration.

It would appear that some enzymes are excellent indicators of metal toxicity, however, when we reduce the 96-hr TL-50 concentrations by one-fourth to one-tenth and look at the enzyme activity after 30 days

and monthly up to 5 months, we find very little if
any change in the activity of most enzymes. Ribonuclease continued to be reduced by Cd and Cu. Catalase
also was slightly reduced by most metals. Most of
the other enzymes looked at in chronic studies were
unaffected. ATPàse was not looked at.

Lead markedly inhibited δamino levulinate dehydrase both in liver and kidney. Figure 4 shows the
effect of lead and mercury on this enzyme in exposed

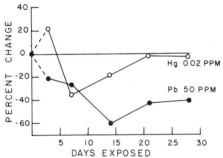

*Fig. 4.   Temporal
changes in δamino
levulinate dehydrase
activity upon exposure
of fish to lead and
mercury.*

fish. Although 50 ppm of lead was added to the fish
tank, only about 1 ppm remained in solution after
1 week. All fish survived the 28-day exposure with
no apparent outward damage. Therefore, this enzyme
is a promising indicator of sublethal lead toxicity.

Recent studies show the importance of salinity
in metals uptake. Table 1 shows that there is a very
significant increase in uptake of Cd with a decrease
in salinity. This should also be taken into consideration when working with other metals.

Changes in enzyme kinetics is another approach
which, although as yet untried, could potentially be
used. Exposure to metals might in some way alter the
enzyme and change its response to cofactors, temperature or pH optimum, or its Michaelis constant.

In our opinion, a practical assay for stress
should be simple enough to be conducted without
sophisticated instrumentation and highly trained
specialists. It should be reproducible under all or
most conditions and be sensitive enough to produce

TABLE 1

Whole Body Burden of Radiocadmium in Fundulus heteroclitus *after an 8-Day Exposure to 3000 CPM per ml of
Cd 115^m at Three Different Salinity Concentrations*

Salinity	Total Body CPM
31.5	1,600
16	33,000
0	50,000

changes of 50% or greater. It should be specific
enough so that minor external or internal changes
will not produce false responses. As yet, we have
not met this criterion.

In conclusion, I feel that enzyme research
should continue with regards to metals toxicology.
At present, however, we do not have a standard method
nor are we likely to get one without much more
research effort and a little serendipity.

LITERATURE CITED

Bunting, S. L. and L. L. Caravaggis. 1963. Studies
    on sodium-potassium-activated adenosinetriphos-
    phatase. V. Correlation of enzymes activity
    with cation flux in six tissues. *Arch. Biochem.
    and Biophysics 101*:37-46.
Cornish, H. H. 1971. Problems posed by observations
    of serum enzyme changes in toxicology. *Critical
    Review in Toxicology 1*:1-32.
Gibbs, R., P. M. Roddy, and E. Titus. 1965. Prepa-
    ration, assay, and properties of an Na$^+$- and K$^+$-
    requiring adenosine triphosphatase from beef
    brain. *J. Biol. Chem. 240*:2181-87.

Jackim, E.   1973.   Influence of lead and other metals
     on fish δamino-livulinate dehydrase activity.
     J. Fish. Res. Bd. Canada 30:560-62.
_____, J. M. Hamlin, and S. Sonis.   1970.
     Effects of metal poisoning on five liver enzymes
     in the killifish *(F. heteroclitus)*. *J. Fish.
     Res. Bd. Canada 27*:383-90.

EFFECTS OF SILVER ON OXYGEN
CONSUMPTION OF BIVALVES
AT VARIOUS SALINITIES

FREDERICK P. THURBERG, ANTHONY CALABRESE,
and MARGARET A. DAWSON

National Marine Fisheries Service
Middle Atlantic Coastal Fisheries Center
Milford Laboratory
Milford, Connecticut 06460

Heavy metals occur naturally in the marine environment and some are essential for normal growth
and development of marine organisms. Industrial pollution, in some cases, has elevated the natural
levels of such metals and a number of studies have
been conducted to assess the effect of such pollutants on living marine resources (Shuster and Pringle,
1968; Bryan, 1971; Ruivo, 1972; Eisler, 1973). Silver
is one of the most toxic heavy metals in the aquatic
environment (Bryan, 1971) and is used in jewelry,
silverware, photographic and food processing indus-
tries, as well as ink and antiseptic manufacture.
From such industrial sources, trace amounts of silver
can be expected to reach coastal waters (McKee and
Wolf, 1971). Long Island Sound, the area of concern
in this report, is fed by a number of rivers flowing
through heavily industrialized areas. Schultz and
Turekian (1965) noted that the silver content of
areas of this body of water is up to ten times higher

than surrounding western North Atlantic waters. Studies to ascertain the effects of silver on aquatic animals are limited and only a few of these have been conducted on molluscs (Eisler, 1973). MacInnes and Thurberg (1973) found that silver depressed whole-animal oxygen consumption in the mud snail, *Nassarius obsoletus*, and Harry and Aldrich (1963) reported behavioral changes in the gastropod, *Taphius glabratus*, after exposure to silver.

The present study was designed to evaluate the sublethal effects of silver on four bivalve species held at various salinities. Oxygen consumption was chosen as a parameter to assess stress because it is a valuable indicator of energy expended to meet the demands of an environmental alteration. A number of recent studies have emphasized the importance of synergistic effects of natural environmental factors and pollutants (Vernberg and O'Hara, 1972; Olson and Harrel, 1973; Thurberg, Dawson, and Collier, 1973). The physiological effect of a pollutant stress may be altered when such factors as temperature and salinity are varied. The bivalves used in this study are frequently found in areas of fluctuating salinity, thus the effects of silver at three different salinity levels were examined.

METHODS AND MATERIALS

The four bivalve species used in this study, the American oyster, *Crassostrea virginica*, the quahog or hard clam, *Mercenaria mercenaria*, the blue mussel, *Mytilus edulis*, and the soft-shell clam, *Mya arenaria*, were collected in Long Island Sound within 10 miles of Milford Harbor, Connecticut. Each species group consisted of animals of uniform size: *C. virginica*, 9.0 to 10.5 cm shell length; *M. mercenaria*, 8.5 to 10.0 cm; *M. edulis*, 5.8 to 6.4 cm; and *M. arenaria*, 6.0 to 7.0 cm. Separate groups of each species were exposed to silver for 96 hrs at salinities of 15, 25, and 35 o/oo. Seawater was obtained from Milford

Harbor and filtered through 36μ screens. Well water
or freeze-concentrated brine was added to this filtered water to attain the desired salinities. All
test animals were acclimated to the proper test salinity for 1 week prior to exposure to silver in 20-
gallon (75.5 liter) all-glass aquaria (four animals
per aquarium). An aqueous solution of silver, as
silver nitrate ($AgNO_3$), was added to each aquarium to
produce the calculated silver concentrations of 1.0,
0.5, 0.1 ppm. All concentrations in this report
refer to calculated ppm of the metal ion. The background level of silver in the seawater used in this
study was 0.0015 ppm; other background metal levels
were negligible. The seawater remained at a pH of
7.7 ± 0.1 and was not altered by the addition of
silver solution. The temperature remained at 20 ±
2°C. Each exposure consisted of 16 animals held at
one salinity: 4 at each silver concentration plus
4 controls. Each exposure, 96 hrs in length, was
repeated four times, for a total of 16 animals of
each species per silver concentration at each salinity. At the end of each exposure period, one valve
was removed from each animal and pieces of gill
tissue were dissected out and placed in 15 ml Warburg-
type flasks. Both gills from one side of each
M. arenaria and M. edulis and a single gill from each
C. virginica and M. mercenaria were used. Each flask
contained 5 ml of Millipore-filtered (0.47μ) silver-
treated water from the aquarium from which the test
animal was removed. Oxygen consumption was monitored
over a 4-hr period in a Gilson Differential Respirometer at 20°C. Oxygen consumption rates were calculated as microliters of oxygen consumed per hour per
milligram dry weight of gill tissue ($\mu l\ O_2/hr/mg$),
corrected to microliters of dry gas at standard temperature and pressure. Selected gill tissues were
analyzed for uptake of silver during the exposure
period. The tissues were dry ashed at 400°C, taken
up in 10% nitric acid and analyzed by atomic absorption spectrophotometry (Perkin-Elmer Model 403).

RESULTS

Crassostrea virginica

Results indicate no significant differences in oxygen consumption between control oysters held at 15, 25, and 35 o/oo salinities (Fig. 1); this is consistent with the finding of VanWinkle (1968).

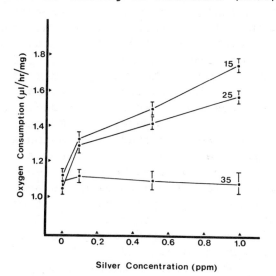

Fig. 1. Crassostrea virginica. *Effects of silver on oxygen consumption. Each point comprises the mean gill tissue oxygen consumption value of 16 animals. Bars: standard errors. Each line represents a series of experiments conducted at the salinity shown thereon.*

Sensitivity to silver did vary, however, with salinity. At 35 o/oo, animals exposed to silver concentrations up to 1.0 ppm respired at the same low rate as the controls. At 25 and 15 o/oo salinities there were significant elevations in the oxygen consumption

rates even at 0.1 ppm silver; such effects increased
with increasing metal concentration. At the two
lower salinities, analysis of variance tests indicated
significant differences in mean respiratory rates between controls and each silver treatment at the 0.01
level of significance.

Mercenaria mercenaria

The oxygen consumption rate of hard clams in the
control tanks increased with decreasing salinity
(Fig. 2), a finding consistent with the observations
of VanWinkle (1968). Oxygen consumption rates increased with increasing silver concentration in both

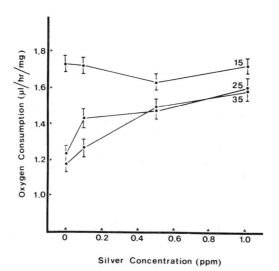

Fig. 2. Mercenaria mercenaria.
Effects of silver on oxygen con-
sumption. Each point comprises
the mean gill tissue oxygen con-
sumption value of 16 animals.
Bars: standard errors. Each line
represents a series of experiments
conducted at the salinity shown
thereon.

25 and 35 o/oo S. Analysis of variance tests at 25 and 35 o/oo salinities indicated significant difference in mean oxygen consumption rates between controls and all silver treatments at the 0.1 level of significance. Silver had no effect on the already highly elevated consumption values in tests conducted at 15 o/oo.

Mytilus edulis

Blue mussels exhibited higher oxygen consumption rates at a salinity of 15 o/oo than at 25 o/oo (Fig. 3). The rates increased with increasing silver concentration, an effect that was quantitatively similar at the two salinities. The animals used in

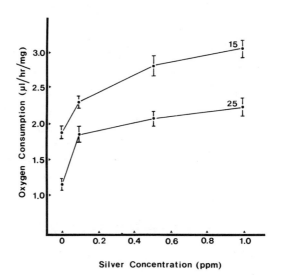

Fig. 3. Mytilus edulis. *Effects of silver on oxygen consumption. Each point comprises the mean gill tissue oxygen consumption value of 16 animals. Bars: standard errors. Each line represents a series of experiments conducted at the salinity shown thereon.*

this study were not able to adjust to 35 o/oo under
the conditions of this experiment and repeated mortality occurred. Analysis of variance tests indicated
significant differences in mean respiratory rates between controls and all treatments at the 0.01 level
of significance.

Mya arenaria

At 25 and 35 o/oo salinity, all of the soft-
shell clams died during exposure to 1.0 ppm silver.
Animals exposed to this concentration at 15 o/oo
showed the highest elevation in oxygen consumption
(Fig. 4). At 0.1 and 0.5 ppm silver, significant
differences (at the 0.01 level) were noted when

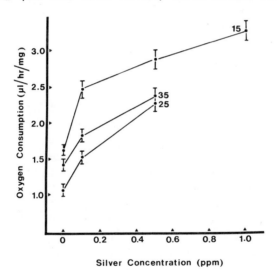

Fig. 4. Mya arenaria. *Effects
of silver on oxygen consumption.
Each point comprises the mean gill
tissue oxygen consumption value of
16 animals. Bars: standard errors.
Each line represents a series of
experiments conducted at the
salinity shown thereon.*

73

comparing these oxygen consumption values with those of the controls. The greatest elevation in oxygen consumption was noted in the experiments conducted at 15 o/oo.

TISSUE UPTAKE STUDIES

Randomly selected animals (12 to 20 per species) were prepared for chemical analysis to determine silver concentration in body tissues after exposure. These specimens were chosen from the 25 o/oo salinity experiments. Because of the large tissue mass in *C. virginica* and *M. mercenaria*, gills were separated from the whole body and analyzed separately. Only whole body analyses were conducted on the considerably smaller *M. edulis* and *M. arenaria*. The results for all four species are shown in Table 1.

DISCUSSION

Although the level of silver in certain areas of Long Island Sound is ten times higher than waters of the surrounding Western North Atlantic, a level which Bryan (1971) contends should stimulate active research on deleterious effects on marine life, literature concerning physiological effects of silver on marine organisms is surprisingly scarce. This study has demonstrated silver-induced alterations of oxygen consumption in four bivalve species.

The interpretation of metal-induced changes in respiration is complicated by the fact that such alterations differ from metal to metal, from species to species, and from one experimental condition to another. The effects of silver noted in this study are in marked contrast to those of copper and zinc. Silver altered the respiration in the majority of cases, and in each alteration an elevation in oxygen consumption was noted. Scott and Major (1972) reported that copper depressed respiration in *Mytilus edulis*. Working with the same species, Brown and

TABLE 1
Uptake of Silver Into Body and Gill Tissue after a 96-Hr Exposure

| Exposure (ppm Ag) | Silver Concentration in ppm Wet Weight | | | | | |
| | C. virginica | | M. mercenaria | | M. edulis | M. arenaria |
	Body	Gills	Body	Gills	Body	Body
Control	6.1	5.9	0.4	1.6	0.2	0.3
0.5	12.4	38.0	0.8	7.6	3.7	10.4
1.0	14.9	33.9	1.0	6.9	5.2	---

Newell (1972) also noted copper-induced depression in oxygen consumption while noting that zinc had no effect. MacInnes and Thurberg (1973) reported that silver depressed oxygen consumption in another mollusc, the gastropod, *Nassarius obsoletus*. The synergistic effects of environmental conditions in combination with a pollutant are also important. The effect of silver on bivalve oxygen consumption varied with salinity. Two species exhibited normal consumption values when exposed to silver at one salinity and highly elevated values when exposed at other salinities. The data concerning uptake of silver into body and gill tissues presented in this report are preliminary and were obtained at only one test salinity. They do, however, indicate how rapidly high levels of silver can accumulate in these animals. It would be of interest in future studies to determine whether uptake varies with salinity and whether such a variation can be related to changes in oxygen consumption.

Although much information is available on the uptake and accumulation of heavy metals by molluscs (Shuster and Pringle, 1968; Pentreath, 1973), very few studies have been conducted in detail on the physiological effects of metals on marine organisms, especially molluscs. Additional studies must be completed, including studies on the synergistic effects of various pollutants and the synergistic effects of pollutants and environmental factors. In addition, long-term studies using low levels of pollutants should be emphasized. These studies must be accomplished before a comprehensive assessment of heavy metal pollution can be obtained.

Note: The use of trade names is to facilitate description and does not imply endorsement by the National Marine Fisheries Service.

ACKNOWLEDGMENT

The authors wish to thank Mr. Richard A. Greig, Chemist, Environmental Microbiology and Chemistry

Investigation, Middle Atlantic Coastal Fisheries
Center, for conducting spectrophotometric analyses of
tissue samples.

LITERATURE CITED

Brown, B. and R. Newell. 1972. The effect of copper
    and zinc on the metabolism of the mussel *Mytilus
    edulis*. *Mar. Biol. 16*:108-18.
Bryan, G. 1971. The effects of heavy metals (other
    than mercury) on marine and estuarine organisms.
    Proc. Roy. Soc. London B 177:389-410.
Eisler, R. 1973. Annotated bibliography on biolog-
    ical effects of metals in aquatic environments.
    Ecological Research Series EPA-R3-73-007.
    Washington, D. C.: U. S. Government Printing
    Office.
Harry, H. and D. Aldrich. 1963. The distress syn-
    drome in *Taphius glabratus* (Say) as a reaction
    to toxic concentrations of inorganic ions.
    Malacologia 1:283-89.
MacInnes, J. and F. Thurberg. 1973. Effects of
    metals on the behaviour and oxygen consumption
    of the mud snail. *Mar. Poll. Bull. 4*:185-86.
McKee, J. and H. Wolf. 1971. Water quality criteria.
    Publs. Calif. St. Wat. Res. Control Bd. 3A:1-548.
Olson, K. and R. Harrel. 1973. Effect of salinity
    on acute toxicity of mercury, copper, and chro-
    mium for *Rangia cuneata* (Pelecypoda, Mactridae).
    Contr. Mar. Sci. 17:9-13.
Pentreath, R. 1973. The accumulation from water of
    ^{65}Zn, ^{54}Mn, ^{58}Co and ^{59}Fe by the mussel *Mytilus
    edulis*. *J. Mar. Biol. Ass. U. K. 53*:127-43.
Ruivo, M. 1972. *Marine Pollution and Sea Life*.
    London: Fishing News (Books) Ltd.
Schultz, D. and K. Turekian. 1965. The investiga-
    tion of the geographical and vertical distribu-
    tion of several trace elements in sea water
    using neutron activation analysis. *Biochim.
    Cosmochim. Acta. 20*:259-313.

Scott, D. and C. Major. 1972. The effect of copper (II) on survival, respiration, and heart rate in the common blue mussel, *Mytilus edulis*. *Biol. Bull.* *143*:679-88.

Shuster, C. and B. Pringle. 1968. Effects of trace metals on estuarine mollusks. *Proc. 1st Mid-Atlantic Industrial Waste Conf. Univ. Delaware, CE-5*:285-304.

Thurberg, F., M. Dawson, and R. Collier. 1973. Effects of copper and cadmium on osmoregulation and oxygen consumption in two species of estuarine crabs. *Mar. Biol.* *23*:171-75.

VanWinkle, W. 1968. The effects of season, temperature, and salinity on the oxygen consumption of bivalve gill tissue. *Comp. Biochem. Physiol.* *26*:69-80.

Vernberg, W. and J. O'Hara. 1972. Temperature-salinity stress and mercury uptake in the fiddler crab, *Uca pugilator*. *J. Fish. Res. Bd. Canada* *29*:1491-94.

THE CYCLING OF ZINC IN THE NEWPORT RIVER ESTUARY, NORTH CAROLINA

DOUGLAS A. WOLFE

National Oceanic and Atmospheric Administration
National Marine Fisheries Service
Atlantic Estuarine Fisheries Center
Beaufort, North Carolina 28516

Most metallic elements occur naturally in estuarine environments, and are generally classified as
pollutants only when added by man in quantities sufficient to produce deleterious effects on some desirable
attribute or product of the ecological system. Although physiological effects of pollutant metals on
estuarine organisms have been repeatedly documented
in acute toxicity experiments, such effects have
rarely been observed in natural systems, and public
concern over metals pollution frequently is not
aroused until the contaminants reach potentially
dangerous levels in organisms consumed as human food.

In the natural environment, organisms are exposed
chronically to sublethal concentrations of several
contaminants simultaneously, and the concentrations
of metals present within the organisms result from the
relative rates of metal accumulation and turnover.
These rates vary among species and also with the concentration and physicochemical form of the metal.
Assessment of potential effects of metal contaminants
on estuarine organisms must consider organismic

response not only to levels of metals in the environment, but also to contained metals. Furthermore, both the cycling of a particular metal and its toxicity characteristics may depend upon the physico-chemical form of the metal introduced into the system.

The effective concentration of a contaminant metal within an organism is affected also by numerous other factors, including ambient metal concentration, salinity, pH, temperature, feeding rate and composition of food. An understanding of the potential physiological effects resulting from contaminative metal additions to estuarine environments, therefore, requires knowledge of those complex physical and biological processes controlling distribution and turnover of metals in the component organisms and in the entire ecosystem. At our Center we have been studying metals cycling in estuaries and in estuarine organisms with the hopeful view of becoming able to predict radionuclide distributions in seafood organisms after contaminative additions of radioactivity. A similar approach will be necessary to understand the physiological effects of contaminant metal additions to estuaries. In this paper, the ecological budget of zinc in the Newport River estuary (Wolfe, 1974) is re-examined in terms of its sensitivity to the various parameters considered in its development and discussed with respect to recent research on other aspects of elemental cycling in estuaries.

THE NEWPORT RIVER ZINC MODEL

In a series of recent papers, we have presented background data on the distribution of trace transition metals in the salt marsh estuaries near Beaufort, North Carolina, and developed a preliminary assessment of the annual fluxes across the system boundaries, and between compartmentalized biological and abiotic components of the system, as typified by the Newport River estuary (Cross et al., 1970; Wolfe and Rice, 1972; Wolfe et al., 1973; Cross et al., 1974; Wolfe,

TABLE 1

Parameters Used in the Development of Zinc Budget of Figure 1

Parameter	Values[a] Used in Budget	Extremes[b] Considered Here	Flux[c] Estimates for Extreme Values		
			PD	PF	DF
Spartina production	5.5	2.7/11	21.1/26.6	32.6/27.1	36.3/41.8
Phytoplankton standing crop	0.1	.05/.2	23.2/22.4	30.5/31.3	38.4/37.6
Phytoplankton production	55	27/110	NEG/115	49.4/NEG	19.5/74.8
Zooplankton standing crop	0.004	.002/.008	23.7/21.3	30.5/31.4	38.5/37.5
Production/standing crop	.05/day	.025/.10	22.4/23.1	30.9/30.7	38.0/38.3
Biological half-life	7.6 days	3.8/15	22.2/23.3	30.8/30.7	38.1/38.2
Assimilation efficiency	20%	10/40	21.8/23.4	30.9/30.8	38.0/38.1
Macrofauna standing crop	0.97	.48/2.0	59.0/NEG	NEG/106	39.2/36.0
Production/standing crop	2/year	1.0/3.0	47.8/NEG	5.9/55.7	38.8/37.5
Biological half-life	300 days	150/600	2.0/33.5	51.7/20.2	37.7/38.5
Number trophic conversions	2 steps	1.25/1.75	74.7/57.5	NEG/NEG	38.2/38.2
Assimilation efficiency	20%	10/40	NEG/74.7	238/NEG	38.3/38.2
Net assimilation-microbiota plus meiofauna	40%	20/80	NEG/214	54.6/NEG	14.3/229
River flushing rate	3/year	1.5/6.0	23.2/22.4	30.7/31.0	38.3/38.1
Tidal flushing	-/year[d]	10/40[e]	21.7/16.5	31.3/33.5	37.7/35.6

[a] All values not otherwise specified are expressed in terms of mg Zn/m^2 or mg Zn/m^2 · yr.

[b] One-half and twice original values, except for macrofaunal production/standing crop ratio and mean number of trophic conversions within macrofauna.

[c] PD = phytoplankton to detritus; PF = phytoplankton to macrofauna; DF = detritus to macrofauna. NEG = negative fluxes and ecologically unfeasible budget.

[d] Not included in original budget.

[e] At P/SC of .05/day, zooplankton productivity cannot sustain this flushing rate.

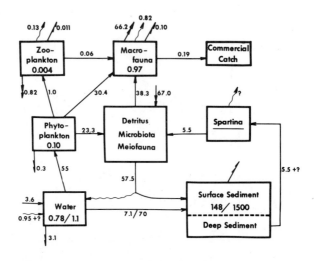

Fig. 1. Annual budget for Zn in the Newport River estuary, based on parameters identified in Table 1 and discussed by Wolfe (1974). Values inside compartments are mg Zn/m^2 for the 31 km^2 estuary at half-tide depth of 1.3 m. Values on arrows represent annual fluxes between compartments mg Zn/m^2 yr. One-sided arrows (⬈) represent flushing in the river volume only plus emigration; squiggly arrows (⬈) represent excretion of soluble metals to water; double-headed arrows (⬈) represent deposition of unassimilated metals to detritus as fecal matter.

1974). The Newport River estuary is a 31 km^2 embay-
ment which receives freshwater runoff from a watershed
of about 340 km^2 consisting principally of pine-
cypress pocosins, low-lying pine forest, and agricultural lands. At the mouth of Newport River estuary
is a complex network of channels, inlets, marshes and
mud flats where Bogue Sound and Back Sound meet just
inside of Beaufort Inlet, the major connection with
the Atlantic Ocean. Tidal amplitude is about 0.8 m
near the ocean inlets and somewhat less at the heads
of the estuaries.

A model was constructed for the zinc flux through
and within this estuary (Fig. 1), based on data gath-
ered at our Center during the past 6 years, and on
numerous specified assumptions concerning elemental
distributions, elemental turnovers, and biological
productivity (Wolfe, 1974). In this model we provided
for: (a) primary production by phytoplankton and
Spartina, (b) conversion of part of this production
to detritus by heterotrophic microorganisms,
(c) direct consumption of a portion of the phytoplankton production by zooplankton and herbivorous macro-
biota, and (d) consumption of detritus either directly
or indirectly after microbial assimilation. The
various parameters discussed by Wolfe (1974) and
summarized in Table 1 were assembled according to the
following series of equations to compute the zinc
fluxes between compartments shown in Figure 1.

Zinc incorporation by primary producers (Spartina
and phytoplankton) was calculated simply:

$$P^* = \text{primary production in g C/m}^2 \cdot \text{yr} \times \text{Zn content in mg Zn/g C} \tag{1}$$

and

$$S^* = \text{primary production in g dry wt/} \text{m}^2 \cdot \text{yr} \times \text{Zn content in mg Zn/g dry wt}$$

where, P^* and S^* are the amount of Zn incorporated
annually by phytoplankton and Spartina, respectively.

The zinc requirements of zooplankton (Z^*) and
macrofauna (F^*) were computed from estimates of production and zinc turnover rates as:

$$Z^* = \frac{(SC)_Z \ (P/SC)_Z + (SC)_Z \ \lambda_Z \ (365)}{A_Z} \qquad (2)$$

and

$$F^* = \frac{(SC)_F \ (P/SC)_F + (SC)_F \ \lambda_F \ (365)}{\overline{A}_F}$$

where:

(SC) = mean zinc content of standing crop of zooplankton or macrofauna

(P/SC) = ratio of annual production to mean standing crop for zooplankton or macrofauna

λ = fractional daily turnover coefficient which is related to the biological half-life by $\lambda = 0.693/T_{b\frac{1}{2}}$

A_Z = zinc assimilation efficiency for zooplankton

$\overline{A}_F$ = net zinc assimilation efficiency for macrofauna, dependent upon number of trophic steps involved within the component.

The annual flushing of phytoplankton (P_f) or zooplankton (Z_f) by the river volume was derived from the annual river flow volume, the volume of the estuary at midtide, and the zinc content of the standing crop of phytoplankton or zooplankton:

$$P_f = \frac{\text{annual river volume}}{\text{midtide volume of estuary}}$$

$$(SC)_P = 2.84 \ (SC)_P \qquad (3)$$

and

$$Z_f = 2.84 \ (SC)_Z$$

The returns of unassimilated zinc to detritus from either zooplankton (ZD) or macrofauna (FD) are available directly from the zooplankton and macrofauna zinc requirements (Z^* and F^*, respectively) and the respective assimilation efficiencies defined above:

$$ZD = Z^* \ (1 - A_Z) \quad \text{and} \quad FD = F^* \ (1 - \overline{A}_F) \qquad (4)$$

The zinc flux from zooplankton to macrofauna (ZF) was presumed to be the excess of the zooplankton zinc requirement above those amounts not assimilated (ZD), flushed from the estuary (Z_f), or returned to the water by biological turnover:

$$ZF = Z^* - ZD - Z_f - (SC)_z \lambda_z (365) \qquad (5)$$

Provided that the system is in steady state, i.e., the compartments show no net change, the three remaining fluxes interconnecting the phytoplankton, detritus, and macrofauna compartments were calculated from the parameters used in equations (1) through (5), by solving the following system of three simultaneous equations:

$$PF + DF = F^* - ZF \qquad (6)$$
$$PD + PF = P^* - Z^* - P_f \qquad (7)$$
$$(-\overline{A}_m)(PD) + DF = (\overline{A}_m)(FD + ZD + S^*) \qquad (8)$$

where:

PF = zinc flux from phytoplankton to macrofauna

DF = zinc flux from detritus to macrofauna

PD = zinc flux from phytoplankton to detritus

$\overline{A}_m$ = net assimilation efficiency of the microbiota and meiofauna inside the detritus compartment or net efficiency with which detrital zinc is available to macrofauna.

Other terms are as previously defined.

The release of zinc unassimilated by microbiota or meiofauna and therefore not transferred from the detritus to the macrofauna remains an ambiguity in the model. It is not known whether, or how much, soluble zinc is released directly to the water through microbial remineralization of detritus, or to what extent physical exchange processes in the interstices of the surface sediments may be involved. Since the intimate association of detritus, sediments and their attendant microflora and fauna has thus far precluded separation of these compartments and meaningful individual chemical analysis, the compartment labeled "surface sediment" includes all these indistinguishable reservoirs to an arbitrary depth of 2 cm. Although the quantity of zinc entering the detritus compartment in excess of the macrofaunal utilization of detrital zinc (DF) is shown as a flux from the detritus to the sediment and water, no attempt has been made to identify the ultimate disposition of this zinc.

This model suggested that zinc is highly conserved in this estuary, with the annual watershed inputs being recycled about 16 times per year through the biological compartments of the system and with very small net losses from the system due to the autumn emigration of migratory species, flushing of planktonic organisms and harvesting of commercial fishery species. Wolfe (1970) observed that fallout Zn-65 is very effectively retained by oysters (Crassostrea virginica) in this ecosystem and suggested that the radio-zinc cycling through the biological compartments of the system was not diluted by the large reservoirs of stable zinc already present in the water and sediments of the estuary or the adjacent ocean. This latter observation led to the conclusion that zinc must occur in more than one physicochemical form in the estuary, and that the form involved in biological cycling is probably distinct from that involved in sediment-water exchange.

Stability may be presumed to be a likely attribute of the zinc cycling system, as it is for many ecological systems, especially those involving detritus-based food webs (Odum, 1969). If this is the case, then the validity of the quantities assigned to the various system parameters (Table 1) can be tested through an analysis of how sensitive are the estimates for selected intercompartmental flows to changes in any of the other parameters. A stable model has at least one attribute of the real ecological system. Such a sensitivity analysis may provide added insight, therefore, on where our current understanding of the real system is most deficient, and thus guide future research priorities.

SENSITIVITY OF ZINC BUDGET TO VARIOUS PARAMETERS

The values assigned to various parameters to develop the Zn budget are summarized in Table 1, where I have also computed the response of the model to independently increasing or decreasing these values, one at a time, by a factor of 2. Exceptions were made in three cases: (a) the production/standing crop ratio for macrofauna where values exceeding 2/yr are rarely observed (Sanders, 1956); (b) the mean number of consecutive trophic conversions assumed to occur within the macrofauna compartment cannot exceed 2, since most of the organisms are filter-feeding or deposit-feeding herbivores (Wolfe *et al.*, 1973); and (c) tidal flushing. No value has been previously considered for (c). The flushing time of the Newport River estuary upstream of the Morehead City-Beaufort causeway has been estimated at 4.5 to 9.6 days, using a modified tidal prism method based on river flow, tidal range, and bathymetry of the estuary (Jennings *et al.*, 1970). Independent estimates, however, based also on movement of bottom drifters, suggested longer flushing times of about 20 days for the Newport River estuary (F. J. Schwartz, personal communication). A range of 9 to 36 days flushing time, or 10 to 40 times per year, was tested in the sensitivity analysis.

In Table 1, values are shown for the predominant fluxes around the phytoplankton and detritus compartments for each of the specified extremes of every parameter. Negative fluxes from phytoplankton to detritus (PD) and from phytoplankton to macrofauna (PF) are designated by NEG in the table and indicate reversal of the directions of the fluxes shown in the flow diagram (Fig. 1). Such reversals are ecologically impossible and, therefore, indicate high sensitivity of the model to change of that particular parameter. Negative values were obtained upon varying the following parameters: phytoplankton

87

production, macrofauna standing crop, macrofauna production/standing crop, mean number of consecutive trophic steps in the macrofauna, macrofaunal assimilation efficiency, and net assimilation in the detrital compartment. The model was also highly sensitive to changes in biological half-life of zinc turnover in macrofauna, although negative values of PD and PF were not obtained when the parameter was doubled or halved. The limiting value of PD and PF was approached particularly when turnover was increased, or when biological half-life was decreased to 150 days. The budget was relatively insensitive to changes in *Spartina* production, phytoplankton standing crop, any features of the zooplankton compartment and flushing rate of the estuary.

Since the parameters associated with the macrofauna compartment proved least flexible and were all interrelated, the quantities assigned to the three parameters were varied simultaneously over a wide range of values, to determine whether a more stable system was produced with different values. In Table 2, resultant values for zinc flux from phytoplankton to detritus (PD) and to macrofauna (PF) are shown over a range of microbiotic and net macrofaunal assimilation efficiencies. At the values used initially (net macrofaunal assimilation = 0.04, and microbiota assimilation = 0.4), both PF and PD are highly sensitive to change in either parameter and the model is feasible only over a narrow range. Blanks in this table represent unfeasible (negative) values. At a net macrofaunal assimilation of 0.02, PD was predicted negative for all values shown for microbiotic assimilation. Maximum stability for PD and PF (the region in which PD and PF changed the least per unit change of either parameter) prevailed at net macrofaunal assimilation of about 0.08 to 0.12 and at low microbiotic assimilation (0.01 to 0.10).

If zinc assimilation efficiency for each macrofaunal consumption is assumed to be 15 to 20% (Wolfe, 1974), then 1.25 to 1.75 consecutive consumption steps within the macrofauna are required to produce a net

TABLE 2

Values for Flux or Zinc from Phytoplankton to Detritus (PD) and to Macrofauna (PF) Predicted for Various Combinations of Effective Assimilation by Macrofauna and Detrital Microbiota. Values are PD/PF

| Assimilation | Net Macrofaunal Assimilation | | | | | |
	.04	.06	.08	.12	.16	.32
.01		8.3/45.4	19.8/33.9	31.3/22.4	37.1/16.6	45.7/8.0
.05		10.8/42.9	22.3/31.4	33.8/19.9	39.5/14.2	48.2/5.5
.10		14.1/39.6	25.6/28.1	37.1/16.6	42.9/10.8	51.5/2.2
.20		22.1/31.6	33.6/20.1	45.1/ 8.6	50.8/ 2.9	
.30	9.3/44.4	32.3/21.4	43.8/ 9.9			
.40	23.3/30.4	45.9/ 7.8				
.50	42.1/11.6					

macrofaunal assimilation of 8 to 12% (Table 3). For example, from Table 3 net macrofaunal assimilation of 0.0863 occurs when there are 1.5 consecutive steps at 0.15 assimilation efficiency. When these values and net microbiotic assimilation of 0.10 are substituted for the values used initially in the model, the total flux of zinc into the macrofauna is reduced considerably and the budget is modified as shown in Figure 2.

TABLE 3
Effective Macrofaunal Assimilation as Affected by Assimilation Efficiency and Mean Number of Trophic Conversions Within the Macrofauna Compartment

| Assimilation | Mean Trophic Level | | | | |
Efficiency (%)	1.0	1.25	1.5	1.75	2.0
5	.0500	.0383	.0263	.0144	.0025
10	.1000	.0775	.0550	.0325	.0100
15	.1500	.1183	.0863	.0544	.0225
20	.2000	.1600	.1200	.0800	.0400
25	.2500	.2032	.1563	.1094	.0625
30	.3000	.2475	.1950	.1425	.0900
35	.3500	.2932	.2363	.1794	.1225

The sensitivity of PD and PF to changes in other parameters was retested under these new conditions, i.e., 15% macrofaunal assimilation efficiency, 1.5 trophic steps in macrofauna, and 0.10 microbiotic assimilation (Table 4). Only those parameters to which the model was previously found highly sensitive

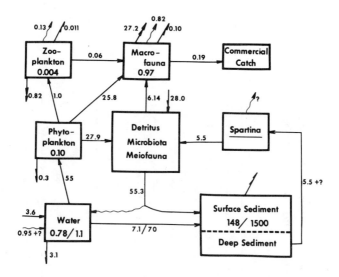

Fig. 2. Annual budget for Zn in the Newport River estuary derived as in Figure 1, except that different estimates of net macrofauna assimilation efficiency (0.086) and microbiotic assimilation (0.10) were used.

are listed in Table 4. Without exception, the sensitivity of PD and PF to changes in these parameters decreased under the new set of net assimilation efficiencies. Unfeasible negative flux estimates were still obtained, however, upon halving or doubling certain parameters. The model remains highly sensitive to phytoplankton production, but is feasible over a wide range of about 35 to more than 100 mg Zn/$m^2 \cdot$ yr (Table 4). This parameter probably is stabilized in the ecosystem by a variable zinc concentration in phytoplankton—i.e., when phytoplankton productivity is large, the zinc content per cell may be reduced and when productivity is small, zinc content of cells may be high (Rice and Ferguson, 1974). Thus, the total amount of zinc "produced" annually probably has a narrower range of homeostasis than does production of fixed carbon by phytoplankton.

91

TABLE 4

Reappraisal of Sensitivity of Model under Revised Conditions of Net Macrofauna Assimilation Efficiency and Net Microbial Assimilation Efficiency

Parameter	Original Values[a]	Revised Values[b]	Extremes[c] Considered Here	Flux[d] Estimates for Extreme Values		
				PD	PF	DF
Phytoplankton production	55	55	27/110	NEG/89	28.9/19.7	3.0/12.2
			35/86	5.9/62.6	27.8/22.1	4.1/9.8
Macrofauna standing crop	0.97	0.97	0.48/2.0	43.2/NEG	10.6/65.4	6.4/5.5
Production/standing crop	2/year	2/year	1.0/3.0	39.5/13.1	14.2/40.6	6.5/6.0
Biological half-life	300 days	300 days	150/600	18.6/32.9	35.1/20.8	6.3/6.4
Number trophic conversions	2 steps	1.5 steps	1.25/1.75	36.8/9.40	16.9/44.3	6.4/6.4
Assimilation efficiency	20%	15%	7.5/30	NEG/46.0	62.0/7.1	6.4/6.4
Net assimilation-microbiota plus meiofauna	40%	10%	5/20	24.8/36.1	38.9/17.6	3.0/14.3

[a] From Table 1; used in Figure 1.

[b] Used in Figure 2.

[c] Approach same as used in Table 1; two ranges tested for phytoplankton production.

[d] PD = phytoplankton to detritus; PF = phytoplankton to macrofauna; DF = detritus to macrofauna; NEG = negative fluxes and ecologically unfeasible system.

When macrofaunal standing crop was halved, the resultant values for PD and PF were feasible, but PD still became negative when this standing crop was doubled (Table 4), suggesting that the estimates of macrofaunal standing crop, or of zinc concentration in these organisms, may already be near the upper limits of feasibility. Halving macrofaunal assimilation efficiency also produced a negative value for PD. It should be emphasized again, however, that the number of macrofaunal trophic conversions is coupled with assimilation efficiency (Table 3), and feasible estimates for PD and PF are produced over a wide range of net macrofaunal assimilation (about 0.05 to 0.20).

If the premise is valid that stability is a likely characteristic of metals flow in the estuarine ecosystem, then those parameters to which the model is most sensitive require further investigation into their accuracy. Of the parameters listed in Table 2, we have attempted so far to gain further insight into microbial transformation of detritus and macrofaunal assimilation. These are discussed in the following sections.

ZINC IN EELGRASS, BENTHIC ALGAE AND DETRITUS

To examine the hypothesized losses of zinc from the detritus compartment to either the sediment or the water compartments due to microbiotic turnover, we analyzed zinc in samples of "detritus," *Zostera*, and various benthic algae, all from a single eelgrass bed in the Newport River estuary (Table 5). "Detritus" was defined as that material separated from sand, silt and clay by sieving through a 1.2 mm screen, after bits of green or red algae, fragments of green eelgrass leaves and live roots and rhizomes were removed by handpicking (Thayer *et al.*, 1974). This "detritus" fraction contained on the average only 60% of the mean zinc level of live eelgrass, and was considerably less variable in zinc content. Smaller particle sizes of detritus, which have undergone more extensive microbial decomposition, cannot readily be separated from contaminating sediment materials with high zinc content, and, therefore, we cannot further test analytically the hypothesis that only 10% of detrital zinc is ultimately assimilated by microbiota and meiofauna and subsequently becomes available to macrofauna.

TABLE 5
Zinc Content of Eelgrass, Various Benthic Algae, and Detritus from the Phillips Island Eelgrass Bed in the Newport River Estuary

Genus	Number of Samples	Zinc Content (mg/g • dry wt.)	
		Range	Mean ± S.E.
Zostera	6	40–190	84 ± 24
Ectocarpus	4	28–210	90 ± 41
Gymnogongus	2	74–140	107 ± 33
Agardiella	3	25–80	47 ± 17
Gracilaria	2	32–34	33 ± 1
Dictyota	1		48
Dasya	2	57–64	60 ± 3
Ulva	2	29–72	50 ± 22
Enteromorpha	2	36–63	50 ± 13
Bryopsis	1		120
Detritus	13	27–96	49 ± 6

MACROFAUNAL ASSIMILATION OF ZINC

By combining data on the zinc content of ingested food, the rate of change in zinc body burden through growth, and the biological turnover of assimilated zinc, Cross *et al.* (1974) were able to estimate the role of juvenile fish in the cycle of zinc within the Newport River estuary. The zinc ingested varied widely among the three species studied (menhaden, spot and pinfish), depending mainly upon the content of inorganic material in the food. The zinc accumulated in body tissues and the estimates of biological turn-over were similar for the three species, however, and thus assimilation efficiency for zinc varied inversely with the zinc content of the food, ranging from 2% for spot to 36% for menhaden with the estimate for pinfish midway between. On the basis of this study, Cross *et al.* estimated the total Zn ingested and assimilated by the population of menhaden, spot and pinfish in the Newport River estuary during the summer months (May 1 to September 1) to be 1065 g/day ingested, 56 g/day assimilated and 1009 g/day defecated. These estimates can be compared to the values shown in Figure 1 or 2, after conversion to units of mg zinc/m^2 and integration over the summer growing period (considered to be 180 days, despite the 2 to 3 months duration for which data were obtained). On this basis the above figures become: 6.2 mg Zn/m^2 ingested, 0.33 mg Zn/m^2 assimilated, and 5.9 mg Zn/m^2 defecated (all per 180 days). This estimate of ingestion for three fish species represents 9% of the total macrofaunal zinc ingestion averaged over the year when calculated as shown in Figure 1, and 19% when calculated as in Figure 2. The correspondence between these values appears more reasonable when one considers that the macrofaunal biomass (and therefore ingestion) is much higher than average during the summer season. These three fish species were estimated (Wolfe *et al.*, 1973; Wolfe, 1974) to constitute more than 40% of the total nekton and epibenthic macrofaunal biomass in the estuary, or about 10% of the total of these plus benthic macrofauna. Insofar as the three species are concerned, agreement between the estimates of Figures 1 and 2 and the data of Cross *et al.* (1974) is thus quite good.

ECOSYSTEM ZINC TURNOVER

Turnover of zinc within the estuary can be discussed in terms of both an open and closed system. Based on the very slow turnover of ^{65}Zn from fallout in oysters, it was concluded (Wolfe, 1974) that the cycling of zinc through the biota of the Newport River estuary represented nearly a closed system, perhaps involving a physicochemical form of zinc distinct from that in the large sediment compartment of the system. In the preceding sensitivity evaluation, additional flushing of dissolved and suspended zinc in the water column was shown to exert little effect on the biological fluxes around phytoplankton and detritus.

If the estuary is considered a closed system, then turnover can be expressed as the faction of the exchangeable sediment zinc which is cycled annually through the biota of the system. For Figure 1, the sediments were considered only to a depth of 2 cm, and on this basis the annual phytoplankton and *Spartina* incorporation of zinc (60.5 mg Zn/m^2 yr) represents about 40% per year turnover. The system might more properly be considered to extend to a depth of 1 meter into the sediments—a depth to which *Spartina* roots penetrate (Pomeroy *et al.*, 1969) and thus the turnover would be about 0.8% per year. Thus, at least 100 years would be required to equilibrate any additions of zinc with the large reservoir represented by the deep sediment compartment. The surface sediments, on the other hand, are probably involved in a more rapid dynamic equilibrium with metals dissolved in the overlying and interstitial waters. At least the superficial sediments are maintained in close physical proximity to the dissolved metals through constant resuspension of fine silt,

clays, and detritus, and through burrowing activities
in the upper several centimeters of sediments. Zinc
turnover was estimated in a similar way for Card
Sound, a mangrove-turtlegrass estuary in southern
Florida (Segar, 1974). There the sediments average
about 15 cm depth over a hard rock substrate, and
annual biological turnover of Zn constituted about
9% of the total element in the ecosystem.

The turnover of Zn in the system can also be
discussed as an open system, in terms of the proportion of the total zinc in the annual biological cycle
which is exported annually from the system. Thus,
the combination of commercial catch (0.19 mg Zn/m^2 yr),
emigration of macrofauna (0.10) and flushing of
zooplankton (.011) and phytoplankton (0.3) accounts
for about 1% of the total Zn incorporated annually by
the primary producers *Spartina* and phytoplankton
(60.5 mg Zn/m^2 yr). This same estimate of biological
export represents about one-sixth of the annual
import in runoff from the watershed. Both considerations support the previous conclusion that the
Newport River estuary has a great retentive capacity
for Zn and approaches a closed system for biological
cycling of this metal.

LITERATURE CITED

Cross, F. A., T. W. Duke, and J. N. Willis. 1970.
    Biogeochemistry of trace elements in a coastal
    plain estuary: Distribution of manganese, iron
    and zinc in sediments, water and polychaetous
    worms. *Chesapeake Sci. 11*:221-34.
_____, J. N. Willis, L. H. Hardy, N. Y. Jones,
    and J. M. Lewis. 1974. Role of juvenile fish
    in cycling of Mn, Fe, Cu and Zn in a coastal
    plain estuary. In: *Proceedings of the Second
    International Estuarine Research Conference.*
    In press.

Jennings, C. D., J. N. Willis, III, and J. M. Lewis. 1970. Hydrography of the Newport River estuary; a preliminary report. In: *Center for Estuarine and Menhaden Research*, Annual Report to the Atomic Energy Commission, pp. 66-70.

Odum, E. P. 1969. The strategy of ecosystem development. *Science* 164:262-70.

Pomeroy, L. R., R. E. Johannes, E. P. Odum, and B. Roffman. 1969. The phosphorus and zinc cycles and productivity of a salt marsh. In: *Symposium on Radioecology*, USAEC CONF-670503, pp. 412-19, ed. by D. J. Nelson and F. C. Evans.

Rice, T. R. and R. L. Ferguson. 1974. Response of estuarine phytoplankton to environmental conditions. In: *Physiological Ecology of Estuarine Organisms*, ed. by F. J. Vernberg. In press.

Sanders, H. L. 1956. Oceanography of Long Island Sound, 1952-1954. X. The biology of marine bottom communities. *Bull. Bingham Oceanogr. Collect. Yale Univ.* 5:345-414.

Segar, D. 1974. Trace metal cycling in sub-tropical estuaries. In: *Proceedings of the Second International Estuarine Research Conference*. In press.

Thayer, G. W., S. M. Adams, and M. W. LaCroix. 1974. Structural and functional aspects of a recently established *Zostera marina* community. In: *Proceedings of the Second International Estuarine Research Conference*. In press.

Wolfe, D. A. 1970. Levels of stable Zn and ^{65}Zn in *Crassostrea virginica* from North Carolina. *J. Fish. Res. Bd. Canada* 27:47-57.

_____. 1974. Modeling the distribution and cycling of metallic elements in estuarine ecosystems. In: *Proceedings of the Second International Estuarine Research Conference*. In press.

_____, F. A. Cross, and C. D. Jennings. 1973. The flux of Mn, Fe, and Zn in an estuarine ecosystem. In: *Radioactive Contamination of the Marine Environment*, pp. 159-75.

_____ and T. R. Rice. 1972. Cycling of elements in estuaries. *Fish. Bull.* *70*:959-72.

METHYL MERCURY AND INORGANIC MERCURY: UPTAKE, DISTRIBUTION, AND EFFECT ON OSMOREGULATORY MECHANISMS IN FISHES

J. LARRY RENFRO, BODIL SCHMIDT-NIELSEN, DAVID MILLER,
DALE BENOS, and JONATHAN ALLEN

Mount Desert Island Biological Laboratory
Salsbury Cove, Maine 04672

Considerable accumulation of mercury has been found in many populations of marine and freshwater organisms. In some populations, the source of the mercury in the animals can be traced directly to industrial pollution of inland and coastal waters, while in others accumulation has taken place through the food chain and cannot easily be traced (Gaskin *et al.*, 1973; Kamps *et al.*, 1972; Rivers *et al.*, 1972). Relatively inert mercury compounds can undergo biotransformation in the bottom strata of streams, lakes, and seas producing methyl mercury (Wood *et al.* 1968; Jensen and Jernelov, 1969) which is readily accumulated by gill-breathing organisms (Hannerz, 1968).

Potentially, this accumulation poses a number of threats to man and to the ecosystem. The direct threats of mercury poisoning through consumption of fish has been highly publicized. The more indirect threats of gradual deterioration of animal life in coastal and marine waters has received much less attention.

Answers to many of the questions that we would like to address ourselves to are, for the most part, unknown. It is known that extremely low concentrations of inorganic mercury (DeCoursey and Vernberg, 1972) are acutely lethal to many marine larvae, but we do not know how many and how widespread are the areas in which such concentrations are encountered. It is also known that the acute lethal concentrations in the water are considerably higher for mature organisms (Vernberg and Vernberg, 1972), but the constant or intermittent exposure to sublethal concentrations of mercury either in water or in food leads to considerable accumulation in the gills, liver, kidney, and muscles. This accumulation, which is due to the fact that excretion of mercury is extremely slow, is deleterious to the animal but to what degree is not known.

It is possible, and even likely, that the steady increase in environmental mercury which has taken place during the last century as indicated from analyses of coastal sediments and arctic ice (Young *et al.*, 1973) will reach a limit that may cause a catastrophic deterioration of all life in the oceans.

"Where that limit is, how close we are to it, whether or not, as the limit is approached, the ecosystem will show a gradual deterioration or will suddenly deteriorate like the collapse of a house of cards, are all unknown" (*Marine Poll. Bull.*, 1973).

One of the numerous questions that needs to be answered in order to assess this potential danger is how the mercurials affect the physiological functions of organisms.

Aquatic organisms, freshwater or marine, maintain body fluid which differs in composition and osmolality from their environment and thus their osmoregulatory mechanisms, which operate primarily by ion transport across epithelial membranes, are essential for their health and survival. Some evidence indicates that mercurial compounds inhibit ion transport. This

evidence stems primarily from our knowledge that mercurial diuretics inhibit sodium reabsorption in the
renal tubules (Cafruny, 1968). Similarly, it has
been shown that the sodium uptake by the gills of
goldfish is inhibited by mercuric chloride (Meyer,
1952).

The present investigations were undertaken:
1. to evaluate the effect of two mercurial compounds,
   methyl mercury and mercuric chloride, upon sodium
   transport in two osmoregulatory organs in fish,
   i.e., gill epithelium and urinary bladder, and
   the effect upon the enzyme ATPase involved in
   sodium transport.
2. to determine distribution of these two mercurial
   compounds in the organism and metabolism of the
   organic form.
3. to determine if the ionic regulation of the
   intracellular compartments was affected by mercury
   accumulation.

METHODS

SODIUM UPTAKE AND PASSIVE EFFLUX BY THE FRESHWATER
ADAPTED TELEOST GILL

Net sodium uptake was determined in *Fundulus
heteroclitus*, weighing 5 to 10 gm, depleted of sodium
for 4 to 6 days in deionized, distilled water. Non-
depleted fish had an average plasma sodium and potassium level of 178 ± 2.9 (S.E.) (n = 10) and 6.60 ±
0.37 (S.E.) (n = 10) mEq/l, respectively. Depleted
fish averaged 126.2 ± 3.9 (S.E.) (n = 10) and 10.18 ±
1.04 (S.E.) (n = 10) mEq/l, respectively. Experimental animals underwent 24-hr exposure to 0.125 ppm
(parts per million; $4.6 \times 10^{-6}M$) concentration of
mercuric chloride or methyl mercury chloride ($4.98 \times
10^{-6}M$) in 700 ml of deionized water just prior to
introduction into the mercury-free uptake medium.
Control animals were treated identically except for
mercury exposure. The uptake medium consisted of

400 ml of a solution containing the following mEq/l:
0.5 Na, 0.1 K, 0.1 Cl, 0.45 HCO_3, and 0.025 H_2PO_4.
This solution was aerated and circulated past a continuously recording sodium electrode (Orion Corp.).
The electrode was periodically calibrated by the use
of flame-photometry (Instrumentation Laboratories,
Inc., Model 343).

To determine if the mercurial compounds increased
sodium loss by causing damage to gills or other epithelia, unidirectional efflux of sodium was determined
on *Fundulus* exposed to 0.18 ppm (6.67×10^{-6}M) of
either mercuric chloride or methyl mercury chloride
(7.17×10^{-6}M) for a total of 8 hrs. Each fish was
injected intraperitoneally with 4.0 μCi (40 μl) of a
solution of ^{22}NaCl and placed in a 50 ml bath. 200 μl
samples of the bath were taken at 2-hr intervals for
24 hrs. The rate constant for sodium turnover and
the efflux rate of sodium were determined by the
method of Motais (1967).

*SODIUM REABSORPTION BY THE ISOLATED TELEOST URINARY
BLADDER*

Marine teleosts, *Pseudopleuronectes americanus*,
weighing about 200 gm, were used in this portion of
the study. Continuous perfusion of the isolated
urinary bladder was employed to maintain a constant
transepithelial electrical potential difference (P.D.).
PE 200 tubing was inserted into each end of the tube-
like bladder and fastened with silk suture. This
preparation was then suspended in a bath as shown in
Figure 1. Perfusion fluid was delivered by gravity
feed at a regulated flow rate. The incubation medium
was Forster's saline (Forster and Hong, 1958) modified
by omission of the biphosphate-bicarbonate buffer
system and substitution of 3 mM imidazole (pH = 7.8).
The external bath volume was exactly 4 ml of aerated
solution maintained at a constant temperature of 10 C.
Net sodium transport ($\mu M/cm^2 \cdot hr^{-1}$) was determined
from the difference in unidirectional influx and
efflux which were performed sequentially rather than

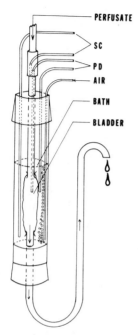

Fig. 1. *Schematic drawings of the apparatus used to perfuse the isolated urinary bladder. The bath volume was exactly 4 ml. SC = short-circuiting electrodes; P.D. = the potential difference electrodes.*

simultaneously with ^{22}Na. The external bath was renewed frequently to avoid back diffusion of the isotope. Two or three determinations of net sodium reabsorption were done on each bladder prior to treatment with one of the mercurial compounds. Because of individual variation among bladders, net sodium transport for each bladder is expressed as a percentage of the values obtained prior to mercurial treatment. Bladders were exposed serosally to either 0.4 mM (108 ppm) $HgCl_2$ or 0.4 mM (100 ppm) methyl mercury chloride for periods up to 110 min. Due to solubility characteristics, it was necessary to dissolve the methyl mercury chloride in 95% ethyl alcohol prior to dissolution in the incubation medium. The final concentration of ethanol in the medium was 1%. The effect of 1% ethanol alone on sodium transport by the isolated bladder was negligible. The average ratio of net sodium transport with ethanol to control fluxes measured on the same bladders was 1.017 ± 0.147 (S.E.) (n = 4). The ATPase activity of the bladders was assayed using the technique of Bonting (1970). The optimum conditions of H^+, Na^+, K^+, Mg^{++}, and ATP concentrations for this particular enzyme have been described (Miller and Renfro, 1973).

DISTRIBUTION OF $HgCl_2$ AND CH_3HgCl, AND METABOLISM OF CH_3HgCl IN THE KILLIFISH, FUNDULUS HETEROCLITUS

One group of fish was exposed to 0.084 ppm ($3.11 \times 10^{-6}M$) ^{203}Hg-labeled mercuric chloride, and another group was exposed to a similar amount of ^{203}Hg-labeled methyl mercury chloride ($3.36 \times 10^{-6}M$) for 24 hrs. At intervals ranging from 4 to 168 hrs, the animals were sacrificed, and the radioactivity of various tissues determined.

Two groups of killifish in identical environments were each exposed for 24 hrs to methyl mercury. One tank contained ^{14}C-labeled methyl mercury, and the other contained ^{203}Hg-labeled methyl mercury. Fish were retrieved from each tank simultaneously at regular intervals up to 117 hrs after initial exposure, and the radioactivity in the whole body of each fish determined. The ratio of the amount of the ^{14}C-label to the amount of ^{203}Hg label was determined for each time period.

DETERMINATION OF INTRACELLULAR ION CONCENTRATIONS AND TISSUE MERCURY LEVELS

American eels, *Anguilla rostrata*, were collected from estuaries of Mount Desert Island, Maine. The animals were maintained in the laboratory at a constant temperature of $10°C$, in either seawater or fresh water for at least one month prior to analysis. The technique of determining the intracellular potassium concentration of epaxial muscle tissue has been described (Schmidt-Nielsen *et al.*, 1973). The total mercury concentration of tissue samples was determined using the technique described by Gaskin *et al.* (1973) on an atomic absorption spectrophotometer (Perkin-Elmer, Model 107).

RESULTS

Na UPTAKE BY FRESHWATER TELEOST GILL

The ability of the sodium-depleted killifish to take up sodium was completely inhibited following 24-hr exposure to $4.6 \times 10^{-6}M$ mercuric chloride (Fig. 2). Exposure to methyl mercury ($5.0 \times 10^{-6}M$) had only a transient inhibitory effect upon sodium uptake (Fig. 2). Animals treated with either compound displayed an initial sodium loss. In mercuric chloride-treated fish, net sodium loss stopped after about

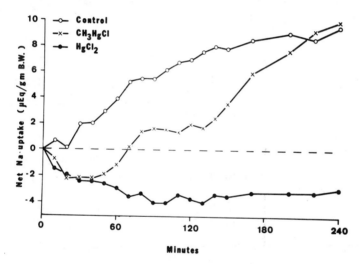

Fig. 2. The effects of methyl mercury and mercuric chloride on net Na uptake by sodium-depleted Fundulus heteroclitus *plotted against time. Negative values on the y-axis indicate Na loss.*

1 hr in mercury-free water, but the fish never showed net sodium uptake. Methyl mercury-treated fish began to take up sodium after about 30 min in mercury-free water and eventually took up as much sodium as the untreated fish.

Measurements of unidirectional sodium efflux in mercury-treated and in control fish showed no significant difference between the experimental groups and the control (Table 1). This finding indicated that the mercuric compounds did not increase the "leakiness" of the gills to sodium and the effect seen on the net sodium uptake must, therefore, have been due to a decrease in unidirectional sodium uptake.

TABLE 1

Rate Constants for Turnover of ^{22}Na, and the Unidirectional Efflux (ϕ_{Na}) of Na in Fundulus heteroclitus *Controls, and after 8 Hrs of $HgCl_2$ Exposure or CH_3HgCl Exposure*

	Control	$HgCl_2$	CH_3HgCl
Rate constant of Na efflux (% hr^{-1})	5.1 (5.0 and 5.1)	4.4 ± 0.8[b] (3)	4.2 ± 1.1 (3)
ϕ_{Na} (mEq/kg[a]· hr^{-1})	2.56 (2.59 and 2.54)	2.44 ± 0.47 (3)	1.99 ± 0.50 (3)

[a]Body wet weight.
[b]Mean ± S.E.M. (n).

PRELIMINARY STUDIES ON THE EFFECT OF MERCURIALS ON Na TRANSPORT BY THE ISOLATED TELEOST URINARY BLADDER

The urinary bladder of the teleost normally transports sodium from the mucosal to the serosal side. The rate of sodium transport does not decrease over a 2-hr period. The average net transport rate for the bladders used in this study prior to mercurial treatment was 1.034 ± 0.108 (S.E.) $\mu M/cm^2 \cdot hr^{-1}$ (n = 4).

However, when $HgCl_2$ $(4.0 \times 10^{-4}M)$ was added to the serosal bathing medium, net sodium transport progressively decreased (Fig. 3). No evidence of recovery

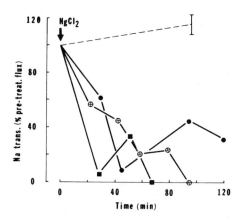

Fig. 3. The relationship of Na transport by the isolated, perfused urinary bladder of Pseudopleuronectes americanus *to time. The dashed line shows the average and standard error of three control bladder fluxes expressed as percentage of time zero flux. The percentage change in flux over a 96-min period with no treatment was negligible. The solid lines represent bladders treated with mercuric chloride at time zero (arrow). Fluxes are expressed as percent of the average of 2 or 3 pretreatment flux measurements on each bladder.*

from the effects of the inorganic mercury was seen up to 110 min after initial exposure. Bladders exposed to methyl mercury chloride $(4.0 \times 10^{-4}M)$ exhibited an immediate marked reduction in sodium reabsorption (Fig. 4). However, between 30 and 60 min after initial exposure each bladder showed a recovery phase. This response of the sodium transport mechanism to methyl mercury appeared to be somewhat similar to that shown by the freshwater teleost gill (Fig. 2).

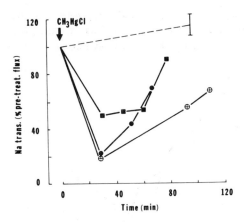

Fig. 4. The effect of methyl mercury chloride on Na transport by the isolated, perfused bladder of Pseudopleuronectes americanus. *See Figure 3 for explanation of graph.*

DISTRIBUTION OF MERCURIC CHLORIDE AND METHYL MERCURY CHLORIDE IN THE KILLIFISH, FUNDULUS HETEROCLITUS

A possible explanation for the different effects initially produced by these two mercurials may be obtained from examination of their distribution within the fish with time. The concentration of the mercury label of inorganic mercury remained highest in the gill tissue as long as 72 hrs after cessation of exposure (Table 2). There was a gradual redistribution of mercuric chloride primarily into liver and kidneys. Methyl mercury, however, redistributed more rapidly into liver and kidney tissues and was higher in these tissues after 48 hrs than in gill tissue (Table 2). Therefore, upon cessation of exposure of killifish to methyl mercury chloride, this substance may be mobilized out of the gills and into other tissues, thus allowing recovery of the sodium uptake mechanism.

In addition to being more mobile, the carbon-mercury bond of methyl mercury was apparently rapidly

TABLE 2

Distribution of $^{203}HgCl_2$ and CH_3 $^{203}HgCl_2$ in Various
Tissues of Fundulus heteroclitus in Fresh Water. The
Distribution Is Shown as the Ratio of the Concentration of ^{203}Hg in Each Tissue and the Concentration in
Gill Tissues

			Time	
HgCl$_2$	5 hrs	24 hrs	48 hrs	72 hrs
Gill	1.00	1.00	1.00	1.00
Liver	0.323	0.394	0.771	0.905
Kidney	0.167	0.493	0.560	0.700
Spleen	---	0.090	0.253	0.221
Muscle	0.015	0.021	0.038	0.038

CH$_3$HgCl	4 hrs	24 hrs	48 hrs	72 hrs	168 hrs
Gill	1.00	1.00	1.00	1.00	1.00
Liver	0.81	0.98	1.89	0.62	1.30
Kidney	0.45	0.17	0.85	2.93	0.70
Spleen	0.73	---	1.06	1.59	1.04
Muscle	0.07	0.20	0.15	0.07	0.18

split after uptake of the compound (Table 3). Comparison of the distribution of the methyl group of
methyl mercury with the mercury moiety of this mercurial showed that the ^{14}C label in the bodies of the
fish decreased with time more rapidly than did the
^{203}Hg label. As seen in Figure 5, the ^{203}Hg label of
methyl mercury remained high in the gill tissue for a

longer time than the ^{14}C label. This was an indication that methyl mercury is probably metabolized
within the animal, possibly resulting in the production of inorganic mercury.

TABLE 3
*The Tendency of the Methyl Group to Leave Methyl Mer-
cury Chloride after Uptake by* Fundulus heteroclitus
Is Shown. The Data Are Expressed as the Ratio of
[14]*C-CH$_3$HgCl to CH$_3$* [203]*HgCl in the Whole Body*

Time (Hrs)	$^{14}C/^{203}Hg$
5	1.72
20	1.31
45	1.26
69	1.35
117	1.07

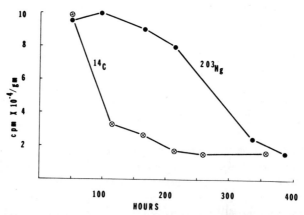

Fig. 5. *Time course for the disappearance
of* [14]*CH$_3$-Hg-Cl and CH$_3$-*[203]*Hg-Cl from the
gill tissue of* Fundulus heteroclitus.

EFFECTS OF MERCURIALS ON ATPase ACTIVITY OF THE
URINARY BLADDER

Attempts to determine the way in which mercurials
inhibit active sodium transport were based on the fact
that several ion transporting systems have displayed
a more or less direct dependence on Na-K-ATPase activity (Bonting, 1970). This has not been shown in the
freshwater fish gill, but sodium transport by the
isolated teleost urinary bladder is directly correlated (r = 0.81) with Na-K-ATPase activity (Miller
and Renfro, 1973). The results of tests on whole
bladder homogenates are shown in Table 4. Mercuric
chloride was given as an aqueous solution and can
thus be compared directly with the "no treatment"
category. This compound totally inhibited Na-K-ATPase
activity at all concentrations used. Since methyl
mercury was given in a 1% ethanol solution, the effect
of methyl mercury on ATPase activity should be compared to the effect of 1% ethanol. Even though
ethanol slightly reduced ATPase activity in homogenates, addition of methyl mercury produced a far
greater effect. About 80% of the Na-K-ATPase was
inhibited by the organic mercurial (Table 4).

MUSCLE INTRACELLULAR K CONCENTRATION IN RELATION TO
TISSUE Hg LEVELS

All cells regulate intracellular concentrations
of K and Na. K is pumped into the cells and Na out,
usually against electrochemical gradients. Na-K-
ATPase is necessary for the pump activity (Bonting,
1970). Because mercurials were shown to inhibit
sodium transport systems, and because both mercurials
have an inhibitory effect on Na-K-ATPase, we would
expect that the intracellular K level of tissues may
be influenced by the mercury concentration of the
tissue. We examined this relationship in epaxial
musculature of American eels, *A. rostrata*. The
results show that in seawater-acclimated eels there
does appear to be a relationship between intracellular

K levels and tissue mercury levels (Fig. 6). The intracellular K concentration decreases with increasing mercury accumulation in the muscle (correlation coefficient r = 0.71). The relationship was not clear in freshwater-acclimated fish (r = 0.42), but the fish containing the highest mercury levels did tend to have lower intracellular K concentrations.

TABLE 4

Effect of Various Treatments on ATPase Activity in Whole Bladder Homogenates of the Winter Flounder, Pseudopleuronectes americanus

	ATPase Activity (μM P_i/mg wet weight$\cdot hr^{-1}$)		
Treatment	Total	Mg	Na-K
None	0.406 ± 0.004[a] (5)	0.248 ± 0.008 (5)	0.152 ± 0.006 (5)
$HgCl_2$			
0.04 mM	0.178	0.181	0
0.20 mM	0.132	0.126	0.006
0.40 mM	0.134	0.152	0
EtOH 0.3%	0.372	0.234	0.138
0.6%	0.380	0.206	0.154
1.3%	0.334 ± 0.010 (3)	0.208 ± 0.010 (3)	0.128 ± 0.004 (3)
EtOH + CH_3HgCl			
0.3% + 0.05 mM	0.174	0.146	0.028
0.6% + 0.27 mM	0.148	0.126	0.018
1.3% + 0.53 mM	0.152	0.136	0.016

[a]Mean ± standard error (n).

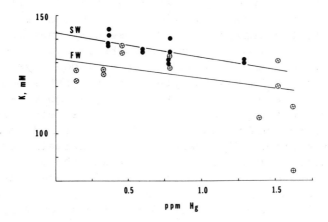

*Fig. 6. The relationship of muscle intra-
cellular K concentration of* Anguilla rostrata
*to endogenous tissue mercury concentrations.
The K concentration is expressed as mM/l cell
water. The correlation coefficient (r) in
seawater (SW) was 0.72. In fresh water,
r = 0.42. The lines through the points are
the regression lines.*

DISCUSSION

Mercuric chloride and methyl mercury chloride
both depressed ion transport in several important
osmoregulatory systems of teleosts. Mercuric chloride
appeared to be a more effective inhibitor of sodium
transport than methyl mercury. Mercuric chloride, at
the concentrations used in this study, totally abol-
ished sodium transport ability possibly, in part, by
direct interference with Na-K-ATPase activity. This
effect persisted during the period of experimentation.
Methyl mercury had a transient inhibitory effect on
sodium transport in both the gill and urinary bladder.
It should be noted that in bladder experiments, the
mercurials were applied during the measurement period
so that we see both initial effects and recovery or

lack of it. In the experiments on gill, mercurials were applied and removed prior to measurements and, therefore, the full effect of these compounds is seen immediately. Also, sodium transport is expressed differently in the gill and urinary bladder systems. The data for gill sodium uptake are not rates but rather amounts of sodium uptake per gram body weight. This was necessary because the sodium-depleted fish were in a nonsteady state and the rate of sodium uptake was continuously changing as the animals replenished their sodium pool. In contrast, the data for sodium transport by the perfused bladder are rates (expressed as percentage). The bladders were in a steady state and sodium flux was affected only by the mercurials. Thus, the similarity between these two systems with regard to methyl mercury inhibition exists only in the fact that subsequent recovery occurred.

Both mercuric chloride and methyl mercury chloride inhibited Na-K-ATPase activity in whole bladder homogenates. Mercuric chloride was somewhat more effective in this regard. Although the perfused bladders and whole bladder homogenates were exposed to similar concentrations of the mercurials, it is not possible to say that they took up similar amounts of mercurial compound. The location of the mercurial within the tissue would be as important as its concentration. The only conclusion that can be drawn from this aspect of the experiment is that, even at the lowest concentrations used, the mercurials greatly reduced ATPase activity. Cafruny (1968) has pointed out the fact that many organic mercurials inhibit Na-K-ATPase in homogenized tissue preparations, but apparently some do not have the same effect within the intact tissue. Jones *et al.* (1965) showed that part of the kidney diuresis produced by mercurial diuretics may be due to inhibition of Na-K-ATPase. However, Burg and Green's (1973) recent work with mersalyl indicated that its primary effect was on chloride transport, not sodium transport. Additionally, they showed that mersalyl was effective in this

way as an intact molecule, not by release of inorganic
mercury. It is likely that mercuric chloride and
methyl mercury chloride act on sodium transport in
quite different ways. There is considerable evidence
to indicate that mercuric chloride and methyl mercury
chloride can cause structural damage to gill epithelia
and kidney tubules (Fowler, 1972; Olson et al., 1973).
If structural damage occurred in the studies reported
here, it did not show up as an increased passive
sodium loss by the fish. The very polar mercuric
chloride is not greatly soluble in the cellular membranes, but has a high affinity for all sulfhydryl
containing molecules which includes most enzymes
(Rothstein, 1970). These very strong bonds would
tend to greatly reduce the mobility of the mercury
within the animal. Therefore, tissues with the highest concentrations of enzymes (liver, gill, and
kidneys) would tend to accumulate greater amounts of
Hg. Also, for this reason many cellular processes in
addition to ion transport are probably inhibited by
Hg (Paterson and Usher, 1971). Nonpolar methyl mercury is highly soluble in the lipid of cellular membranes. However, the carbon-mercury bond of methyl
mercury is apparently slowly split by cellular processes (Clarkson, 1968). The charged mercury would
then prefer to bind with sulfhydryl groups. We did
not determine the concentration or location of mercury within the transporting epithelia.

Teleosts, in their natural environment, encoun-
tering methyl mercury either in solution or via their
food in a sublethal amount may experience a decrease
in sodium transport efficiency. The mercurial contaminant may express itself initially by decreasing
gill and renal sodium transport. This would lead to
some degree of disturbance of the constancy of the
composition of the animal's body fluids. The result
is a weakened animal, less able to compete.

Continuous accumulation of mercury may reduce
the ability of every cell in the body of the fish to
maintain proper intracellular ionic composition. We
have shown a possible correlation between maintenance

of cellular ion levels and endogenous mercury concentration in the American eel, and this gives weight to the argument that mercury at "sublethal" levels may be debilitating to wild animals. We measured only tissue mercury in our animals. The possibility exists that eels having high Hg levels had a past history of living in polluted waters. Thus, substances other than Hg may also be high in these animals, and it may be these substances which caused the decrease in intracellular K concentrations. However, mercury is a likely possibility in light of its effect on Na-K-ATPase. Regulation of intracellular ion concentrations places one of the largest demands for energy on the metabolic processes of the animal. Thus, the fitness of an animal will inevitably be affected by any increase in the burden on the ion-regulating system.

It should be noted that most of the mercury found in fishes is in the form of methyl mercury (Kamps *et al.*, 1972; Rivers *et al.*, 1972). However, this study, as well as others (Clarkson, 1968), indicated that transformation of organic mercurials could take place within animals and inorganic mercury may be one of the many possible byproducts of the metabolism of methyl mercury. Therefore, the danger of methyl mercury to osmoregulatory systems may not be its initial effects, which appear to be rapidly reversed, but rather the slow release of inorganic mercury which has a longer lasting effect on sodium transport processes.

Comparison of muscle mercury levels of eels caught in Northeast Creek, Mt. Desert Island, Maine, with levels in several freshwater teleosts (Table 5) shows that the levels in the eels are as high as those caught in heavily polluted lakes in Sweden and the United States. If mercury accumulation was, indeed, the cause of the lowered intracellular K concentration in the eels, we could expect that these other fishes may also experience a disruption of cellular ionic regulation. Even higher accumulations of mercury have been found in the muscle tissue of several marine species. However, in view of the

TABLE 5

Comparison of Tissue Mercury Levels of the Eel, Anguilla rostrata, with Several Freshwater Fishes

Animal	Location	Tissue	Total Hg ppm (Range and Average)	Authors
American Eel (SW)	Mt. Desert Island	Muscle	0.36 – 1.29 0.618	Present study
(FW)		Muscle	0.13 – 1.61 1.28	
Northern Pike (FW)	Sweden	Muscle	1.10 – 1.94 1.38	Kamps *et al.* (1972)
Northern Pike (FW)	Clay Lake	Whole fish	7.03 – 8.71 8.00	Uthe *et al.* (1971)
Northern Pike (FW)	Lake St. Clair	Muscle	0.85	Kamps *et al.* (1972)
Walleye Pike (FW)	Lake St. Clair	Muscle	2.51	Greig and Seagran (1971)
White Sucker (FW)	Lake St. Clair	Muscle	1.44	Greig and Seagran (1971)
Yellow Perch (FW)	Lake St. Clair	Muscle	1.50	Greig and Seagran (1971)

recent finding that selenium may have a protective effect against mercury in porpoises and seals (Koeman *et al.*, 1973), it will be difficult to assess these effects without a more thorough analysis of these tissues.

ACKNOWLEDGMENTS

This study was supported in part by NIH Grant AM15972 to Dr. Bodil Schmidt-Nielsen, USPHS Grant AM15973 to Dr. William B. Kinter, and NSF Grant GB28139 to the Mount Desert Island Biological Laboratory.

LITERATURE CITED

Anonymous. 1973. Mercury hazard. *Marine Poll. Bull.* *4*:129-30.

Bonting, S. L. 1970. Sodium-potassium activated adenosinetriphosphatase and cation transport. In: *Membranes and Ion Transport*, pp. 257-363, ed. by E. E. Bittar. New York: Wiley.

Burg, M. and N. Green. 1973. Effect of mersalyl on the thick ascending limb of Henle's loop. *Kidney Internat.* *4*:245-51.

Cafruny, E. J. 1968. The site and mechanism of action of mercurial diuretics. *Pharmacol. Rev.* *20*:89-116.

Clarkson, T. W. 1968. Biochemical aspects of Hg poisoning. *J. Occupational Med.* *10*:351.

DeCoursey, P. J. and W. B. Vernberg. 1972. Effect of mercury on survival, metabolism and behavior of larval *Uca pugilator* (Brachyura). *Oikos 23*: 241-47.

Forster, R. P. and S. K. Hong. 1958. *In vitro* transport of dyes by isolated renal tubules of the flounder as disclosed by direct visualization. Intracellular accumulation and transcellular movement. *J. Cell. Comp. Physiol.* *51*:259-27.

Fowler, B. A. 1972. Ultrastructural evidence for nephropathy induced by long-term exposure to small amounts of methyl mercury. *Science 175*: 780-81.

Gaskin, D. E., K. Ishida, and R. Frank. 1973. Mercury in harbour porpoises *(Phocoena phocoena)* from the Bay of Fundy region. *J. Fish. Res. Bd. Canada 29*:1644-46.

Greig, R. A. and H. L. Seagran. 1971. Survey of mercury concentrations in fish of Lakes St. Clair, Erie and Huron. Prepublication manuscript.

Hannerz, L. 1968. Experimental investigations on the accumulation of Hg in H_2O organisms. *Fish. Bd. of Sweden, Inst. of Freshwater Res., Drottningholm 48*:120-76.

Jensen, S. and A. Jernelov. 1969. Biological methylation of Hg in aquatic organisms. *Nature 223*:753-54.

Jones, V. D., G. Lockett, and E. J. Landon. 1965. A cellular action of mercurial diuretics. *J. Pharmacol. Exp. Therap. 147*:23-31.

Kamps, L. R., R. Carr, and H. Miller. 1972. Total mercury-monomethyl-mercury content of several species of fish. *Bull. Environ. Contam. Tox. 8*: 273-79.

Koeman, J. H., W. H. M. Peeters, C. H. M. Koudstaal-Hol, P. S. Tjioe, J. J. M. DeGoeij. 1973. Mercury-selenium correlations in marine mammals. *Nature 245*:385-86.

Meyer, D. K. 1952. Effects of mercuric ion on sodium movement through the gills of goldfish. *Fed. Proc 11*:107.

Miller, D. and J. L. Renfro. 1973. Preliminary characterization of Na-K-activated ATPase in the urinary bladder of *Pseudopleuronectes americanus* and its relation to sodium transport. Mt. Desert Island Biological Laboratory, *The Bulletin*. In press.

Motais, R. 1967. Les mecanismes d'échanges ioniques branchiaux chez les teleostéens. *Ann. Inst. Ocean 45*:1-83.

Olson, K. R., P. O. Fromm, and W. L. Frantz. 1973. Ultrastructural changes of rainbow trout gills exposed to methyl mercury or mercuric chloride. *Fed. Proc. 32*:261.

Paterson, R. A. and D. R. Usher. 1971. Acute toxicity of methyl mercury on glycolytic intermediates and adenine nucleotides of rat brain. *Life Sci. 10*:121-28.

Rivers, J. B., J. E. Pearson, and C. D. Shultz. 1972. Total and organic mercury in marine fish. *Bull. Environ. Contam. Tox. 5*:257-66.

Rothstein, A. 1970. Sulfhydryl groups in membrane structure and function. In: *Current Topics in Membranes and Transport*, pp. 135-76, ed. by Felix Bronner and Arnost Kleinzeller. New York: Academic Press.

Schmidt-Nielsen, B., J. L. Renfro, and B. J. Benos. 1973. Estimation of extracellular space and intracellular ion concentrations in osmoconformers, hypo- and hyperosmoregulators. Mt. Desert Island Biological Laboratory, *The Bulletin 12*: 99-104.

Uthe, J. F., F. A. J. Armstrong, and K. C. Tam. 1971. Determination of trace amounts of mercury in fish tissues: Results of a North American check sample study. *J. AOAC 54*:866-69.

Vernberg, W. B. and F. J. Vernberg. 1972. The synergistic effects of temperature, salinity, and mercury on survival and metabolism of the adult fiddler crab, *Uca pugilator*. *Fish. Bull. 70*:415-20.

Wood, J. M., F. F. Kennedy, and C. G. Rosen. 1968. Synthesis of methyl-Hg compounds by extracts of a methanogenic bacterium. *Nature 220*:173-74.

Young, D. R., J. N. Johnson, A. Soutar, and J. D. Isaacs. 1973. Mercury concentrations in dated varved marine sediments collected off southern California. *Nature 244*:273-75.

ADAPTATION OF AN ESTUARINE POLYCHAETE TO SEDIMENTS CONTAINING HIGH CONCENTRATIONS OF HEAVY METALS

G. W. BRYAN

Marine Biological Association Laboratory
Citadel Hill
Plymouth, England

The existence of metal-resistant strains of organisms is now fairly well known and has, for example, been described in land plants (Bradshaw, 1970), marine algae (Russell and Morris, 1970), and bacteria (Beppu and Arima, 1964). Recently, Bryan and Hummerstone (1971) discovered a copper-resistant population of the polychaete *Nereis diversicolor* O. F. Müller in an estuary contamination with mining wastes. There is reason to suppose that other species from the same area are also resistant. However, this paper will be confined to *Nereis diversicolor* and describes both the conditions under which metal-tolerant animals are found and some of the ways in which tolerance is achieved.

Nereis diversicolor is extremely common in the estuaries of Southwest England and lives in burrows in the intertidal sediments. It can tolerate a very wide range of salinities and is usually distributed from near the mouth of an estuary into regions where fresh water is sometimes encountered.

The geographical area from which the animals
were collected is shown in Figure 1 and is drained by
the estuaries of more than 20 small rivers. It is an

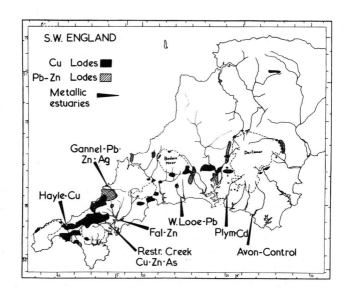

*Fig. 1. Map of Southwest England showing
estuaries referred to in text. Information
on lodes from Dines (1956).*

extremely metalliferous area and Figure 1 shows the
general positions of the main copper and lead-zinc
lodes. Other widely distributed metals are tin,
arsenic, and silver. Of more than 1000 mines in this
area which have been described by Dines (1956), only
a handful of tin mines and tin streaming operations
exist at the present time. But in its heyday in the
nineteenth century, the production of copper, tin,
and arsenic in this area sometimes represented 50% of
the world supply. Production declined rapidly toward
the end of the nineteenth century, but erosion of the
old mine dumps and drainage from the adits continues
at the present time. This is reflected in the concentrations of metals in the sediments of different
estuaries, which vary by as much as two orders of

magnitude. The highest concentrations we have encountered are about 3000 ppm of copper, 3000 ppm of
zinc, and over 5000 ppm of arsenic in Restronguet
Creek; over 8000 ppm of lead and about 20 ppm of
silver in the Gannel Estuary; and up to 20 ppm of
cadmium in the Plym Estuary. *Nereis diversicolor*
thrives in these metallic sediments and how this may
be achieved is the subject of this paper.

METHODS

The methods involved in this work have been described by Bryan and Hummerstone (1971, 1973a, 1973b)
and will not be repeated. Before analysis of the
whole worms they were kept in acid-washed sand covered with 50% seawater for 6 days and in 50% seawater
for 1 day further. With the exception of manganese
(Bryan and Hummerstone, 1973b), appreciable losses of
metals from the tissues do not appear to occur. This
cleaning process and all experimental work was carried
out at 13°C.

The dry weight of *Nereis* is 14 to 15% of the wet
weight and results on a dry weight basis may be converted to a wet basis by dividing by seven.

RESULTS

COPPER

Concentrations of copper in whole animals and
sediments from different estuaries are summarized in
Figure 2. Each point is an average value from an
individual estuary or part of an estuary and the line
drawn through the points is that of direct proportionality. The values range over two orders of magnitude
and concentrations in the worms are roughly proportional to those of the sediments.

Bryan and Hummerstone (1971) showed that if
animals were transferred from a low copper sediment

125

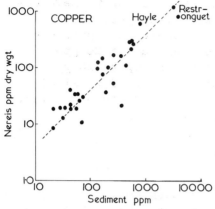

Fig. 2. *Relationship between concentrations of metal in whole* Nereis *and total concentrations in sediments. Broken line shows direct proportionality.*

to the high copper sediment from Restronguet Creek, they tended to die. Analyses of the dead worms showed that, although they had absorbed copper, they still contained far less than animals which normally live in the high copper sediment. It was suspected that animals from the high copper sediment were especially tolerant to copper, and this was confirmed by toxicity experiments. Table 1 shows that, in terms of the 96-hr LC50 concentrations, the tolerant worms are about four times more resistant than the nontolerant individuals. Since the sediments from Restronguet Creek also contain about 3000 ppm of

TABLE 1
Toxicity of Metals to Tolerant and Nontolerant Worms. Results Expressed as 96-Hr LC50 Concentrations in ppm as Determined in 50% Seawater at 13°C

Collection Site	Copper	Zinc
Restronguet Creek (tolerant)	2.3	94
Avon (nontolerant)	0.54	55

126

zinc, it is not unexpected that the copper-tolerant worms are also more resistant to zinc. However, tolerance to both metals need not occur simultaneously, and, in the Hayle Estuary, the worms are tolerant to copper only.

Analyses of interstitial water from the high copper sediments in Restronguet Creek showed that sometimes levels of copper were of the same order (~ 0.1 ppm) as the experimental threshold for toxicity in nontolerant worms. Therefore, it was not surprising that normal animals could be affected by high copper sediments. The presence of copper in the interstitial water suggested that in high copper sediments an appreciable amount is probably absorbed from solution across the body surface. This tends to be confirmed by histochemical staining with rubeanic acid, which showed that in high copper worms the metal is confined mainly to the epidermis and to parts of the nephridia.

From their early work, Bryan and Hummerstone (1971) concluded that copper-tolerant worms can withstand a greater intake into the tissues than nontolerant worms, although uncertainty remained about the relative permeabilities of the two types of animal. Using copper-64, rates of absorption of copper from media containing different concentrations have been measured. It might be expected that the tolerant worms would be less permeable than the nontolerant, but Figure 3 shows that the tolerant worms absorb copper more rapidly. The broken lines in Figure 3 show the external concentrations at which toxic effects subsequently occur and indicates that one of the toxic effects is to increase the permeability of the body surface to copper.

We are not yet certain why the tolerant animals are tolerant. Clearly they withstand much greater amounts of copper than normal worms and we think that they are better able to chelate and detoxify the metal as it enters the body. The more rapid absorption of copper by the tolerant worms suggests that the chelating capacity of the body surface may be

127

higher, and the fact that they can store so much copper in the epidermis tends to support this idea.

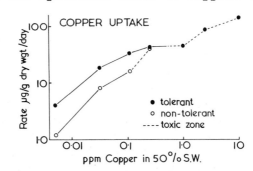

Fig. 3. *Rates of absorption of copper (as citrate) by tolerant (Restronguet Creek) and nontolerant (Avon) animals from different concentrations in 50% seawater. Broken lines represent external concentrations at which toxic effects occur.*

ZINC

The relationship between concentrations of zinc in the worms and those in the sediments is shown in Figure 4.

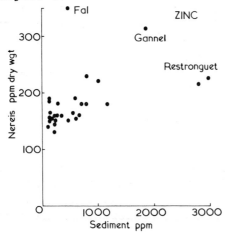

Fig. 4. *Relationship between concentrations of metal in whole* Nereis *and total concentrations in sediments.*

Concentrations in the worms increase with decreasing size, and the results in Figure 4 are for animals having an average dry weight of 0.05 g. Although the concentration of zinc in the sediment varies by a factor of 30 from 100 to 3000 ppm, that in the worms varies by less than a factor of 3. This suggests that its level of zinc is fairly accurately controlled by the animal. Three estuaries differ from the remainder in that the amount of zinc in interstitial water from the surface sediment exceeded 0.2 ppm, whereas in other areas it did not reach 0.1 ppm. In these three estuaries the concentrations in the interstitial water increased in the same order as the total concentration (Fal < Gannel < Restronguet Creek), but concentrations in the worms are in the reverse order (Fig. 4).

Worms from the most heavily contaminated sediments in Restronguet Creek contain little more zinc than those from uncontaminated sediments. Bryan and Hummerstone (1973a) have shown that this is because animals from the high zinc sediments are better able to handle zinc. First, they are about 30% less permeable to the metal than normal worms and, secondly, they are probably better able to excrete zinc. Because of these characteristics they are able to retain a relatively normal zinc concentration and are more resistant to the toxic effects of zinc than "normal" worms (Table 1). Evidence of adaptation also has been found in worms from the Gannel Estuary, but those from the Fal, which contain the highest concentration of zinc, do not appear to be adapted. Possibly contamination of the Fal Estuary has never been sufficiently high to promote the development of adapted animals.

CADMIUM

Bryan and Hummerstone (1973a) found that cadmium was not very toxic to *Nereis* and this may explain why in toxicity experiments no clear differences have been detected to date between different populations.

129

Animals from Restronguet Creek absorb zinc more
slowly than other populations and absorb cadmium
about 15% more slowly. This might confer on them
a slightly greater resistance to cadmium, but the
concentrations to which they are exposed in the field
are fairly low (Fig. 5).

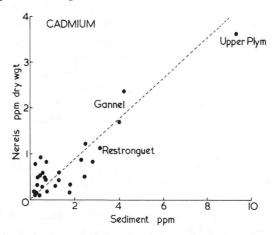

*Fig. 5. Relationship between con-
centrations of metal in whole
Nereis and total concentrations
in sediments. Broken line shows
direct proportionality.*

LEAD

Like copper, the amount of lead in the animals
is roughly proportional to that in the sediment
(Fig. 6). Although worms are found in sediments containing more than 8000 ppm of lead and themselves
contain about 1000 ppm, we have been unable to demonstrate that they are especially adapted to handle the
metal. For example, animals containing only a few
ppm of lead initially were able to absorb almost 2000
ppm from high lead sediments without being killed.

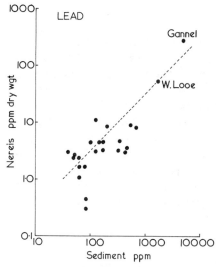

Fig. 6. *Relationship between concentrations of metal in whole* Nereis *and total concentrations in sediments. Broken line shows direct proportionality.*

SILVER

Concentrations of silver in the worms are roughly proportional to those in the sediments (Fig. 7), and the highest concentrations in the sediments are found in the Gannel Estuary and Restronguet Creek.

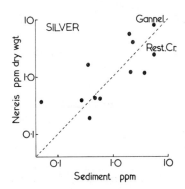

Fig. 7. *Relationship between concentrations of metal in whole* Nereis *and total concentrations in sediments. Broken line shows direct proportionality.*

Although problems of precipitation were encountered at concentrations lower than 1 ppm, preliminary toxicity experiments have indicated that the most resistant animals come from high silver sediments.

ARSENIC

Few analyses for arsenic have been made in either worms or sediments, but in Restronguet Creek more than 5000 ppm have been found in the sediment. Experiments on the toxicity of arsenate to worms from Restronguet Creek and the Avon Estuary failed to demonstrate any adaptation, since even at 100 ppm the worms were unaffected over a period of 17 days. However, experiments showed that arsenic-74 was absorbed more rapidly and to a greater degree by the worms from Restronguet Creek (Table 2). Whether or not, as in the case of copper, this is a sign of adaptation is not yet known.

TABLE 2
Rates of Absorption of Arsenic as Arsenate. Results Expressed as µg/g Dry Weight/Day (Mean of 4 Animals)

ppm As in 50% S.W.	Avon Worms	Restronguet Creek Worms
~ 0.001	0.0052	0.0090
1.0	1.76	4.96
10.0	11.6	17.2

DISCUSSION

Tolerance to heavy metals can be acquired during the life of an organism as has been demonstrated in fish (Lloyd, 1960; Pickering and Vigor, 1965; Edwards and Brown, 1967). This could account for some of the differences which we have observed between different populations, but, certainly in the case of copper, we

favor the idea that tolerant and nontolerant populations differ genetically. Attempts to change the
resistance of tolerant and nontolerant worms by exposing them to sediments of the opposite type were
unsuccessful (Bryan and Hummerstone, 1971), suggesting
that resistance is built into the animal. In land
plants, those which are tolerant to conditions on old
mine dumps differ genetically from nontolerant
individuals (Bradshaw, 1970).

Animals from Restronguet Creek, which is contaminated with all three metals, are resistant to
copper, zinc, and possibly arsenic. Adaptations to
copper and zinc appear to have developed separately
based on two reasons: First, they have different
characteristics and, secondly, worms from the Hayle
Estuary were resistant to copper but not to zinc.
Tolerance to zinc can be explained by a decrease in
the permeability of the body surface and probably to
improvements in the excretion of zinc. On the other
hand, tolerance to copper, and possibly silver and
arsenic, appears to depend on the presence of a better
detoxification system in the body surface and possibly in the excretory organs. Both types of processes
have been observed previously in metal-tolerance
studies. Decreased permeability has been described
as the mechanism of arsenite resistance in the
bacterium *Pseudomonas* (Beppu and Arima, 1964), and,
in land plants, zinc is much more strongly bound by
the cell walls in tolerant than in nontolerant individuals (Turner and Marshall, 1971).

If the tolerant populations of *Nereis* are
genetically different, we should like to know how
long it takes for resistance to evolve. Evidence
from land plants suggests that it may occur very
rapidly (Bradshaw, 1970), but in *Nereis* we can only
set the maximum limit. There has been plenty of time
for tolerant populations to develop, since mining
pollution has occurred for at least 200 years and was
probably at its worst about 100 years ago. In a contaminated estuary, the pressure to evolve almost
certainly develops in the less saline areas, because

metals are more toxic to the worms at low salinities when processes, such as those of ionic and osmotic regulation, become most vulnerable.

As a result of the development of metal tolerance in organisms, it is possible, at least after a long period of time, to have a quite heavily contaminated estuary where the distribution and diversity of species is fairly normal. However, in such an estuary, apparently normal organisms may contain two to three orders of magnitude higher concentrations of metals and these could be transmitted to man.

SUMMARY

The burrowing polychaete *Nereis diversicolor* O. F. Müller thrives in the sediments of estuaries in Southwest England, although some are contaminated with metalliferous mining wastes. Concentrations of Cu, Zn, Cd, Pb, Ag and As in the sediments vary by as much as two orders of magnitude. Relationships between the concentrations in the sediments and animals are described and related to the results of experimental work. In animals from the more metallic sediments, evidence of greater tolerance to the toxic effects of copper and zinc, and possibly silver and arsenic, has been found. Physiological differences between tolerant and nontolerant worms are described and the development of tolerant populations is discussed in relation to tolerance in other organisms.

LITERATURE CITED

Beppu, M. and K. Arima. 1964. Decreased permeability as the mechanism of arsenite resistance in *Pseudomonas pseudomallei*. *J. Bacteriol. 88*: 151-57.
Bradshaw, A. 1970. Pollution and plant evolution. *New Scientist 17*:497-500.

Bryan, G. W. and L. G. Hummerstone. 1971. Adaptation of the polychaete *Nereis diversicolor* to sediments containing high concentrations of heavy metals. I. General observations and adaptation to copper. *J. Mar. Biol. Ass. U. K.* *51*:845-63.

_____ and _____. 1973a. Adaptation of the polychaete *Nereis diversicolor* to estuarine sediments containing high concentrations of zinc and cadmium. *J. Mar. Biol. Ass. U. K. 53*:839-57.

_____ and _____. 1973b. Adaptation of the polychaete *Nereis diversicolor* to manganese in estuarine sediments. *J. Mar. Biol. Ass. U. K. 53*:859-72.

Dines, H. G. 1956. *The Metalliferous Mining Region of South-West England.* Vols. 1 and 2. London: H. M. Stationery Office.

Edwards, R. W. and V. M. Brown. 1967. Pollution and fisheries: a progress report. *Wat. Pollut. Control 66*:63-78.

Lloyd, R. 1960. The toxicity of zinc sulphate to rainbow trout. *Ann. Appl. Biol. 48*:84-94.

Pickering, Q. H. and W. N. Vigor. 1965. The acute toxicity of zinc to eggs and fry of the fathead minnow. *Progve Fish Cult. 27*:153-57.

Russell, G. and O. P. Morris. 1970. Copper tolerance in the marine fouling alga *Ectocarpus siliculosus*. *Nature, London 228*:288-89.

Turner, R. G. and C. Marshall. 1971. The accumulation of ^{65}Zn by root homogenates of zinc-tolerant and non-tolerant clones of *Agrostis tenuis* Sibth. *New Phytol. 70*:539-45.

IMPLICATIONS OF PESTICIDE RESIDUES
IN THE COASTAL ENVIRONMENT

THOMAS W. DUKE and DAVID P. DUMAS

U. S. Environmental Protection Agency
Gulf Breeze Environmental Research Laboratory
Sabine Island, Gulf Breeze, Florida   32561

Residues of pesticides occur in biological and
physical components of coastal and oceanic environments and some of the residues have been implicated
in degradation of portions of these environments.
The presence of many pesticides can be detected at
the parts-per-trillion level, but the effects of
such levels of pesticides on the organisms and systems in which they occur are not clear in many instances. Knowledge of these effects is especially
important when the residues occur in the coastal environment—a dynamic, highly productive system where
fresh water from rivers mingles with salt water from
the sea. The coastal zone interfaces with man's
activities on land and, therefore, is especially susceptible to exposure to acute doses of degradable
pesticides, as well as chronic doses of persistent
ones.
     This paper briefly reports the state-of-the-art
of research on the effects of pesticides on coastal
aquatic organisms. For a comprehensive review of
recent literature in this field, see Walsh (1972b);

for a compilation of data, see the EPA Report to the
States (1973).

Patterns of pesticide usage are changing in this
country and these changes are reflected in amounts of
various pesticides produced annually. Smaller amounts
of the organochlorine pesticides are being applied
because of their persistence in the environment, the
capability of organisms to concentrate them (bioconcentration) and their adverse effects on nontarget
organisms. For many uses, organophosphates and
carbamates have replaced organochlorines because
organophosphates and carbamates hydrolyze rapidly in
water and, therefore, are not accumulated to the same
extent as organochlorines. Some of the organophosphates, however, are extremely toxic to aquatic
organisms on a short-time basis (Coppage, 1972).
Much effort is being devoted to developing biological
control measures that will introduce viruses and
juvenile insect hormones into the environment as part
of an integrated pest control program. The integrated
pest control approach combines biological and chemical
methods to control pests in an effort to reduce the
amount of synthetic chemicals being added to the
environment. A list of several important pesticides
that are used currently or appear as residues in
marine organisms or both is presented in Table 1.

Samples collected in the National Estuarine
Monitoring Program and in other programs show that a
variety of pesticides occur in biota and nonliving
components of the marine environment. Pesticide
residues have been reported in whales from the Pacific
Ocean (Wolman and Wilson, 1970), fish from southern
California (Modin, 1969), invertebrates and fish from
the Gulf of Mexico (Giam et al., 1972), fish from
estuaries along the Gulf of Mexico (Hansen and Wilson,
1970), fauna in an Atlantic coast estuary (Woodwell
et al., 1967), zooplankton from the Atlantic Ocean
(Harvey et al., 1972), and shellfish from all three
coasts (Butler, 1973). These residues indicate that
pesticides can reach nontarget organisms in the
marine environment and give some indications of

TABLE 1

Toxic Organics Used as Pesticides or Appearing as Residues in Marine Organisms or Both

Organochlorines (Insecticides)	Organophosphates (Insecticides)	Carbamates (Insecticides)	Herbicides
Chlordane	Diazinon	Carbaryl	2,4-D
DDT	Guthion	Carbofuran	Picloram
Dieldrin	Malathion		Triazines
Endrin	Naled		Urea
Methoxychlor	Parathion		
Mirex	Phorate		
Toxaphene			

biological reservoirs of pesticides in this environment. The information obtained in these monitoring
programs is invaluable to those interested in manag-
ing our natural resources, but care must be exercised
in interpreting monitoring data.

Biological problems that affect the interpretation of monitoring data were discussed recently by
Butler (1974). Factors affecting persistent organochlorine residues include kind of species sampled,
age of individuals monitored, natural variations in
individuals, seasonal variation, and selection of
tissues to be analyzed. Laboratory experiments and
observations in the field have shown that filter-
feeding mollusks are good indicators of the presence
of organochlorine pesticides in estuarine waters.
These animals are sedentary, have the capacity to
concentrate the chemicals in their soft tissues many
times the concentration in the water and lose the
chemicals rather quickly when exposed to clean water.
Obviously, mollusks would be helpful in locating the
source of a particular organochlorine. Conversely,
pelagic fish might not be useful in locating a particular source because they could have accumulated a
residue some distance from the point of collection.

As patterns of pesticide usage change, techniques
for monitoring the occurrence of the pesticides also
must change. Occurrences of organophosphates, carbamates and biological control agents cannot be monitored in the same manner as occurrences of organochlorine and other more persistent chemicals. To
help identify the presence of a pesticide it may be
necessary to utilize changes in biological systems,
as opposed to routine chemical analyses of organisms
or other components of the environment. Also required
is a concomitant effort to understand the effect of
residues on the organisms and systems in which they
occur.

CONCEPT OF EFFECTS

The implications of pesticide residues in the marine and other environments depends upon the effect of the chemicals on the component in which they occur. A conceptual model of possible effects of pesticides and other toxic substances on biological systems is shown in Figure 1 (Dr. John Couch, Gulf Breeze Environmental Research Laboratory, Gulf Breeze, Florida, unpublished personal communication). The possible impact of a stressor on a biological system is described as the system changes from (1) a normal steady-state to (2) one of compensation to (3) decompensation to death. Accordingly, a pesticide could be considered to have an adverse effect if it temporarily or permanently altered the normal steady-state of a particular biological system to such a degree as to render the homeostatic (compensating) mechanism incapable of maintaining an acceptable altered steady-state.

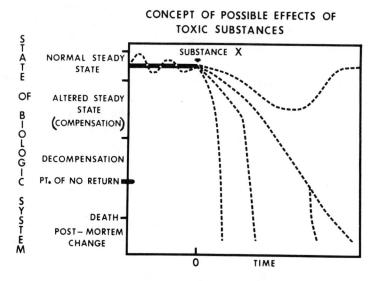

Fig. 1. *Concept of possible effects of toxic substances.*

NORMAL STEADY-STATE

It has been said that the most consistent trait of biological systems is their inconsistency. The normal steady-state of a particular biological system, therefore, is difficult to define. Each system, from an estuarine ecosystem to a system within individual organisms, has a natural range of variability in such factors as population density, species diversity, community metabolism, oxygen consumption, enzyme production, avoidance mechanisms, osmotic regulation, natural pathogens, and others. Obviously, much must be known about the normal or healthy system before an evaluation can be made of the effect of a pesticide on the system.

In relation to this, the impact of pesticides on ecosystems is poorly understood because often the "normal" system itself is poorly understood. An ecosystem can be considered a biological component that consists of all of the plants and animals interacting in a complex manner with their physical environment. The "normal" state of a dynamic coastal ecosystem no doubt depends upon the characteristics of a particular ecosystem, and changes as the system matures. The importance of symbiosis, nutrient conservation, and stability as a result of biological action in an estuarine ecosystem is pointed out by Odum (1969). According to Odum, in many instances, biological control of population and nutrient cycles prevents destructive oscillations within the system. Therefore, a pollutant that interferes with these biological actions could adversely affect the ecosystem.

ALTERED STEADY-STATE (COMPENSATION)

An acute dose of a pesticide could cause a biological system to oscillate outside its normal range of variation, yet with time, the system could return to the normal state without suffering lasting effects. An example of this phenomenon at the ecosystem level was demonstrated by Walsh, Miller, and Heitmuller

(1971), who introduced the herbicide dichlobenil into
a small pond on Santa Rosa Island. Applied as a wet-
table powder at a concentration of one part per
million, the herbicide eliminated the rooted plants
in the pond. As the benthic plants died, blooms of
phytoplankton and zooplankton occurred and a normal
oxygen regime was maintained. As benthic plants
returned, the number of plankters dropped. The pond
returned to a "normal" state in reference to the primary producers approximately 3 months after treatment.
A possible example of such compensation in an individual organism was shown recently when spot,
Leiostomus xanthurus, were exposed to Aroclor ®ª 1254
under laboratory conditions (Couch, 1974). Even
though in many fish no outward signs of stress were
present, the livers of the fish accumulated excess
fat during the tests. For a period of time, the
liver evidently was able to contend with excessive
fat accumulation, but eventually chronic damage leading to necrosis occurred; therefore, the fish entered
another biological state.

DECOMPENSATION TO DEATH

The effect of a stress can eventually reach the
point where the biological system can no longer compensate and death results. In the instance in which
Aroclor 1254 was related to fat globules in the liver
of fish, continued exposure to the chemical caused a
necrotic liver. Eventually, the test organisms died
as a result of the exposure. In the past, most of
the data upon which criteria and standards were based
used death as the criterion for effect. Much time
and effort now are being devoted to developing other
criteria, such as effects of relative concentrations
of the chemicals on tissue and cell structure, enzyme

ª ® Registered trademark: Aroclor, Monsanto Co.
Mention of commercial products does not constitute endorsement by the U. S. Environmental Protection Agency.

reaction, osmotic regulation, behavioral patterns, growth and reproduction.

ASSESSMENT OF EFFECTS

The concept just presented is helpful in visualizing the manner in which pesticides can affect coastal organisms and systems. However, quantitative information must be developed in order to assess the effect of a particular pesticide on the environment or on a component of the environment. For example, it is not enough to know that a pesticide causes an altered steady-state in a fish and eventually causes death. The level of pesticide in the environment that causes the effect must be known and, perhaps even more important, the level at which no effect occurs must be known.

Much of the quantitative information available on effects of pesticides on marine organisms is in terms of acute mortality of individual organisms. In many instances, these data were obtained through routine bioassay tests in which known amounts of pesticides are administered to test organisms for a given period of time. In routine bioassays, the test organisms are examined periodically and compared with control organisms. If conducted for a short time in relation to the life span of the organisms, usually 96 hrs, the tests are considered acute. Longer tests over some developmental stage or reproductive cycles are termed chronic. (An excellent discussion of bioassays and their usefulness is presented by Sprague (1969, 1970).)

Often, it is necessary to estimate the effect of a pesticide on the coastal environment from only a minimum amount of data. Interim guidelines sometimes must be issued on the basis of a few acute bioassays while more meaningful data are being obtained. An application factor is helpful in these instances. This factor is a numerical ratio of a safe concentration of a pesticide to the acutely lethal concentration

(LC$_{50}$). An estimate can be made for an "acceptable" level of a pesticide in marine waters by multiplying the LC$_{50}$ determined in acute bioassays by the appropriate application factor. In many instances, an arbitrary application factor of 0.01 is used when necessary scientific data have not yet been developed. For a discussion on obtaining the application factor experimentally, see Mount (1968) and Brungs (1969).

Information obtained by various bioassay tests on some toxic organics of current interest is shown in Table 2. These results were compiled from the literature and indicate the most sensitive organisms tested against these pesticides and organochlorines. The data give a general idea of the relative toxicity of the various pollutants.

During the past few years, the need for data on chronic or partial chronic exposures and on sublethal effects of pesticides on marine organisms has become evident. Chronic studies involve the exposure of organisms to a pesticide over an entire life cycle, and often are referred to as "egg-to-egg" studies. A subacute chronic is conducted over part of a life cycle. Sublethal studies are designed to determine if a pesticide has an effect at concentrations less than those that are lethal to the organisms and utilize such criteria as growth, function of enzyme systems, and behavior of populations of organisms.

EFFECT OF PESTICIDES ON GROWTH OF ORGANISMS

The effects of pesticides on marine phytoplankton are often related to growth of the organisms. The effects often vary according to the pesticide and to the species of phytoplankton. For example, Menzel *et al*. (1970) found that growth in cultures of marine phytoplankton was affected by DDT, dieldrin and endrin. *Dunaliella* apparently was not affected by concentrations up to 1000 parts per billion. In *Cyclotella*, cell division was completely inhibited by dieldrin and endrin and DDT slowed division of the cells. The authors suggested that estuarine species,

TABLE 2
Toxicity of Selected Pesticides to Marine Organisms

Substance Tested	Formulation	Organism Tested	Common Name	Conc. (ppb Act. Ingred.) in Water	Method of Assessment	Test Procedure	Reference
Insecticides Organochlorines:							
Chlordane	100%	Palaemon macrodactylus	Korean shrimp	18 (10-38)	TL-50	96-hr static lab bioassay	Earnest, unpublished
DDT Compounds							
p,p'-DDT(1,1,1-Trichloro-2,2-bis(p-chlorophenyl)ethane	Technical 77%	Penaeus duorarum	Pink shrimp	0.12 0.17 (0.09-0.32)	TL-50 TL-50	28-day bioassay 96-hr intermittent flow lab bioassay	Nimmo et al., 1971 Earnest, unpublished
p,p'-DDD(p,p'-TDE) (1,1-Dichloro-2,2-bis(p-chlorophenyl)ethane	99%	Palaemon macrodactylus	Korean shrimp	2.5 (1.6-4.0)	TL-50	96-hr intermittent flow lab bioassay	Earnest, unpublished
p,p'-DDE (1,1-Dichloro-2,2-bis(p-chlorophenyl)ethylene	---	Falco peregrinus	Peregrine falcon	---	eggshell thinning	DDE in eggs highly correlated with shell thinning	Cade et al., 1971
Dieldrin	100%	Anguilla rostrata	American eel	0.9	LC-50	96-hr static lab bioassay	Eisler, 1970
Endrin	100%	Mugil cephalus	Striped mullet	0.3	LC-50	96-hr static lab bioassay	Eisler, 1970
	100%	Menidia menidia	Atlantic silverside	0.05	LC-50	96-hr static lab bioassay	Eisler, 1970
Methoxychlor	89.5%	Palaemon macrodactylus	Korean shrimp	0.44 (0.21-0.93)	TL-50	96-hr static lab bioassay	Earnest, unpublished
Mirex	Technical	Penaeus duorarum	Pink shrimp	1.0	100% paralysis/ death in 11 days	Flowing water bioassay	Lowe et al., 1971
Toxaphene	100%	Gasterosteus aculeatus	Threespine stickleback	7.8	TLM	96-hr static lab bioassay	Katz, 1961

TABLE 2—Continued
Toxicity of Selected Pesticides to Marine Organisms

Substance Tested	Formulation	Organism Tested	Common Name	Conc. (ppb Act. Ingred.) in Water	Method of Assessment	Test Procedure	Reference
Insecticides Organophosphates: Diazinon	Technical Grade	Cyprinodon variegatus	Sheepshead minnow	100	Mean inhibition of brain AChE (Result: >84%)	Static bioassay, 48-hr LC 40-60	Coppage, 1972
Guthion	93%	Gasterosteus aculeatus	Threespine stickleback	4.8	TLM	96-hr static lab bioassay	Katz, 1961
	Technical Grade	Cyprinodon variegatus	Sheepshead minnow	3	Mean inhibition of brain AChE (Result: 84%)	Static bioassay, 72-hr LC 40-60	Coppage, 1972
	Technical Grade	Lagodon rhomboides	Pinfish	10	Mean inhibition of brain AChE (Result: 80%)	Flowing seawater bio-assay, 24-hr LC 40-60	Coppage and Matthews, 1974
	Technical Grade	Leiostomus xanthurus	Spot	20	Mean inhibition of brain AChE (Result: 96%)	Flowing seawater bio-assay, 24-hr LC 40-60	Coppage and Matthews, 1974
Malathion	Technical Grade	Cyprinodon variegatus	Sheepshead minnow	190	Mean inhibition of brain AChE (Result: >84%)	Static bioassay, 24-hr LC 40-60	Coppage, 1972
	Technical Grade	Lagodon rhomboides	Pinfish	238	Mean inhibition of brain AChE (Result: 88%)	Flowing seawater bio-assay, 24-hr LC 40-60	Coppage and Matthews, 1974
	Technical Grade	Leiostomus xanthurus	Spot	238	Mean inhibition of brain AChE (Result: 70%)	Flowing seawater bio-assay, 24-hr LC 40-60	Coppage and Matthews, 1974
	100%	Thalassoma bifasciatum	Bluehead	27	LC-50	96-hr static lab bioassay	Eisler, 1970

TABLE 2—Continued
Toxicity of Selected Pesticides to Marine Organisms

Substance Tested	Formulation	Organism Tested	Common Name	Conc. (ppb Act. Ingred.) in Water	Method of Assessment	Test Procedure	Reference
Insecticides							
Organophosphates:							
Naled	Technical Grade	*Lagodon rhomboides*	Pinfish	23	Mean inhibition of brain AChE (Result: 89%)	Flowing seawater bio-assay, 72-hr LC 40-60	Coppage and Matthews, 1974
	Technical Grade	*Leiostomus xanthurus*	Spot	70	Mean inhibition of brain AChE (Result: 85%)	Flowing seawater bio-assay, 24-hr LC 40-60	Coppage and Matthews, 1974
Parathion	Technical Grade	*Cyprinodon variegatus*	Sheepshead minnow	10	Mean inhibition of brain AChE (Result: 84%)	Static bioassay, 72-hr LC 40-60	Coppage, 1972
	Technical Grade	*Lagodon rhomboides*	Pinfish	10	Mean inhibition of brain AChE	Flowing seawater bio-assay, 24-hr LC 40-60	Coppage and Matthews, 1974
	Technical Grade	*Leiostomus xanthurus*	Spot	10	Mean inhibition of brain AChE (Result: 90%)	Flowing seawater bio-assay, 24-hr LC 40-60	Coppage and Matthews, 1974
Methyl Parathion	100%	*Crangon septemspinosa*	Sand shrimp	2	LC-50	96-hr static lab bioassay	Eisler, 1969
Phorate	Technical Grade	*Cyprinodon variegatus*	Sheepshead minnow	5	Mean inhibition of brain AChE (Result: >84%)	Static bioassay, 72-hr LC 40	Coppage, 1972
Carbamates:							
Carbaryl	100%	*Palaemon macrodactylus*	Korean shrimp	7.0 (1.5-28)	TL-50	96-hr intermittent flow lab bioassay	Earnest, unpublished
	Technical Grade	*Lagodon rhomboides*	Pinfish	1333	Mean inhibition of brain AChE (Result: 81%)	Flowing seawater bio-assay, 24-hr LC 40-60	Coppage, unpublished

TABLE 2—Continued
Toxicity of Selected Pesticides to Marine Organisms

Substance Tested	Formulation	Organism Tested	Common Name	Conc. (ppb Act. Ingred.) in Water	Method of Assessment	Test Procedure	Reference
Insecticides							
Carbamates:							
Carbofuran	Acetone wash from sand-coated particle formulation	Cyprinodon variegatus	Sheepshead minnow	Unknown	Mean inhibition of brain AChE (Result: 84%)	Static bioassay 48-hr LC 40-60	Coppage, unpublished
Herbicides							
2,4-D and derivatives							
Picloram	Ester	Crassostrea virginica	American oyster	740	TLM	14-day static lab bioassay	Davis and Hidu, 1969
Tordon ® 101 (39.6% 2,4-D 14.3% picloram)	---	Isochrysis galbana	---	5×10^5	50% decrease in O_2 evolution[a]	---	Walsh, 1972a
Triazines:							
Ametryne	Technical acid	Chlorococcum sp.	---	10	50% decrease in growth	Measured as ABS. (525mμ) after 10 days[b]	Walsh, 1972a
	Technical acid	Isochrysis galbana	---	10	50% decrease in O_2 evolution[a]	---	Walsh, 1972a
	Technical acid	Monochrysis lutheri	---	10	50% decrease in O_2 evolution[a]	---	Walsh, 1972a
	Technical acid	Phaeodactylum tricornutum	---	10	50% decrease in O_2 evolution[a]	---	Walsh, 1972a
Atrazine	Technical acid	Chlorococcum sp.	---	100	50% decrease in growth	Measured as ABS. (525mμ) after 10 days[b]	Hollister and Walsh, 1973
	Technical acid	Chlamydomonas sp.	---	60	50% decrease in O_2 evolution[a]	---	Hollister and Walsh, 1973
	Technical acid	Monochrysis lutheri	---	77	50% decrease in O_2 evolution[a]	---	

TABLE 2—Continued
Toxicity of Selected Pesticides to Marine Organisms

Substance Tested	Formulation	Organisms Tested	Common Name	Conc. (ppb Act. Ingred.) in Water	Method of Assessment	Test Procedure	Reference
Herbicides: Triazines:							
Atrazine	Technical acid	Isochrysis galbana	---	100	50% decrease in O_2 evolution[a]	---	Walsh, 1972a
	Technical acid	Phaeodactylum tricornutum	---	100	50% decrease in O_2 evolution[a]	---	Walsh, 1972a
Simazine	Technical acid	Isochrysis galbana	---	500	50% decrease in growth	Measured as ABS. (525mμ) after 10 days[b]	Walsh, 1972a
	Technical acid	Phaeodactylum tricornutum	---	500	50% decrease in growth	Measured as ABS. (525mμ) after 10 days[b]	Walsh, 1972a
Urea:							
Diuron	---	Protococcus sp.	---	0.02	0.52 OPT. DEN. expt/OPT. DEN. control[b]	10-day growth	Ukeles, 1962
	---	Monochrysis lutheri	---	0.02	0.00 OPT. DEN. expt/OPT. DEN. control[b]	10-day growth	Ukeles, 1962
	Technical	Chlorococcum sp.	---	10	50% decrease in growth	10-day growth	Walsh, 1972a
	Technical acid	Isochrysis galbana	---	10	50% decrease in growth	10-day growth	Walsh, 1972a
	---	Monochrysis lutheri	---	290	0.67 OPT. DEN. expt/OPT. DEN. control[b]	10-day growth	Ukeles, 1962

[a] O_2 evolution measured by Gilson differential respirometer on 4 mℓ of culture in log phase. Length of test 90 min.

[b] ABS. (525mμ) = Absorbance at 525 millimicrons wavelength. OPT. DEN. expt/OPT. DEN. control = Optical density of experimental culture/optical density of control culture.

such as *Dunaliella*, are perhaps less susceptible than
are open ocean forms, such as *Cyclotella*. Similar
studies on phytoplankton and PCBs by Fisher *et al.*
(1972) also suggested that coastal phytoplankton may
be more resistant to organochlorines than are those
found in open ocean. Isolates of diatoms from the
Sargasso Sea were more sensitive than clones from
estuaries and the continental shelf. Herbicides applied to four species of marine unicellular algae
adversely affected their growth (Walsh, 1972a). Urea
and triazine herbicides were the most toxic of the
formulations tested. In some instances, smaller
amounts of herbicides were required to inhibit growth
than to inhibit oxygen evolution. Interestingly,
Dunaliella was most resistant of the four species
tested, as occurred in Menzel's *et al.* studies (1970).

The effect of mirex and a PCB, Aroclor 1254, on
growth of ciliate, *Tetrahymena pyriformis*, was studied by Cooley *et al.* (1972). Both chemicals caused
significant reductions in growth rate and population
density and the ciliate accumulated both toxicants
from the culture media, concentrating mirex up to
193 times and Aroclor to approximately 60 times the
nominal concentration in the media. The authors postulate that if this ciliate encountered similar concentrations of these materials in nature, the results
would be a reduction of their availability as food
organisms and nutrient regenerators. Also, the
capacity of the organisms to concentrate mirex and
Aroclor could provide a pathway for entry of these
chemicals into the food web.

Growth rates of young oysters, *Crassostrea*
virginica, as indicated by height and in-water weight,
was significantly reduced in individuals exposed to
5 micrograms of Aroclor 1254 per liter (ppb) for
24 weeks, but growth rate was not affected in individuals exposed to 1 part per billion for 30 weeks
(Lowe *et al.*, 1972). Oysters exposed to 1 part per
billion concentrated the chemical 101,000 times, but
less than 0.2 part per million remained after 12
weeks of depuration. The growth rate of the oyster

was a much more sensitive indicator, since no significant mortality occurred in oysters exposed to
5 ppb.

The effects of mirex on growth of crabs, as
measured by the duration of developmental stages of
crabs as an indicator of their growth, is illustrated
by the work of Bookhout et al. (1972). The duration
of developmental stages of zoea and the total time of
development was generally lengthened with an increase
in concentration of mirex from 0.01 to 10.0 parts per
billion. Menippe did not demonstrate this effect,
but the percentage of the extra 6th zoeal stage increased as concentrations of mirex increased. This
method of determining the effect of mirex on crabs
appears to be more sensitive than previous tests with
juvenile blue crabs reported by McKenzie (1970) and
Lowe et al. (1971).

EFFECTS OF PESTICIDES ON BEHAVIOR OF ORGANISMS

The behavioral activity of organisms is a sensitive criterion for determining the effect of pesticides on marine organisms. Dr. H. G. Kleerekoper has
successfully studied the interactions of temperature
and a heavy metal on the locomotor behavior of fish
in the laboratory (Kleerekoper and Waxman, 1973) and
will present data on the effect of pesticides on
marine fish later in this volume. Hansen (1969)
showed that the estuarine fish, *Cyprinodon variegatus*,
avoided water containing DDT, endrin, Dursban ®[b] or
2,4-D in controlled laboratory experiments, but the
fish did not avoid test concentrations of malathion
or Sevin ®.[c] Likewise, grass shrimp, *Palaemonetes
pugio*, an important forage food for estuarine organisms, avoided 1.0 and 10.0 ppm of 2,4-D by seeking

---

[b] ® Registered trademark: Dursban, Dow Chemical
Company.
[c] ® Registered trademark: Sevin, Union Carbide
Company.

water free of this herbicide, but did not avoid the
five insecticides tested (Hansen *et al.*, 1973). The
capacity of coastal organisms to avoid water containing pesticides may enhance their survival by causing
them to move to an area free of pesticides. Avoidance could be disastrous to a population if, by
avoiding the pesticides, the population is unable to
reach an area where spawning normally occurs.

EFFECTS OF PESTICIDES ON ENZYME SYSTEMS

Inhibition of the hydrolyzing enzyme, acetylcholinesterase (AChE), by organophosphate and carba-
mate pesticides can be used as an indication of the
effect of these chemicals on estuarine fish (Coppage,
1972). Evidently, esterase-inhibiting pesticides
bind active sites of the enzyme and block the breakdown of acetylcholine, which causes toxic accumulation of acetylcholine. As a result, nerve impulse
transfers can be disrupted. Laboratory bioassays
with estuarine fish spot, *Leiostomus xanthurus*,
showed that lethal exposures of this fish to malathion
reduced the AChE activity level by 81%. Such information developed in the laboratory is useful in evaluating effects of pesticides applied in the field.

EFFECTS OF PESTICIDES ON ECOSYSTEMS AND COMMUNITIES

Few data are available concerning the effects of
pesticides at the ecosystem or community level of
organization. This is not surprising considering the
complexities of ecosystems and our lack of knowledge
of the structure and function of coastal zones.
Effects of pesticides could be masked by variations
in population densities and it would require several
years to evaluate such variations. However, it is
possible to design laboratory and field experiments
to yield information on this complex system.
An experimental community that received 10
micrograms per liter of a polychlorinated biphenyl,
Aroclor 1254, did not recover to a "normal" state in

terms of numbers of phyla and species after 4 months (Hansen, 1974). Communities of planktonic larvae were allowed to develop in "control" aquaria and aquaria that received the Aroclor 1254. Communities that received 10 micrograms per liter of the chemical were dominated by tunicates, whereas controls were dominated by arthropods. The Shannon-Weaver species diversity index was not altered by Aroclor 1254, but numbers of phyla, species and individuals decreased.

The capacity of a fish population to compensate for the effect of a pesticide was suggested in a recent study made in Louisiana, where malathion was applied aerially to control mosquito vectors of Venezuelan equine encephalomyelitis (Coppage and Duke, 1972). Fish were collected from the coastal area before, during and after the application of malathion. Acetylcholinesterase (AChE) activity in the brains of fish were used as an indicator of the effect of malathion on the community of fish. Levels of inhibition during and soon after spraying in one lake approached levels that were associated with death of fish in laboratory bioassay studies. The AChE activity of the fish population returned to normal within 40 days after application of the chemical.

CONCENTRATION FACTORS

The capacity of organisms to concentrate a pesticide is another factor that must be considered when evaluating the impact of these chemicals on a coastal system. Many of the persistent pesticides are passed through the food web through accumulation and bioconcentration. Some question exists about the mechanisms involved in trophic accumulation of fat-soluble hydrocarbons from water by aquatic organisms (Hamelink *et al.*, 1971). Whatever the mechanisms for accumulation, many coastal organisms have the capacity to concentrate pesticides many times more than the concentration occurring in the water around them. Concentration factors, the ratio of the amount of

pesticide in the animal to that in the water, for
some specific organisms and pesticides determined by
investigators at the Gulf Breeze Environmental
Research Laboratory are shown in Table 3.

STATE OF THE ART

Concern about the occurrence of pesticides in
the marine environment is continually emphasized
because surveillance and research on these chemicals
are given high priority by knowledgeable scientists.
The analytical capability for determining residues of
some pesticides in the parts per trillion range is
available, but we often do not understand the biological or ecological significance of these residues.
We need more information on chronic exposures of
sensitive marine organisms during complete reproductive cycles and on effects of sublethal levels of
exposure. Also, information is required on the
structure and function of coastal ecosystems and
criteria for evaluating the stress of pesticides on
these systems. Laboratory microcosms and other kinds
of experimental environments no doubt will be useful
in this evaluation.

As mentioned previously, use-patterns of pesticides in this country are changing. We must be prepared to evaluate possible effects, on the environment, of integrated pest control procedures, whereby
biological control may be just important as chemical
control of pests. Viruses and juvenile-hormone
mimics are being tested for use as pesticides and
could inadvertently reach the coastal zone. The
research effort to evaluate the impact of these new
agents must take into account that the coastal
environment already contains residues of pesticides,
persistent organochlorines, and other pollutants.

TABLE 3

Accumulation of Pesticides from Water by Marine Organisms[a]

Substance Tested	Organism Tested	Common Name	Exp. Conc.	Conc. Factor	Time	Special Details	Reference
Insecticides Organochlorines:							
Chlordane	Pseudomonas spp.	---	10 ppm	0.83	10 days	Mixed culture of four species	Bourquin, unpublished
DDT	Brachidontes recurvus	Hooked mussel	1 ppb	24,000	1 week	Whole body residues (Meats)	Butler, 1966
	Mercenaria mercenaria	Hard-shell clam	1 ppb	6,000	1 week	Whole body residues (Meats)	Butler, 1966
	Mya arenaria	Soft-shell clam	0.1 ppb	8,800	5 days	Whole body residues (Meats)	Butler, 1971
	Crassostrea gigas	Pacific oyster	1.0 ppb	20,000	7 days	Whole body residues (Meats)	Butler, 1966
	Penaeus duorarum	Pink shrimp	0.14 ppb	1,500	3 weeks	Whole body residues	Nimmo et al., 1970
	Lagodon rhomboides	Pinfish	0.1, 1.0 ppb	10,600 38,000	2 weeks	Whole body residues	Hansen and Wilson, 1970
Dieldrin	Mercenaria mercenaria	Hard-shell clam	0.5 ppb	760	5 days	Whole body residues (Meats)	Butler, 1971
Endrin	Mercenaria mercenaria	Hard-shell clam	0.5 ppb	480	5 days	Whole body residues (Meats)	Butler, 1971
Methoxychlor	Mercenaria mercenaria	Hard-shell clam	1.0 ppb	470	5 days	Whole body residues (Meats)	Butler, 1971
Mirex	Tetrahymena pyriformis W	---	0.9 ppb	193	1 week	Axenic cultures incubated at 26°C; concentration factor on dry weight basis	Cooley et al., 1972
	Penaeus duorarum	Pink shrimp	0.1 ppb	2,600	3 weeks	Whole body residues	Lowe et al., 1971

156

TABLE 3—Continued

Accumulation of Pesticides from Water by Marine Organisms[a]

Substance Tested	Organism Tested	Common Name	Exp. Conc.	Conc. Factor	Time	Special Details	Reference
Insecticides Organochlorines: Mirex	*Rhithropanopeus harrisii*	Mud crab (larvae)	0.1 ppb	1,000	7 weeks	Static culture bowl method with a change to fresh medium + chemical each day	Bookhout *et al.*, 1972
	Callinectes sapidus	Blue crab (juveniles)	0.1 ppb	1,100–5,200	3 weeks	Whole body residues	Lowe, unpublished
	Thalassia testudinum	Turtle grass	0.1 ppb	0 leaves 0.36 rhizomes	10 days	Plants exposed to chemical through rhizomes; concentration factor on wet weight basis	Walsh and Hollister, unpublished
Halogenated hydrocarbon: Polychlorinated biphenyl (PCB) Aroclor ® 1254	*Tetrahymena pyriformis* W	---	1 ppm	60	1 week	Axenic cultures incubated at 26°C; concentration factor on dry weight basis	Cooley *et al.*, 1972
	Palaemonetes pugio	Grass shrimp	0.62 ppb	2,069	1 week	Whole body residues (Meats)	Nimmo and Heitmuller, unpublished
				26,580	5 weeks	Whole body residues (Meats)	Nimmo and Heitmuller, unpublished

157

TABLE 3—Continued
Accumulation of Pesticides from Water by Marine Organisms[a]

Substance Tested	Organism Tested	Common Name	Exp. Conc.	Conc. Factor	Time	Special Details	Reference
Insecticides Halogenated Hydrocarbon: Polychlorinated biphenyl (PCB) Aroclor ® 1254							
	Penaeus duorarum	Pink shrimp	2.5 ppb	1,800	2 days	Whole body residues	Nimmo et al., 1971
				7,600	9 days	Whole body residues	Nimmo et al., 1971
	Lagodon rhomboides	Pinfish	5 ppb	2,800–21,800	2–15 weeks	Whole body residues	Hansen et al., 1971
	Leiostomus xanthurus	Spot	1 ppb	17,000–27,000	4–8 weeks	Whole body residues	Hansen et al., 1971
			5 ppb	9,200–30,400	3–6 weeks	Whole body residues	Hansen et al., 1971
	Thalassia testudinum	Turtle grass	5820 ppb	0 leaves 0 rhizomes	10 days	Plants exposed to chemical through rhizomes; concentration factor on wet weight basis	Walsh and Hollister, unpublished
Herbicide Tordon ® 101 (39.6% 2,4-D; 14.3% picloram)							
	Rhizophora mangle	Red mangrove	14.4 ppb	Stems 1.28 (2,4-D) 0.64 (picolinic acid)	20 days	Seedlings treated when two pairs of leaves were present; concentration factor on wet weight basis	Walsh et al., 1973
	Thalassia testudinum	Turtle grass	5 ppm	Leaves 0 (2,4-D) 0 (picolinic acid)	10 days	Plants exposed to chemical through rhizomes; concentration factor on wet weight basis	Walsh and Hollister, unpublished

[a]Information developed at the Environmental Protection Agency's Gulf Breeze Environmental Research Laboratory.

ACKNOWLEDGMENTS

We thank C. M. Herndon for assistance in preparing the manuscript; N. R. Cooley, D. J. Hansen and M. E. Tagatz for their constructive criticism of the manuscript; and the investigators at this laboratory for freely supplying data (some unpublished) for the tables.

LITERATURE CITED

Bookhout, C. G., A. J. Wilson, Jr., T. W. Duke, and J. I. Lowe. 1972. Effects of mirex on the larval development of two crabs. *Water, Air and Soil Pollution 1*:165-80.

Bourquin, A. W. Unpublished data. Environmental Protection Agency, Gulf Breeze Environmental Research Laboratory, Gulf Breeze, Florida.

Brungs, W. A. 1969. Chronic toxicity of zinc to the fathead minnow, *Pimephales promelas* Rafinesque. *Trans. Am. Fish. Soc. 98*:272-79.

Butler, P. A. 1966. Pesticides in the marine environment. *J. Appl. Ecol. 3*:253-59.

_____. 1971. Influence of pesticides on marine ecosystems. *Proc. Roy. Soc. London B 177*:321-29.

_____. 1973. Organochlorine residues in estuarine mollusks, 1965-1972—National Pesticide Monitoring Program. *Pestic. Monitor. J. 6*:238-62.

_____. 1974. Biological problems in estuarine monitoring. Proceedings of Seminar on Standardization of Methodology in Monitoring the Marine Environment held in Seattle, Washington, October 16-18, 1973. In press.

Cade, T. J., J. L. Lincer, C. M. White, D. G. Roseneau, and L. G. Swartz. 1971. DDE residues and eggshell changes in Alaskan falcons and hawks. *Science 172*:955-57.

Cooley, N. R., J. M. Keltner, Jr., and J. Forester. 1972. Mirex and Aroclor ® 1254: effect and accumulation by *Tetrahymena pyriformis* Strain W. *J. Protozool. 19*:636-38.

Coppage, D. L. Unpublished data. Environmental Protection Agency, Gulf Breeze Environmental Research Laboratory, Gulf Breeze, Florida.

_____. 1972. Organophosphate pesticides: specific level of brain AChE inhibition related to death in sheepshead minnows. *Trans. Am. Fish. Soc. 101*:534-36.

_____ and T. W. Duke. 1972. Effect of pesticides in estuaries along the Gulf and southeast Atlantic Coasts. In: *Proc. Second Gulf Conf. on Mosquito Suppression and Wildl. Manage.*, pp. 24-31.

_____ and E. Matthews. 1974. Short-term effects of organophosphate pesticides on cholinesterases of estuarine fishes and pink shrimp. In press.

Couch, J. 1974. Histopathologic effects of pesticides and related chemicals on the livers of fishes. Proc. of Symposium on Research of Fishes sponsored by the Registry of Comparative Pathology. University of Wisconsin Press. In press.

Davis, H. C. and H. Hidu. 1969. Effects of pesticides development of clams and oysters and on survival and growth of the larvae. *Fish. Bull. 67*:393-404.

Earnest, R. Unpublished data. Effects of pesticides on aquatic animals in the estuarine and marine environment. In: *Annual Progress Report 1970*. Columbia, Mo.: Fish-Pesticide Research Laboratory, Bur. Sport Fish. Wildl., U. S. Dept. Interior.

Eisler, R. 1969. Acute toxicities of insecticides to marine decapod crustaceans. *Crustaceana 16*: 302-10.

_____. 1970. Acute toxicities of organo-
chlorine and organophosphorus insecticides to
estuarine fishes. *Bur. Sport Fish. Wildl.
Tech. Paper* No. 46.

Environmental Protection Agency. 1973. Effects of
pesticides in water—a report to the states.

Fisher, N. S., L. B. Graham, E. J. Carpenter, and
C. F. Wurster. 1973. Geographic differences in
phytoplankton sensitivity to PCBs. *Nature 241*:
548-49.

Giam, C. S., A. R. Hanks, R. L. Richardson, W. M.
Sackett, and M. K. Wong. 1972. DDT, DDE, and
polychlorinated biphenyls in biota from the Gulf
of Mexico and the Caribbean Sea. *Pestic.
Monitor. J.* 6:139-43.

Hamelink, J. L., R. C. Waybrant, and R. C. Ball.
1971. A proposal: exchange equilibria control
the degree chlorinated hydrocarbons are biologically magnified in lentic environments. *Trans.
Am. Fish. Soc.* 100:207-14.

Hansen, D. J. 1969. Avoidance of pesticides by
untrained sheepshead minnows. *Trans. Am. Fish.
Soc.* 98:426-29.

_____. 1974. Aroclor ® 1254: effect on com-
position of developing estuarine animal communities in the laboratory. Contributions in Marine
Science, University of Texas. In press.

_____, P. R. Parrish, J. I. Lowe, A. J. Wilson,
Jr., and P. D. Wilson. 1971. Chronic toxicity,
uptake, and retention of Aroclor 1254 in two
estuarine fishes. *Bull. Environ. Contam. and
Toxicol.* 6:113-19.

_____, S. C. Schimmel, and J. M. Keltner, Jr.
1973. Avoidance of pesticides by grass shrimp
(*Palaemonetes pugio*). *Bull. Environ. Contam.
and Toxicol.* 9:129-33.

_____ and A. J. Wilson, Jr. 1970. Significant
of DDT residues in the estuary near Pensacola,
Florida. *Pestic. Monitor. J.* 4:51-56.

Harvey, G. R., V. T. Bowen, R. H. Backus, and G. D. Grice. 1972. Chlorinated hydrocarbons in open-ocean Atlantic organisms. In: *The Changing Chemistry of the Oceans*, pp. 177-88, ed. by D. Dyrssen and D. Jagner. Stockholm: Almquist and Wiksell.

Hollister, T. A. and G. E. Walsh. 1973. Differential responses of marine phytoplankton to herbicides: oxygen evolution. *Bull. Environ. Contam. and Toxicol. 9*:291-95.

Katz, M. 1961. Acute toxicity of some organic insecticides to three species of salmonids and to the threespine stickleback. *Trans. Am. Fish. Soc. 90*:264-68.

Kleerekoper, H. and J. B. Waxman. 1973. Interaction of temperature and copper ions as orienting stimuli in the locomotion behavior of the goldfish *(Carassius auratus)*. *J. Res. Bd. Canada 30*:725-28.

Lowe, J. I. Unpublished data. Environmental Protection Agency, Gulf Breeze Environmental Research Laboratory, Gulf Breeze, Florida.

_____, P. R. Parrish, J. M. Patrick, Jr., and J. Forester. 1972. Effects of the poly-chlorinated biphenyl Aroclor ® on the American oyster *Crassostrea virginica*. *Mar. Biol. 17*: 209-14.

_____, _____, A. J. Wilson, Jr., P. D. Wilson, and T. W. Duke. 1971. Effects of mirex on selected estuarine organisms. *Trans. 36th N. Am. Wildl. and Natural Resources Conf.*, March 7-10, 1971, Portland, Oregon, 171-86.

McKenzie, M. D. 1970. Fluctuations in abundance of the blue crab and factors affecting mortalities. *South Carolina Wildlife Resources Dept., Marine Resources Division Technical Report 1*.

Menzel, D. W., J. Anderson, and A. Randike. 1970. Marine phytoplankton vary in their response to chlorinated hydrocarbons. *Science 167*:1724-26.

Modin, J. C. 1969. Chlorinated hydrocarbon pesticides in California bays and estuaries. *Pestic.
Monitor. J. 3*:1-7.

Mount, D. I. 1968. Chronic toxicity of copper to
fathead minnows (*Pimephales promelas*, Rafinesque).
Water Research 2:215-23.

Nimmo, D. R., R. R. Blackman, A. J. Wilson, Jr., and
J. Forester. 1971. Toxicity and distribution
of Aroclor ® 1254 in the pink shrimp, *Penaeus
duorarum. Mar. Biol. 11*:191-97.

_____ and P. T. Heitmuller. Unpublished data.
Environmental Protection Agency, Gulf Breeze
Environmental Research Laboratory, Gulf Breeze,
Florida.

_____, A. J. Wilson, Jr., and R. R. Blackman.
1970. Localization of DDT in the body organs of
pink and white shrimp. *Bull. Environ. Contam.
and Toxicol. 5*:333-41.

Odum, E. P. 1969. The strategy of ecosystem
development. *Science 164*:262-70.

Sprague, J. B. 1969. Measurement of pollutant
toxicity to fish. I. Bioassay methods for acute
toxicity. *Water Research 3*:793-821.

_____. 1970. Measurement of pollutant toxicity
to fish. II. Utilizing and applying bioassay
results. *Water Research 4*:3-32.

Ukeles, R. 1962. Growth of pure cultures of marine
phytoplankton in the presence of toxicants.
Appl. Microbiol. 10:532-37.

Walsh, G. E. 1972a. Effects of herbicides on photo-
synthesis and growth of marine unicellular algae.
Hyacinth Control J. 10:45-48.

_____. 1972b. Insecticides, herbicides and
polychlorinated biphenyls in estuaries. *J. Wash.
Acad. Sci. 62*:122-39.

_____, R. Barrett, G. H. Cook, and T. A.
Hollister. 1973. Effects of herbicides on
seedlings of the red mangrove, *Rhizophora mangle*
L. *BioScience 23*:361-64.

_____ and T. A. Hollister. Unpublished data. Environmental Protection Agency, Gulf Breeze Environmental Research Laboratory, Gulf Breeze, Florida.

_____, C. W. Miller, and P. T. Heitmuller. 1971. Uptake and effects of dichlobenil in a small pond. *Bull. Environ. Contam. and Toxicol. 6*: 279-88.

Wolman, A. A. and A. J. Wilson, Jr. 1970. Occurrence of pesticides in whales. *Pestic. Monitor. J. 4*: 8-10.

Woodwell, G. M., C. F. Wurster, and P. A. Isaacson. 1967. DDT residues in an east coast estuary: a case of biological concentration of a persistent insecticide. *Science 156*:821-24.

RENAL HANDLING OF DDA
BY THE SOUTHERN FLOUNDER

JOHN B. PRITCHARD

Department of Physiology
Medical University of South Carolina
Charleston, South Carolina 29401

The problems associated with the presence of
chlorinated hydrocarbons in the environment arise
particularly from their physical and chemical stability (Gustafson, 1970; O'Brien, 1967). Because of
this stability, they are extremely persistent and
widely distributed in the environment (Risebrough,
1969; Sladen *et al.*, 1966). Furthermore, Woodwell
et al. (1967), Rudd and Herman (1972), and others
have documented extensive accumulation of DDT (1,1,1-
trichloro-2,2-bis(p-chlorophenyl)ethane) and metabolites by living organisms to levels far exceeding
that of their environment and additional accumulation
at successive trophic levels in the food chain. Such
accumulation, or bioconcentration, is determined by
several factors including: (a) the availability of
the pesticide, (b) the mode of entry, and particularly, (c) the rate of elimination. The work described below examines one of the mechanisms which
may control the rate of excretion of DDT and metabolites, specifically, the renal handling of the polar
DDT metabolite, DDA (2,2-bis(p-chlorophenyl)acetic
acid).

DDA is particularly of interest since it is the terminal and most polar DDT metabolite produced by most higher organisms (O'Brien, 1967; Peterson and Robison, 1964). DDA is the major form excreted in the urine and feces of mammals (Judah, 1949; White and Sweeney, 1945) and at least some fish (Pritchard *et al.*, 1973). In addition, Wolfe *et al.* (1970) have shown that DDA excretion is a very good indicator of the extent of DDT exposure in man. Since DDA is the major form excreted, a better understanding of the mechanism by which it is eliminated should be valuable in establishing the basis for the observed limited excretion of DDT and metabolites. A likely possibility is that the excretion of DDA, itself an organic acid, may be controlled by the same system which transports phenol red and p-aminohippurate (organic acid system). Preliminary support for this hypothesis is provided by our previous observation that DDA competitively inhibits organic acid transport in flounder kidney (Pritchard and Kinter, 1970). The present study examines renal handling of DDA directly using ring-labeled ^{14}C-DDA.

METHODS

ANIMALS

Southern flounder *(Paralichthys lethostigma)* were collected locally in a small otter trawl. Fish, usually 100-300 gms, were maintained in a 100-gallon Frigid Units holding tank using Instant Ocean artificial seawater (35 o/oo) at 20°C. They were fed live shrimp or minnows.

CHEMICALS

All ^{14}C-labeled compounds used in this study were purchased from Mallinckrodt Nuclear. Unlabeled pesticides were obtained from Aldrich Chemical Company. Unlabeled organic acids were from Eastman Chemicals.

PROCEDURES

The isolated renal tubule preparation developed by Forster (1948) was used for this study. The fish were killed by decapitation. The kidney was rapidly removed and transferred to iced, oxygenated saline where small masses of tubules (1-2 mg each) were teased from its caudal portion with sharp forceps. Three to five such masses were transferred to each incubation vial where they were incubated in 1 ml of medium for from 0.5 to 60 min at $20°C$ with gentle agitation. Each vial was gassed with 100% O_2. The incubation medium was Forster's (1948) saltwater fish medium at pH 7.8 containing ^{14}C-labeled organic acids or DDA. Unless otherwise specified, the concentration of labeled compound was always $10^{-5}M$. Metabolic inhibitors, if any, were added at $10^{-3}M$ final concentration. At the conclusion of the incubation period, the tissue was removed, blotted, and weighed. Tissue and medium samples were solublized (Soluene, Packard Instrument Company) and analyzed by liquid scintillation counting. Results were then expressed as tissue to medium ratios (T/M), i.e., $\dfrac{ng/g\ tissue}{ng/ml\ medium}$.

Binding of DDA to tubular tissue was measured by a modified version of the method of Gigon and Guarino (1970). Fifty or 100 mg of the tubules were homogenized per 1 ml of Forster's medium (5 and 10% homogenates w/v). ^{14}C-DDA was added (final concentration of $10^{-5}M$) to the homogenate, and the homogenate was centrifuged at $0°C$ for 24 hrs at 30,000 $\times$ G. Samples of the resulting pellet and supernate were analyzed by liquid scintillation counting as above. Results were expressed as a binding ratio (i.e., $\dfrac{DPM/gm\ pellet}{DPM/ml\ supernate}$). The organic acids chlorophenol red (CPR) and p-aminohippuric acid (PAH) were added at a final concentration of $10^{-3}M$ to some samples before addition of the ^{14}C-DDA.

RESULTS

Since uptake of organic acids by southern flounder in this *in vitro* system had not previously been
examined, accumulation of the well-studied organic
acid, PAH, was examined first (Fig. 1). Tissue to
medium ratios (T/M) for PAH reached a maximum by
1 hr at about 12. Cyanide (10^{-3}M), which inhibits
the production of metabolic energy by blocking electron transport (Racker, 1965), reduced the T/M to 1,
the distribution ratio predicted for simple diffusion.
A preincubation with cyanide of at least 5 min was
necessary to demonstrate the full cyanide effect.

Like PAH, DDA is concentrated by flounder kidney
tubules. After 60 min, a T/M of approximately 10 had
been reached; longer incubation increased accumulation very little. Unlike PAH, cyanide did not return
the T/M for DDA to 1, but did reduce maximum uptake
by 30%. This result has been normalized so that 60-
min control uptake equals 100, thus permitting direct
visual comparison of the effects of cyanide and the
organic acid, chlorophenol red (CPR), on DDA accumulation (Fig. 2). CPR also reduces DDA uptake by the
isolated renal tubules (15% reduction of maximal
uptake at 10^{-4}M, 50% at 10^{-3}M). Thus, a portion of
DDA uptake requires metabolic energy and a portion
is inhibited by another organic acid as would be
predicted if accumulation were on the organic acid
transport system.

Accumulation of DDT itself in this system
reached a T/M of 8 by 60 min; however, no reduction
of uptake was produced by cyanide or CPR.

To establish whether the inhibition produced by
the organic acid, CPR, was on the portion of the DDA
uptake requiring metabolic energy or on the cyanide-
insensitive portion, 60-min DDA accumulation was
measured in the presence of CPR both with and without
10^{-3}M cyanide. Figure 3 depicts the results of one
such experiment. Increasing the concentration of the
organic acid, CPR, inhibits both total uptake and
cyanide-insensitive uptake. However, the bulk of the

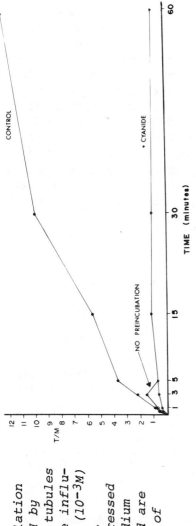

Fig. 1. Accumulation of 10⁻⁵M 14C-PAH by flounder kidney tubules in vitro and the influence of cyanide (10⁻³M) on this process. Results are expressed as tissue to medium ratios (T/M) and are plotted vs time of incubation. →

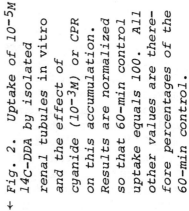

← *Fig. 2. Uptake of 10⁻⁵M 14C-DDA by isolated renal tubules in vitro and the effect of cyanide (10⁻³M) or CPR on this accumulation. Results are normalized so that 60-min control uptake equals 100. All other values are therefore percentages of the 60-min control.*

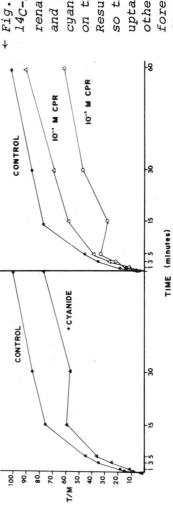

reduction occurs in the fraction of the uptake requiring metabolic energy, i.e., the cyanide-sensitive
component depicted here as the increment between
total uptake and uptake in the presence of cyanide.
In the absence of CPR, the 60-min uptake of 10^{-5}M DDA
gives a T/M of 9.7 and cyanide reduces uptake to a
T/M of 6.2, an increment of 3.5. This increment was
reduced to 1.4 at 10^{-4}M CPR (60% reduction) and 0.9
at 10^{-3}M CPR (73% inhibition), while the cyanide-
insensitive uptake was reduced only 14% by 10^{-3}M CPR.
The results of all experiments of this type and the
statistical significance of the results are summarized in Table 1, which again emphasizes the much
greater sensitivity of energy-dependent uptake to
organic acid inhibition.

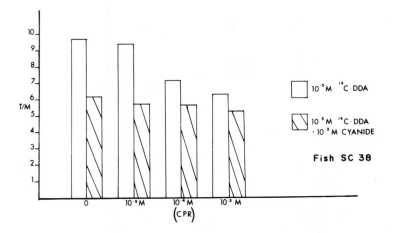

Fig. 3. Effects of the organic acid, CPR,
on total ^{14}C-DDA uptake and on cyanide
inhibited ^{14}C-DDA uptake. [DDA] = 10^{-5}M.
Results are expressed as the T/M achieved
by the tubules of a single fish after 60-
min incubation under the conditions
indicated above.

Chlorophenol red is heavily bound (70%) to
plasma proteins in fish, while PAH shows only 5%

TABLE 1

Effects of the Organic Acid, Chlorophenol Red, on Tissue to Medium Ratios (T/M) of ^{14}C-DDA Achieved by Flounder Kidney Tubules in the Presence or Absence of $10^{-3}M$ Cyanide. Data Presented as Mean T/M ± Standard Error of the Mean (N's range from 2 to 8)

Substance	Concentration of			
	0	$10^{-5}M$	$10^{-4}M$	$10^{-3}M$
$10^{-5}M$ DDA	10.21 ± 0.55	8.60 ± 0.80	$7.56 \pm 0.49^{*}$	$6.07 \pm 0.13^{*}$
$10^{-5}M$ DDA + $10^{-3}M$ KCN	$6.96 \pm 0.34^{*}$	$5.76 \pm 0.04^{*}$	$6.17 \pm 0.26^{*}$	$5.04 \pm 0.24^{\dagger *}$

*$p < 0.01$ vs. $10^{-5}M$ DDA alone.

†$p < 0.05$ vs. $10^{-5}M$ DDA plus $10^{-3}M$ KCN.

171

binding (Maack *et al.*, 1969). Since the small but significant reduction of cyanide-insensitive uptake might arise from a competition between DDA and CPR for such nonspecific binding sites, the above experiments were repeated using PAH, which should not so effectively compete for binding sites. Indeed, PAH produced no significant depression of cyanide-insensitive uptake (Fig. 4). Nevertheless, this organic acid was at least as effective as CPR in reducing the energy-dependent uptake (76% inhibition at 10^{-4}M PAH; 86% at 10^{-3}M PAH).

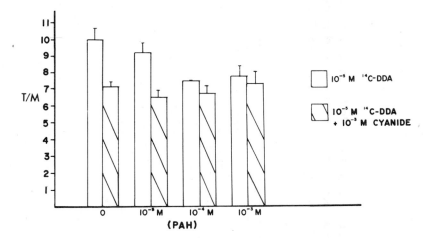

Fig. 4. Summary of the effects of the organic acid, PAH, on total ^{14}C-DDA uptake and cyanide inhibited uptake. [DDA] = 10^{-5}M. Results are expressed as 60-min T/M. Standard error of the mean is indicated by the line extended from each bar. N's range from 2 to 8.

Inhibition of energy-dependent and cyanide-insensitive uptake by the two organic acids (CPR and PAH) is summarized in Table 2. Note the marked differential in sensitivity between the energy-dependent, or active, uptake and the cyanide-insensitive uptake. The active accumulation was greatly reduced by both

TABLE 2

Percent Inhibition of Energy-Dependent Uptake and Cyanide-Insensitive Uptake of 14C-DDA in Flounder Kidney Tubules by the Organic Acids, Chlorophenol Red (CPR) and P-Aminohippuric Acid (PAH)

Concentration of Inhibitor	Energy-Dependent Uptake % Inhibition by		Cyanide-Insensitive Uptake % Inhibition by	
	CPR	PAH	CPR	PAH
10^{-5}M	11.1	16.6	4.3	-6.3
10^{-4}M	75.5*	75.6*	1.6	-11.3
10^{-3}M	65.7*	85.7*	16.24*	-6.6

*$p < 0.05$ vs. Control.

organic acids, while only CPR, at the highest concentration used, significantly decreased cyanide-
insensitive uptake.

Nonspecific tissue binding could account for DDA
accumulation while energy production was inhibited.
The parent pesticide, DDT, is known to be heavily
bound to plasma proteins, lecithin, and many cellular
fractions (Dvorchik and Maren, 1972; Tinsley et al.,
1971; Wells and Yarbrough, 1972). In addition, as
indicated above, we have measured DDT accumulation by
renal tubules in this system and found that it was
not sensitive to cyanide or organic acids, and thus
might be tissue binding. To test the possibility
that cyanide-insensitive uptake might be tissue binding, the binding of $10^{-5}M$ DDA to renal tubule homogenates was measured both alone and in the presence of
the organic acids, CPR, and PAH. As indicated in
Table 3, 13.6 times as much DDA was associated with
the particulate material in the pellet as was found
in the supernatant after ^{14}C-DDA (final concentration
$10^{-5}M$) was mixed with a 5% homogenate of renal tubules
and centrifuged. CPR ($10^{-3}M$) significantly reduced
the binding ratio (19%), while PAH ($10^{-3}M$) did not
change the binding ratio. This result was identical
to the effect of these two compounds on the cyanide-
insensitive uptake.

DISCUSSION

The results presented above are consistent with
two simultaneously operating mechanisms of DDA accumulation. The first is an energy-dependent process
capable of producing a concentration of DDA in the
tissue exceeding that in the medium, i.e., active
transport. Furthermore, the energy-dependent uptake
is strongly inhibited by the presence of other organic acids known to be actively transported on the
renal organic acid system (Weiner, 1967). Therefore,
pending more complete kinetic examination, it is
suggested that the active accumulation of DDA seen

174

TABLE 3

Effect of Chlorophenol Red (CPR) and P-Aminohippuric Acid (PAH) on the Binding Ratio* of DDA with Flounder Kidney Tubule Homogenates. Values for 5% Homogenate Are Mean ± Standard Error of the Mean (N = 3)

	Concentration of Inhibitor		
% Homogenate	0	10^{-3}M CPR	10^{-3}M PAH
5	13.59 ± 0.19	10.95 ± 0.11[†]	13.68 ± 0.09[†]
10	13.3	—	—

*Calculated as $\dfrac{\text{DPM/gram Pellet}}{\text{DPM/ml Supernatant}}$.

[†]p < 0.01 for CPR vs. Control; PAH treatment gave a binding ratio not signifi-cantly different for the control.

175

here occurs via the organic acid system. The second mode of DDA accumulation is apparently a nonspecific binding to the tissue masses. This hypothesis is supported by the lack of energy dependence of this component, the well-documented binding of DDT and metabolites to plasma proteins and cellular components mentioned above, and the identical responses of cyanide-insensitive uptake and tissue binding to chlorophenol red and p-aminohippuric acid.

The value of such a mechanistic or physiologic approach to environmental problems lies not only in the results *per se*, but also in predicting further specific consequences of exposure to environmental agents, means of investigating these effects, and methods for changing the handling of environmental agents. For example, excretion of DDA via the organic acid transport system could explain why urinary DDA excretion in man is such a good indicator of exposure to DDT (Wolfe *et al.*, 1970). Similarly, active intracellular accumulation of DDA could have severe consequences for renal cells, since DDT and related compounds have been shown to inhibit enzymes involved in oxidative phosphorylation (Cutkomp *et al.*, 1971; Yarbrough and Wells, 1971). Such a sequence of events is not unlike that seen with dinitrophenol (DNP), which is also actively accumulated by the organic acid system, and subsequently exerts its effects on oxidative phosphorylation (Haung and Lin, 1965; Ross and Weiner, 1972). In fact, such a process could explain the cellular damage seen by Gruppuso and Kinter (1973) in goldfish renal tubules following DDA exposure *in vitro*. A third potential consequence of active DDA uptake arises from the observation that DDA and other DDT metabolites inhibit the Na^+, K^+-ATPase. If DDA has a similar effect in the kidney, it could reduce sodium transport as well as transport of glucose, amino acids, and organic acids which may be coupled to the sodium gradient (Curran *et al.*, 1967; Robinson, 1972; Vogel and Kröger, 1966). A fourth possibility is that DDA may competitively decrease the transport of

constituents normally carried by this system, such
as uric acid or the bile acids (Weiner, 1967; Weiner
et al., 1964). Our own earlier work (Pritchard and
Kinter, 1970) presented evidence that such competitive inhibition does occur at least in vitro.

Finally, the observation that DDA is transported
on the organic acid system suggests that it may be
possible to increase DDA excretion pharmacologically,
just as urate excretion via this system is enhanced
by probenecid or salicylate (Zins and Weiner, 1968).
DDA excretion might also be increased by decreasing
its binding, increasing the fraction of total DDA
available for excretion via the organic acid transport system.

ACKNOWLEDGMENTS

This work was supported in part by USPHS Grant
RR-0542, EPA Contract 68-02-0577, and South Carolina
State Appropriation.

I wish to thank the South Carolina Department
of Wildlife and Marine Resources, particularly
Mr. Charles Farmer, for supplying many of the flounder used in this study. I also thank Ms. Margaret
Cauthen for her able technical assistance.

LITERATURE CITED

Curran, P. F., S. G. Shultz, R. A. Chez, and R. E.
    Fuisz. 1967. Kinetic relations of the Na-amino
    acid interaction at the mucosal border of
    intestine. J. Gen. Physiol. 50:1261-86.
Cutkomp, L. K., H. H. Yap, E. Y. Cheng, and R. B.
    Koch. 1971. ATPase activity in fish tissue
    homogenates and inhibitory effects of DDT and
    related compounds. Chem. Biol. Interactions 3:
    439-47.

Dvorchik, B. H. and T. H. Maren. 1972. The fate
p,p'-DDT [2,2-bis(P-chlorophenyl)-1,1,1-
trichloroethane] in the dogfish, *Squalus
acanthias*. *Comp. Biochem. Physiol. 42A*:205-11.

Forster, R. P. 1948. Use of thin kidney slices and
isolated renal tubules for direct study of
cellular transport kinetics. *Science 108*:65-67.

Gigon, P. L. and A. M. Guarino. 1970. Uptake of
probenecid by rat liver slices. *Biochem.
Pharmacol. 19*:2653-62.

Gruppuso, P. A. and L. B. Kinter. 1973. DDT inhibition of active chlorophenol red transport in
goldfish *(Crassius auratus)* renal tubules.
Bull. Environ. Contam. Toxicol. 10:181-86.

Gustafson, C. G. 1970. PCB's—prevalent and persistent. *Environ. Sci. Tech. 4*:814-19.

Haung, K. C. and D. S. T. Lin. 1965. Kinetic studies on transport of PAH and other organic acids
in isolated renal tubules. *Amer. J. Physiol.
208*:391-96.

Judah, J. D. 1949. Studies on the metabolism and
mode of action of DDT. *Brit. J. Pharmacol.
Chemother. 4*:120-31.

Maack, T., D. D. S. Mackenzie, and W. B. Kinter.
1969. Renal excretion of chlorophenol red, PAH,
and Diodrast by intact flounder *(Pseudopleuronectes americanus)*. *Bull. Mt. Desert
Island Biol. Lab. 9*:29-30.

O'Brien, R. D. 1967. *Insecticides: Action and
Metabolism*. New York: Academic Press.

Peterson, J. E. and W. H. Robison. 1964. Metabolic
products of p,p'-DDT in the rat. *Toxicol. Appl.
Pharmacol. 6*:321-27.

Pritchard, J. B., A. M. Guarino, and W. B. Kinter.
1973. Distribution, metabolism, and excretion
of DDT and Mirex in a marine teleost. *Environ.
Health Perspect. 4*:45-54.

_____ and W. B. Kinter. 1970. DDA: An
inhibitor of chlorophenol red transport by
flounder kidney tubules *in vitro*. *Bull. Mt.
Desert Island Biol. Lab. 10*:67-68.

Racker, E. 1965. *Mechanisms in Bioenergetics*. New York: Academic Press.

Risebrough, R. W. 1969. Chlorinated hydrocarbons in marine ecosystems. In: *Chemical Fallout*, pp. 5-23, ed. by M. W. Miller and G. G. Berg. Springfield, Ill.: Thomas.

Robinson, J. W. L. 1972. The inhibition of glycine and B-methylglucoside transport in dog kidney cortex slices by ouabain and ethacrynic acid: Contribution to the understanding of sodium-pumping mechanisms. *Comp. Gen. Physiol. 3*: 145-59.

Ross, C. R. and I. M. Weiner. 1972. Adenine nucleotides and PAH transport in slices of renal cortex: Effects of DNP and CN⁻. *Amer. J. Physiol. 222*:356-59.

Rudd, R. L. and S. G. Herman. 1972. Toxic effect of pesticide residues on wildlife. In: *Environmental Toxicology of Pesticides*, pp. 471-85, ed. by F. Matsumura, G. M. Boush, and T. Misata. New York: Academic Press.

Sladen, W. J. L., C. M. Menzie, and W. L. Reichel. 1966. DDT residues in Adelie penguins and a crabeater seal from Antartica. *Nature 210*: 670-73.

Tinsley, I. J., R. Hague, and D. Schmedding. 1971. Binding of DDT to lecithin. *Science 174*:145-47.

Vogel, G. and W. Kröger. 1966. Die Bedeutung des Transports, der Konzentration und der Darbietungsrichtung von Na⁺ fur den tubulären Glucose-und PAH-transport. *Arch. Ges. Physiol. 288*:342-58.

Weiner, I. M. 1967. Mechanisms of drug absorption and excretion. *Ann. Rev. Pharmacol. 7*:39-56.

_____, J. E. Glasser, and L. Lack. 1964. Renal excretion of bile acids: taurocholic, glycocholic, and cholic acids. *Amer. J. Physiol. 207*:964-70.

Wells, M. R. and J. E. Yarbrough. 1972. Retention of ^{14}C-DDT in cellular fractions of vertebrate insecticide-resistance and susceptible fish. *Toxicol. Appl. Pharmacol. 22*:409-14.

White, W. C. and T. R. Sweeney. 1945. The metabolism of 2,2-bis(P-chlorophenyl)-1,1,1-trichloroethane. I. A metabolite from rabbit urine, di(P-chlorophenyl)acetic acid; its isolation, identification, and synthesis. *Public Health Rept.* (U. S.) *60*:66-71.

Wolfe, H. R., W. F. Durham, and J. F. Armstrong. 1970. Urinary excretion of insecticide metabolites. *Arch. Environ. Health 21*:411-16.

Woodwell, G. M., C. F. Wurster, and P. A. Isaccson. 1967. DDT residues in an east coast estuary: a case of biological concentration of a persistent insecticide. *Science 156*:821-24.

Yarbrough, J. D. and M. R. Wells. 1971. Vertebrate insecticide resistance: the *in vitro* endrin effect on succinic dehydrogenase activity of endrin-resistance and susceptible mosquitofish. *Bull. Environ. Contam. Toxicol. 6*:171-76.

Zins, G. R. and I. M. Weiner. 1968. Bidirectional urate transport limited to the proximal tubule in dogs. *Amer. J. Physiol. 215*:411-22.

BEHAVIORAL EFFECTS OF DIELDRIN
UPON THE FIDDLER CRAB, *UCA PUGILATOR*

MARIAN L. KLEIN and JEFFREY L. LINCER

Mote Marine Laboratory
Sarasota, Florida   33581

Long-term sublethal effects resulting from chronic exposure of organisms to low levels of toxicants probably have the most important impact on the estuarine community (Butler, 1966). Odum *et al.* (1969) reported that, after exposure to DDT, fiddler crabs exhibited maladaptive behavior which probably resulted from abnormal neurophysiological functions. These findings suggest that gross effects on behavior requiring total coordination should receive careful attention. *Uca pugnax*, as well as other estuarine organisms, showed disturbed motor coordination after their habitat was sprayed with BHC at 0.1 pound per acre (George *et al.*, 1957). These changes resulted in the crab's inability to escape predators, which was reflected by unusual numbers of egrets and raccoons feeding on the crabs.

Fiddler crabs are an important intermediate step in the food web of the estuaries. The organic detritus on which fiddlers feed has often carried chemical pollutants into the estuarine habitat from connected watersheds (Butler, 1971; Hansen and Wilson, 1970). These crabs, in turn, serve as food for innumerable predators, like marine birds, raccoons, skunks, bobcats, larger crustacea, fish, etc. (Rathbun, 1918; Springer, 1963).

Uca pugilator was selected for this investigation because (1) it is the most abundant local fiddler crab species; (2) it adapts well to the laboratory environment; (3) it is at the summit of crustacean development in nervous system organization and complexity of display (Crane, 1943); and (4) extensive literature describing normal behavior in this species provides a basis for identifying aberrant behavior.

METHODS

On June 22, 1973, approximately 300 *Uca pugilator* (males and females) were collected on a sandy beach at the north end of Casey Key (Florida). Crabs were placed in a 160 cm × 120 cm × 60 cm high fiberglass-lined wood box outfitted with a wire mesh cover and movable sunshade. Sand was placed in this crabbery sloping from 15 cm at one end to 0 cm at the opposite end where there was an intake hose and drain. Crabs were maintained with unfiltered seawater pumped from Little Sarasota Bay daily to maintain a water depth of about 5 cm. Crabs were fed frozen smelts daily and *ad libidum*.

Twelve 15 cm × 30 cm × 20 cm high all-glass aquaria were used for the bioassay and were prepared by thorough washing and rinsing. Sifted and sterilized (2 hrs at 177°C) sand was placed in each aquarium and sloped from 5 cm deep to 0 cm at the drain end. Once a day, fresh, sand-filtered seawater was supplied to each aquarium to the top of a 2½ cm overflow for 1 min. The results of a chemical analysis of this water can be found in Table 1. Salinity of incoming water and pH of water standing in the aquaria were measured twice a week throughout the experiment. Salinity of incoming water ranged from 36.8 to 37.6 o/oo. pH ranged between 6.8 and 8.0. Water temperature in each aquarium was taken at every testing session and ranged from 28 to 30°C.

TABLE 1
Chemical Analysis[a] of Filtered Seawater

Chemical Compound	Concentration
PO_4-P	.074 ppm
NO_3-N	.0085 ppm
NH_3-N	.032 ppm
SiO_2-Si	.172 ppm
NO_2-N	.0012 ppm

[a]Samples analyzed on Technicon Auto-analyzer II using standard techniques of the Environmental Protection Agency (1971) with slight modifications as described by Humm (1971).

Approximately 1 month after capture, 72 male crabs were randomly selected and assigned in groups of six to the 12 small aquaria. At this time, 10 additional male crabs were removed from the crabbery and frozen for later dieldrin analysis to establish a preexperiment baseline. Only male crabs were used in order to eliminate sex differences in reaction to dieldrin, a difference observed by the authors in an earlier pilot study (unpublished data). Crabs were maintained for 1 week and *ad lib.* on "Tetramenu 4 in 1 ®" brand tetramin fish flakes which contained a minimum of 42% crude protein, 4% crude fat, and 9% crude fiber according to the container label.

Behavioral patterns chosen for observation were:

(1) Righting Response—Each crab was placed on its back, held there for 3 sec, released and time to return to normal stance recorded. A maximum of 60 sec was allowed as crabs that did not succeed in

turning over in that time had to be put in an upright position by the experimenter. Time to right was measured for one trial only and not repeated at that testing, in order to eliminate observed fatigue in higher dosed groups as a factor in the response.

(2) Runway Behavior—Each crab was placed in a wooden runway (15 cm × 60 cm × 5 cm deep) and subjectively scored on a 1 to 5 scale as to its escape ability. Taken into consideration were (a) escape attempts, such as running to end of runway and climbing over side, (b) running speed and agility in the escape attempts, (c) reaction to "threat" of experimenter's finger placed about 7 to 10 cm from the front of crab, and (d) position of major cheliped and ambulatories.

(3) Aquarium Behavior—Unquantified observations were made on moving about in the aquarium, burrowing, feeding, response to presence of experimenter, and response to "threat" of experimenter's finger.

After the week of acclimatization, all crabs were weighed and observed for the above behavioral patterns. Each aquarium was randomly assigned to a treatment level, two aquaria to each of six dietary groups. Food was prepared by macerating tetramin flakes with a glass mortar and pestle, then adding appropriate amounts of dieldrin dissolved in 1 ml pesticide grade acetone to achieve 0.1, 1, 10, and 50 ppm based on the dry weight of food. Food for the control groups was prepared as above by adding an equal amount of acetone. The acetone was allowed to evaporate before feeding commenced. Between feedings, food was stored in the dark to reduce the possibility of dieldrin photoisomerization. In view of previously observed maladaptive feeding behavior, which was possibly in response to dieldrin (unpublished data), two aquaria were designated extreme stress controls (ESC). Sand was removed from these aquaria and crabs maintained in 2½ cm of filtered seawater with clean rocks to provide shelter. This group received no food during the experiment (except for one crab which was cannibalized in each replicate aquarium) to

elucidate behavioral changes that might result from
starvation.

Behavior of crabs during 1 week acclimatization,
as well as continuing behavior of control groups, was
considered as "normal" for crabs confined under conditions of the experiment. That is: (1) all crabs
righted too quickly for time to be measured; (2) when
experimenter approached the aquarium table, crabs
scurried to the farthest corner; (3) when experimenter
put a finger in aquarium (about 5 to 7 cm from crab),
the crab made thrusting motions with major cheliped
and attempted to pinch the experimenter's finger;
(4) when crabs were observed from a distance, they
moved about the aquarium, feeding, burrowing in the
sand and interacting with other crabs, the major
cheliped being carried flexed horizontally across the
front of, and close to, the body except when used to
ward off threat; and (5) crabs in the runway frequently eluded experimenter, often climbing over the
side of the runway and scurrying to any available
shelter.

For the first 2 weeks (Phase I) crabs were fed
daily 0.57 g of dry, treated food per aquarium.
Behavior, as described above, was observed daily.
Although social activity in *Uca* generally is most in
evidence before and after low tide, there is a peak
of activity occurring between 8 A.M. and noon that is
somewhat independent of the tide. However, tides in
the Sarasota area do not follow a regular temporal
pattern and, therefore for consistency, daily behavioral observations were made between 10 A.M. and noon.

Crabs were weighed by group at the beginning of
the experiment and on days 8 and 15. At the end of
2 weeks, two crabs from each aquarium, i.e., four
from each treatment level, were removed and frozen
for later dieldrin analysis to determine rate of
toxicant uptake. Since there was significant variation in crabs selected for the experiment and in
those removed thereafter, weighing was discontinued
after 2 weeks since mean weight per aquarium was no
longer meaningful. Overall, the average weight of

the 72 crabs used was 4.23 g at the beginning of the experiment.

After 2 weeks on treated food, crabs were fed nontreated tetramin flakes for 4 more weeks (Phase II); during this period behavioral parameters were observed every other day. At the end of the experiment, all crabs were frozen for dieldrin analysis.

RESULTS

BEHAVIOR

Observed behavior of crabs in the treatment groups showed early and marked changes following dieldrin feedings. After 1 day on treated food, crabs in the 50 ppm groups had difficulty in righting themselves (Fig. 1). Crabs in the 10 ppm groups showed the same responses as the 50 ppm groups, but to a lesser degree. Their delay in righting was evident by day 2, and continued in time at one-third to one-half the impairment of the 50 ppm groups. By day 11, crabs in the 1 ppm groups were measurably slow in righting, reaching a maximum average delay in righting of 15 sec on day 23. Righting response delay in the 0.1 ppm group was only great enough to be measured on day 11 and from day 21 to day 25. At no time was the speed of righting delayed in either control or extreme stress control groups.

In interpreting Figure 1, it should be noted that in one 50 ppm aquarium, because of mortality and planned removal, only one crab remained on day 17. This was a small crab and, like most small crabs, it was consistently better able to right itself than larger crabs. Thus, the apparent brief recovery on day 33 in the 50 ppm groups is due to this remaining small crab righting in 14 sec.

Changes in runway behavior (Table 2) occurred immediately in the 50 ppm groups and persisted throughout the experiment. The posture of these crabs when placed in the runway was rigid with fully

186

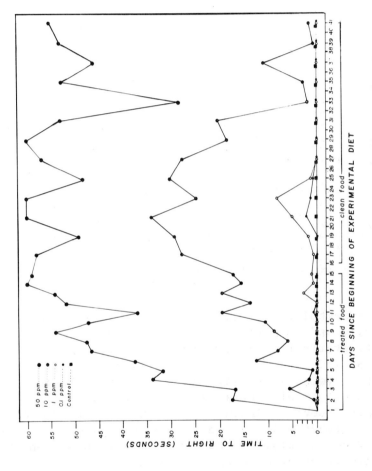

Fig. 1. Righting response time of fiddler crabs as affected by various levels of dietary dieldrin. Each point is the mean of two replicate means.

187

TABLE 2

Behavioral Responses of Fiddler Crabs on Various Dietary Dieldrin Doses to "Threat" of Investigator

Treatment	Day																			
	1	3	5	7	9	11	13	15	17	19	21	23	25	27	29	31	33	35	37	39
C	1	1	1	1	1	1	1	1	1	1	1	1	1	1	1	1	1	1	1	1
ESC	1	1	1	1	1	1	1	1	1	1	1	1	1	1	1	2	2	2	2	2
0.1 ppm	1	1	1	1	1	1	1	1	1	1	2	2	2	2	2	2	2	2	2	2
1.0 ppm	1	1	2	2	2	2	3	3	3	3	3	3	3	3	3	2	2	2	2	2
10 ppm	2	2	4	4	4	4	4	5	5	5	5	5	5	5	5	5	5	5	5	5
50 ppm	5	5	5	5	5	5	5	5	5	5	5	5	5	5	5	5	5	5	5	5

Key to runway behavior:

1—"Normal" behavior: very active, constant, and usually successful escape attempts. Movements of major chela in response to finger "threat" well coordinated.

2—Movements parallel "normal" but show sluggish responses.

3—Running is slow and awkward. Escape over runway side not always successful. Major chela movement shows moderate loss of coordination. Occasionally "freeze" rather than seeking escape.

4—Running very awkward with frequent stumbling. Frequently "freeze." Tremor in major chela which is held rigidly extended dorso-laterally. Movements of major chela in response to threat lack direction, are slow and jerky.

5—Posture is motionless with ambulatories so rigidly extended that several do not touch runway floor. No movement unless prodded. Major chela extended rigidly dorso-laterally. Any movement of major chela causes crab to fall over backwards.

extended ambulatories and major cheliped extended laterally and dorsally. In response to "threat" of investigator's finger, the crabs raised the major cheliped and in so doing usually fell over backwards. Walking movements, if any, were slow with frequent stumbling. Ambulatories were so rigidly extended that often several did not touch the ground. There was no attempt made to escape from the runway. Crabs maintained the rigid posture for 1 to 2 min, at which time ambulatories collapsed and the body rested on the substrate. Movements of the 10 ppm groups were slow after 1 day of the experimental diet, and by day 3 followed the same pattern as described for the 50 ppm group crabs. Both of these groups showed no recovery in runway behavior after being placed on clean food. Tremor was present in both 50 and 10 ppm groups from day 1 and day 5, respectively.

The first changes in runway behavior in the 1 ppm groups occurred on day 4 when sluggish movements were observed. The major chela was carried, away from the body, although not rigidly extended as in the higher dosage groups. Throughout the experiment, the 1 ppm groups showed delayed and awkward responses in the runway, ranging from almost normal to "freezing" with rigidly extended major chela. They were able to run, but seldom succeeded in getting over the side of the runway. Occasional tremor of the major chela was observed.

Runway behavior of the 0.1 groups appeared almost normal throughout the experiment. It should be noted that on day 21, uncoordinated running in the runway was observed which paralleled ephemeral righting response effect (Fig. 1). These crabs were able to escape fairly consistently from the investigator in the runway tests. However, for the last 3 weeks of the experiment, during which period they were on clean food, the 0.1 ppm groups appeared lethargic in their movements.

Crabs in the control and extreme stress control groups showed no changes in coordination of runway behavior, running swiftly, climbing out of the runway easily, and responding to "threat" with well-coordinated gestures. ESC crabs displayed, however, a slight overall slowing down of movements during the final week.

Behavioral changes in the aquaria occurred on day 1 in the 50 ppm groups, and by day 3 in the 10 ppm groups. These crabs maintained a rigid posture in the aquarium, moving only when prodded, and holding the major cheliped in an awkward position, fully extended laterally and dorsally. Tremor in this cheliped was frequent. There was no burrowing in the sand by these groups. When approached by investigator's finger, crabs raised their major chelae dorsally, falling over backwards in so doing. Crabs in the 1 and the 0.1 ppm groups were less active in the aquaria than the control groups, but did move about, feeding, burrowing and retreating to a far corner of the aquarium in response to approach by the investigator. The behavior in the aquaria of the control groups was consistently "normal" (as described previously) throughout the experiment. Although runway behavior of the extreme stress control groups remained normal throughout the experiment, generalized activity increased in the aquaria on day 11 until day 14. After this spurt of activity, these crabs became increasingly less active than controls, although response to disturbance of any kind remained normal.

When food was placed on the sand, the 50 ppm groups seemed unable to feed. If the crab was placed on the food, it made feeding motions with the small cheliped but seldom succeeded in getting food to its mouth. When the crab was placed a few centimeters away from the food, it made feeding movements but did not move close enough to be able to pick up any food. Some food remained on the sand most of the time between feedings. Crabs in the 10 ppm groups were able to feed, although with limited efficiency since some food often remained in the sand in these aquaria. The 1, 0.1 ppm and control groups consumed all the food given them.

WEIGHT CHANGE

No significant weight changes were observed in any group under the weighing schedule previously described. Since no change occurred in the ESC groups, which were not fed, nor in the 50 and 10 ppm groups which displayed feeding difficulty, the possible effects of starvation are obviously not readily discernable by simply measuring weight changes. Crabs not feeding might maintain their weight by increasing their water content but changes in protein, carbohydrates, fats, etc., would go unnoticed.

MORTALITY

The first mortality occurred in a 0.1 ppm group after 1 week of dieldrin feeding (Fig. 2). A large puncture wound in the carapace, which appeared to have been inflicted by the cheliped of another crab, was presumed to be the cause of death. The next mortality occurred in a 50 ppm group the next day. One more crab in the 50 ppm group died before the end of Phase I. During Phase II, deaths were observed in the 50 ppm groups at 1, 8, and 22 days (i.e., after the start of the clean food diet). Delayed mortality also occurred in the 10 ppm groups at 9 and 14 days after the start of Phase II. There were no deaths in the 1 ppm nor the control groups during the experiment. One ESC group cannibalized one crab after 16 days fasting and the other group cannibalized another crab after 27 days fasting. One crab in the latter ESC group also died from unknown causes after 39 days fasting.

DISCUSSION AND CONCLUSIONS

It appears that small amounts of dieldrin ingested over a short period of time by fiddler crabs cause behavioral changes that would interfere with their survival. Of special interest is that some of these behavioral changes occurred within 24 hrs of the first dieldrin feeding. Both behavioral and lethal effects showed a delayed response which continued into the period of clean food, and both severity of behavioral changes and mortality are dose-related (Figs. 1 and 2 and Table 2). Future analysis for dieldrin in these specimens should provide a meaningful comparison with levels in field-collected specimens.

As with most studies, more questions were created than were answered. Figure 1 indicates that

191

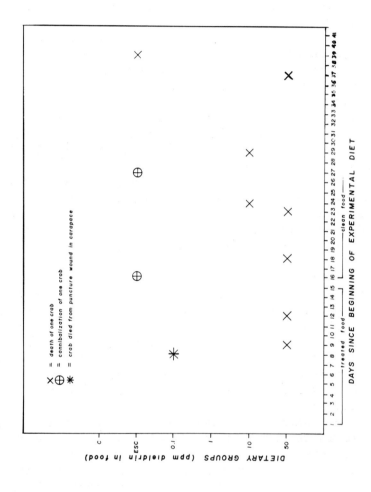

Fig. 2. Mortality response of fiddler crabs to various levels of dietary dieldrin.

some recovery from the toxic effects of dieldrin *may* have occurred during the last 2 weeks of the experiment. The extension of the experiment for several more weeks should distinguish if this is a true recovery or if it is an artifact of averaging the scores of a diminishing population, a population represented by only those individuals fit enough to survive the experimental insults.

It would be of value to repeat this study with more standardization of the size of crabs selected. We observed that the smaller males, whose major cheliped was smaller, relative to body size, than that of the large males, had less difficulty in righting themselves. This suggests that claw weight is a factor in the righting response. In a previous study a very apparent sex difference was noted, with females less deleteriously affected than males by dieldrin (unpublished data). Interestingly, there were no mortalities among females, but 50% of the males died in this pilot study in which 1 to 100 ppm dieldrin was fed for only 1 week. The obvious questions are whether this sex difference is physiological in origin or possibly the result of the males consuming more dieldrin. Were the observed differences in behavioral responses merely due to the mechanical difference caused by the males' large cheliped or is there a physiological explanation, as has been reported for the effect of mercury on this species (Vernberg and Vernberg, 1972; Vernberg *et al.*, 1974)? Is the female fiddler, with her unique set of enzymes, more able to induce the metabolic breakdown of exogenous toxicants?

Once again, it becomes obvious that sublethal effects in response to low level, chronic exposure to an environmental toxicant can be quite significant. The integrity of the righting response and coordinated escape behavior in this species is essential for survival. Crabs unable to carry out complex coordination would be unable to escape predation and/or avoid dehydration. An extremely important pressure on any population can be that of predation. We know that organochlorines applied under field conditions can, in fact, bring about behavior so maladaptive that extreme predation results (George *et al.*, 1957). Other aberrant and maladaptive behavior, such as the inability to burrow in the sand, or feed properly, would prove equally hazardous for this and related estuarine species.

In order to assess the ecological significance of the dieldrin-induced mortality and aberrant behavior which we have reported, it is necessary to relate our experimental dietary levels to available field levels.

Deubert and Zuckerman (1969) found dieldrin to remain primarily in the upper 2 inches of the soil where it averaged 1.2 ppm (oven dry weight basis) and was, thereby, presumably more available to be transferred downstream by erosional forces. That dieldrin residues do find their way to the sediments of estuaries is exemplified by reports of 0.5 to almost 4 ppb (Fay and Newland, 1972; Rowe *et al.*, 1971). Marine algae have been shown to concentrate dieldrin to well above dieldrin's solubility in water (Rice and Sikka, 1973) with especially high concentrations associated with runoff from heavily farmed areas (Seba and Corcoran, 1969). Fiddler crabs are opportunistic in their feeding, often scavaging and consuming dead fish that wash up on the shore. In this light, it is especially interesting that dieldrin applications as low as 0.2 lb/A killed many fish (Ginsburg and Sobbins, 1954) and dieldrin levels in apparently healthy marine fish have been reported at 3-13 ppb (Fay and Newland, 1972).

Whether the fiddler feeds on the diatoms on the sediment or the fishes and plankton washed up on the beach, the literature indicates that dieldrin residues approaching our lower dietary levels *are* present in some estuaries. We fed low levels of dieldrin for only 2 weeks. The results indicate that the effects of even the lower dietary levels were on the rise as time progressed (Figs. 1 and 2 and Table 2). What effects even lower dietary levels would have over a longer period of time, such as a year, awaits further experimentation.

ACKNOWLEDGMENTS

We gratefully acknowledge the William G. Selby and Marie Selby Foundation for support of this research. Special thanks go to Dr. William Tavolga for commenting on the manuscript and Ms. Val Maynard for analyzing the seawater.

LITERATURE CITED

Butler, P. A. 1966. Fixation of DDT in estuaries. Trans. North Amer. Wildlife and Nat. Res. Conf., pp. 184-89.
_____. 1971. Influence of pesticides on marine ecosystems. *Proc. Roy. Soc. London B 177*:321-29.
Crane, J. 1943. Display, breeding and relationships of fiddler crabs (Brachyura, genus *Uca*) in the Northeastern United States. *Zoologica 28*:217-23.
Deubert, K. H. and B. M. Zuckerman. 1969. Distribution of dieldrin and DDT in Cranberry Bog Soil. *Pestic. Monit. J. 2*:172-75.
Environmental Protection Agency. 1971. Methods for chemical analysis of water and wastes. U. S. Government Printing Office, Washington, D. C.
Fay, R. R. and L. E. Newland. 1972. Organochlorine insecticide residues in water, sediment and organisms, Aransas Bay, Texas, September, 1969- June, 1970. *Pestic. Monitor. J. 6*:97-102.
George, J. L., R. F. Darsie, Jr., and P. F. Springer. 1957. Effects on wildlife of aerial applications of Strobane, DDT, and BHC to tidal marshes in Delaware. *J. Wildlife Mgmt. 21*:42-53.

Ginsburg, J. M. and D. M. Jobbins. 1954. Research activities in chemical control of mosquitoes in New Jersey in 1953. Proc. of the 41st Annual Meeting, New Jersey Mosquito Extermination Assoc., and 10th Annual Meeting, American Mosquito Control Assoc., pp. 279-91.

Hansen, D. G. and A. T. Wilson, Jr. 1970. Residues in fish, wildlife, and estuaries. Significance of DDT residues from the estuary near Pensacola, Florida. *Pestic. Monit. J.* 4:51-56.

Humm, H. (Ed.). 1971. Anclote Environmental Project. Report. Submitted to Florida Power Corp. University of South Florida, Department of Marine Biology.

Odum, W. E., G. M. Woodwell, and C. F. Wurster. 1969. DDT residues absorbed from organic detritus by fiddler crabs. *Science 1964*:576-77.

Rathbun, M. J. 1918. The Grapsoid crabs of America. *Smith. Inst. Bull.* 97:401.

Rice, C. P. and H. C. Sikka. 1973. Fate of dieldrin in selected species of marine algae. *Bull. of Environ. Contam. and Toxicol.* 9:116-23.

Rowe, D. R., L. W. Canter, P. J. Snyder, and J. W. Mason. 1971. Dieldrin and endrin concentrations in a Louisiana estuary. *Pestic. Monit. J.* 4: 177-83.

Seba, D. and E. F. Corcoran. 1969. Surface slicks as concentrations of pesticides in the marine environment. *Pestic. Monit. J.* 3:190-93.

Springer, P. F. 1963. Fish and wildlife aspects of chemical mosquito control. Proc. 50th Annual Meeting of New Jersey Mosquito Extermination Assoc., pp. 194-203.

Vernberg, W. B., P. J. DeCoursey, and J. O'Hara. 1974. Multiple environmental factor effects on physiology and behavior of the fiddler crab, *Uca pugilator*. This volume.

_____ and J. Vernberg. 1972. The synergistic effects of temperature, salinity, and mercury on survival and metabolism of the adult fiddler crab, *Uca pugilator*. *Fishery Bull. 70*:415-20.

OSMOTIC AND IONIC REGULATION IN DECAPOD CRUSTACEA EXPOSED TO METHOXYCHLOR

RICHARD S. CALDWELL

Department of Fisheries and Wildlife
Marine Science Center
Oregon State University
Newport, Oregon 97365

Crustaceans exhibit a wide diversity of response with respect to osmotic and ionic regulatory abilities. Many forms such as brine shrimp, prawns and certain shore crabs are able to maintain a degree of osmotic balance in both hypo- and hypersaline environments. Other species regulate blood hyperosmotic to the medium at low salinities, but conform at higher salinities while many of the true marine crabs are completely deficient in osmoregulatory ability (Lockwood, 1967). All crustaceans are capable of some degree of ionic regulation of the blood. The general pattern in normal seawater is toward elevated blood K^+ and Ca^{++}, a depression of the blood Mg^{++} and SO_4^{--}, while blood Na^+ and Cl^- are frequently unchanged (Prosser, 1973). The regulation of the ionic composition of body fluids in animals is presumed to have adaptive significance (Burton, 1973).

Features important to osmotic and ionic regulation in decapod crustaceans are the permeability of the integument to salts (Gross, 1957; Nagel, 1934) and water (Rudy, 1967), active ion uptake or excretion in the gills (Shaw, 1961), and occasionally the gut (Green *et al.*, 1959), and ion transport in the excretory organ (Dehnel and Carefoot, 1965). It is widely believed that active transport processes in biological membranes are driven by the energy stored in ATP and released by the activity of ATPases. The Na^+, K^+ activated, Mg^{++} dependent ATPases (NaKMg ATPases) are central to most mechanisms of active membrane transport and are believed to be involved not only in Na^+ and K^+ transport, but also in the transport of other chemical or ionic species (Whittam and Wheeler, 1970).

Recently, several studies have shown that DDT and related pesticides and the polychlorinated biphenyls (PCB's) inhibit the activity of NaKMg-ATPases (Cutkomp *et al.*, 1971; Davis *et al.*, 1972; Desaiah *et al.*, 1972; Koch, 1969/70; Yap *et al.*, 1971). It has been postulated that ATPase inhibition by these compounds may impair the osmoregulatory ability of eels *(Anguilla rostrata)* and killifish *(Fundulus heteroclitus)* (Janicki and Kinter, 1971; Kinter *et al.*, 1972). In earlier work, Eisler and Edmunds (1966) have shown that alterations of blood and tissue ions occur in puffers *(Sphaeroides maculatus)* exposed to the organochlorine pesticide endrin. However, puffers exposed to sublethal concentrations of methoxychlor did not differ from controls in any of these same parameters (Eisler, 1967). In a similar study Nimmo and Blackman (1972) reported that Na^+, K^+ and possibly Mg^{++} levels of shrimp *(Penaeus aztecus* and *Penaeus duorarum)* hepatopancreas were reduced following exposure to sublethal concentrations of DDT in seawater. These observed effects on shrimp may have been caused by an alteration of ion transport processes in the cell membrane.

In the present study, we have examined the resistance to osmotic stress of adult *Cancer magister* and *Hemigrapsus nudus* exposed to the insecticide methoxychlor [2,2-bis(p-methoxyphenyl)-1,1,1-trichloroethane] in seawater. In addition, we have

studied the physiological consequences of such exposure on the blood osmotic regulation of both of these
species in normal and dilute seawater, and on the
blood ionic regulation and gill ATPase activity of
C. magister.

MATERIALS AND METHODS

Adult male *Cancer magister*, 80-100 mm carapace
width, were collected by trawl from Yaquina Bay,
Oregon. Male *Hemigrapsus nudus*, 15-30 mm carapace
width, were collected from the intertidal zone of an
exposed rocky shore. Both species were laboratory
acclimated for 2 to 4 days in large polyethylene
tanks continually supplied with fresh filtered seawater at a rate of 4 l/min. Subsequently, both
species of crabs were exposed to 0.01 mg/l of methoxychlor (administered as the technical product) in the
same tanks for from 7 to 14 days in different experiments. Methoxychlor was metered into seawater as an
acetone solution at such a rate as to give a final
acetone concentration of 0.1 ml/l. The control tanks
received acetone only. Crabs were exposed to constant
laboratory illumination at all times. Seawater temperatures during the acclimation period and during
later pesticide exposures in these tanks ranged from
12 to 14°C, salinities were 30 to 33 o/oo and dissolved oxygen was maintained at saturation levels by
vigorous aeration. Crabs were fed pieces of frozen
flounder every 2 to 3 days during the acclimation and
pesticide exposure periods, but were not fed during
the tolerance experiments or during the low salinity
exposure periods described later.

The lethal concentrations of methoxychlor for
H. nudus were determined in static acute toxicity
bioassay tests following laboratory acclimation. The
concentrations of methoxychlor tested ranged from 0.1
to 50 ug/l. Lethality bioassays were conducted in
1-gallon glass jars containing 3 liters of seawater
and 10 crabs per jar. Seawater temperatures during

the 4-day test period ranged from 17 to 18°C, the salinity was 30 o/oo, and the dissolved oxygen levels were maintained at saturation by aeration. Deaths were recorded at 24-hr intervals, and the criterion of death used was lack of visible appendage or mouth part movement during a 1-min observation period.

Salinity tolerance of *H. nudus* was determined after a 7-day preexposure of crabs to 0.01 mg/l methoxychlor in flowing seawater as previously described. A control group was exposed to acetone alone. At the start of the tolerance test, 7 to 10 crabs were placed in each of several 3 liter bioassay jars containing various seawater dilutions and either 0.1 ml/l acetone or acetone plus 0.01 mg/l methoxychlor. Mortalities were recorded daily for up to 6 days. The dissolved oxygen level in the jars was maintained at saturation by aeration, and the temperature was 17 to 18°C.

In salinity tolerance tests with *C. magister*, crabs were preexposed in the flowing water system for 10 days to 0.01 mg/l methoxychlor or acetone only as previously described. They were then transferred to glass troughs containing various seawater dilutions and 0.01 mg/l methoxychlor or acetone supplied in a flowing bioassay system, and subsequent mortalities were recorded daily during a 7-day exposure period. Temperature and dissolved oxygen were the same as during the prebioassay period. Frequent analysis of the seawater in these and later tests showed that the methoxychlor concentration was maintained in the flowing water systems to within ± 10% of the desired value.

In order to examine the effects of methoxychlor on osmotic and ionic regulation, *C. magister* were preexposed for 14 days to 0.01 mg/l of methoxychlor or to acetone alone, as described previously. The crabs were then transferred to the glass troughs used in the salinity tolerance experiment and exposed for 48 hrs to various seawater dilutions with continued pesticide exposure. After 48 hrs blood and urine samples were collected for determination of the osmotic concentrations and the concentrations of Na$^+$, K$^+$ and Mg^{++} ions. Following the 14-day exposure period, two other groups of crabs were taken for determination of whole body and gill methoxychlor residues, and for determination of gill ATPase activities, respectively.

In osmoregulatory studies with *H. nudus*, crabs were preexposed to 0.01 mg/l of methoxychlor in the flowing seawater system for a period of 7 days. Groups of treated crabs were then transferred into 3 liters of various seawater dilutions in glass jars as in the salinity tolerance tests with this species, and, after 48 hrs, blood samples were taken for the determination of osmotic concentration. The seawater was not changed during this period, but analysis of the water bathing the crabs exposed to pesticide showed that methoxychlor levels were not reduced below 0.005 mg/l during this period. All other conditions of exposure were the same as for the salinity tolerance tests. After the 7-day preexposure period, additional crabs were frozen and stored for subsequent whole body and gill tissue residue analysis as for *C. magister*.

Blood samples were obtained from both species of crabs by puncture of the thin membrane at the base of the second or third walking leg. Blood was taken from *C. magister* using Pasteur pipettes, transferred to polyethylene vials, and frozen for storage. Upon thawing, the clots were broken up with a glass stirring rod and aliquots of serum were used undiluted for measurements of osmotic concentration or after dilution for the analysis of Na$^+$, K$^+$ and Mg^{++} ions. Samples of blood from *H. nudus* were collected in capillary tubes and immediately used for the determination of osmotic concentration. Urine samples were collected from *C. magister* into 0.25 ml syringes by careful insertion of a blunted 24 gauge needle into the nephropore, and were stored frozen in small polyethylene vials. Osmotic concentrations were determined using a Wescor Model 5100 vapor pressure osmometer. A Perkin-Elmer Model 403 atomic absorption spectrophotometer was used in the determination of Na^+, K^+ and Mg^{++} in blood and urine samples.

ATPase activity was determined on whole homogenates of *C. magister* gill. Gills from freshly killed crabs were homogenized in 0.05 M glycylglycine buffer, pH 7.4, at high speed in a Sorvall omnimixer. The whole homogenate was then centrifuged for 10 min at $30 \times g$ to sediment course materials. Total ATPase activity was measured in a 2 ml reaction medium containing 100 mM NaCl, 15 mM KCl, 5 mM $MgCl_2$, 6 mM ATP and 50 mM glycylglycine buffer, pH 7.4. In some assays, Mg ATPase was determined by adding 1 mM ouabain to the reaction medium to inhibit the NaK activated enzyme. NaK ATPase was calculated as the difference between total ATPase and Mg ATPase. The reaction was allowed to proceed for 30 min at $15^{\circ}C$ and was terminated by pipetting 0.5 ml aliquots of the reaction medium into 0.5 ml 10% TCA. Phosphate was determined by the method of Fiske and SubbaRow (1925). Protein was estimated by the Lowry procedure (Lowry *et al.*, 1951). Specific activity was expressed as µmoles P_i hydrolysed per mg protein per hour.

Methoxychlor residues in whole crab and excised gills were determined by gas liquid chromatography after solvent extraction of the homogenized tissue. The procedure involved a 15-min extraction of 2 g of homogenized tissue with 10 ml of glass redistilled acetone followed by an additional 15-min extraction with shaking after the addition of 20 ml of glass redistilled hexane. The extract was then shaken with 20 ml of hexane extracted, glass distilled water. Two ml of the hexane layer were then taken for further clean-up on a florisil column (3 g activated florisil) contained in a 10 mm I.D. glass chromatography tube. Methoxychlor was eluted from the column with 40 ml of 5% ether in hexane and the eluate was chromatographed without further treatment. Gas chromatographic analyses were performed on a Hewlett-Packard Model 5140 chromatograph equipped with a ^{63}Ni electron capture detector, and a 6 ft $\times$ 4 mm I.D. glass column packed with a 3.8% UCW-98 on 80/100 mesh chromosorb

W-HP. The flow rate was 60 ml per minute with an oven temperature of 220°C and a detector temperature of 300°C.

RESULTS AND DISCUSSION

In earlier work we had determined that the 96-hr LC_{50} of technical methoxychlor for adult *C. magister* was 0.13 mg/l in static acute bioassay tests, and that a 96-hr EC_{50} was 0.046 mg/l (Armstrong *et al.*, 1974). The toxic criterion used in the EC_{50} tests was inability to regain an upright posture after being placed on the dorsal surface. In flowing water tests lasting 85 days, we found that 50% of the crabs exposed to 0.04 mg/l methoxychlor survived for 20 days with complete mortality occurring in 48 days (Armstrong *et al.*, 1974). In crabs exposed to only 0.004 mg/l methoxychlor, survival exceeded 50% until day 80. On the basis of these values, a methoxychlor concentration of 0.01 mg/l was selected for use in the present studies with *C. magister* to test the hypothesis that osmotic or ionic imbalance was a con-tributing factor in methoxychlor-induced mortality. Because this concentration would ultimately be lethal, it was assumed that blood osmotic or ionic imbalance would be detected if these factors were contributory. Since we had not previously determined the lethal level of methoxychlor for *H. nudus*, 96-hr static acute bioassays were conducted in order to estimate a suitable exposure concentration for the osmotic ef-fects experiments with this species. The mortality data, listed in Table 1, show the normal dose-response relationship. At 96 hrs the estimated LC_{50} was 0.017 mg/l. Crabs exposed to 0.005 mg/l and higher were clearly hyperactive while the threshold of hyper-activity appeared to occur between 0.0005 and 0.001 mg/l. From these data it was apparent that *H. nudus* was nearly an order of magnitude more sensitive to the pesticide than was *C. magister*. Nevertheless, for uniformity, we also selected an exposure

TABLE 1
Acute Toxicity of Methoxychlor to Hemigrapsus nudus

Initial Methoxychlor Concentration in Seawater (mg/l)	Total Number of Crabs	% Mortality (Cumulative)			
		24 Hrs	48 Hrs	72 Hrs	96 Hrs
0.005	10	0	0	0	0
0.010	10	30	30	40	40
0.017	10	30	30	30	50
0.025	10	20	30	40	60
0.050	10	80	90	90	90

concentration of 0.01 mg/l methoxychlor for osmotic studies with *H. nudus*.

An indication of whether exposure of crabs to methoxychlor was seriously affecting osmotic responses could be obtained by studying the relative survival of treated and control crabs exposed to low salinities. Tables 2 and 3 give the results of two such experiments using *C. magister* and *H. nudus*, respectively. For both species there was a clear indication that the crabs treated with pesticide were less resistant to the low salinity exposures. Although no mortalities occurred at the higher two salinities tested, within 2 days mortalities were occurring in *C. magister* at 9.7 and 5.5 o/oo salinity (Table 2).

TABLE 2
Percent Mortality of Methoxychlor-Exposed
Cancer magister *in Dilute Seawater*[a]

| Treatment | Salinity Exposure (o/oo) | | | |
	26.5	17.1	9.7	5.5
2 days				
Control	0	0	29	43
Methoxychlor,				
0.01 mg/liter	0	0	47	73
7 days				
Control	17	7	64	100
Methoxychlor,				
0.01 mg/liter	13	7	100	100

[a]Prior to the initiation of low salinity exposures, crabs had been preexposed to 0.01 mg/l methoxychlor or acetone only for 10 days at 12-14°C and 30-33 o/oo salinity.

The percentage mortality of crabs exposed to methoxychlor at this time was nearly twice that for the control group of crabs. By 7 days complete mortalities
were noted in both groups of crabs exposed to 5.5 o/oo
salinity and complete mortality was found in the methoxychlor treated group at 9.7 o/oo, but only 64% mortality was noted in the control group at this salinity. Alspach (1972) reported that adult *C. magister*
held in 25% seawater were unable to survive for longer
than 48 hrs, and that crabs exposed to 50% seawater
survived for only 3 days before becoming moribund.
Survival of *H. nudus* in dilute seawater was better
than that of *C. magister* with no mortalities occurring within 2 days in the control group exposed to
7.4 o/oo salinity. After 6 days no mortalities were

TABLE 3
Percent Mortality of Methoxychlor-Exposed Hemigrapsus
nudus *in Dilute Seawater*[a]

Treatment	Salinity Exposure (o/oo)			
	24.6	17.9	12.8	7.4
2 days				
Control	0	0	0	0
Methoxychlor,				
0.01 mg/liter	0	0	14	29
6 days				
Control	0	0	0	50
Methoxychlor,				
0.01 mg/liter	0	14	29	86

[a]Prior to the initiation of low salinity exposures, crabs had been preexposed to 0.01 mg/liter
methoxychlor or acetone only for 7 days at 12-14°C
and 30-33 o/oo salinity.

observed in the control group at 12.8 o/oo salinity, but 50% mortality occurred at 7.4 o/oo salinity. Survival of the crabs treated with methoxychlor, however, was considerably poorer. After 2 days exposure to reduced salinity, mortalities reached 14 and 29% in 12.8 and 7.4 o/oo, respectively. After 6 days 14% mortalities were observed even at 17.9 o/oo salinity, while mortalities reached 29 and 86% at the two lower salinities tested.

These results show that pesticide-exposed crabs of both species are less resistant than control crabs to periods of low salinity exposure. Decreased survival at low salinities could be caused by a reduced osmoregulatory response in the pesticide-treated crabs. Kinter *et al.* (1972) have shown that the osmolarity and the Na^+ and K^+ concentrations in *Anguilla rostrata* serum are significantly elevated following a 6-hr exposure to 1 ppm DDT in seawater, a concentration of the pesticide which is fatal in 10 hrs. In additional studies with killifish, *Fundulus heteroclitus*, both DDT and Aroclor 1221 significantly elevated serum osmolarity but only Aroclor 1221 exposure caused elevation of serum Na^+ levels (Kinter *et al.*, 1972). To determine whether a similar effect on osmotic and ionic regulation was occurring in crabs, we exposed adult *C. magister* and *H. nudus* to 0.01 mg/l methoxychlor in seawater for 14 and 7 days, respectively, and subsequently held the crabs for 2 more days in a series of reduced salinities. The osmoregulatory responses for both species are illustrated in Figures 1 and 2. Blood osmotic concentrations were not affected at any of the salinities examined in either species. The osmotic concentrations of the blood of *H. nudus* exposed to the pesticide appeared instead to be slightly higher than found in the control animals, but the differences were not found to be statistically significant (P > 0.05). The results confirm the observations of Alspach (1972) that *C. magister* is a weak hyperosmoregulator at low seawater salinities. The results for *H. nudus* also confirm the superior

207

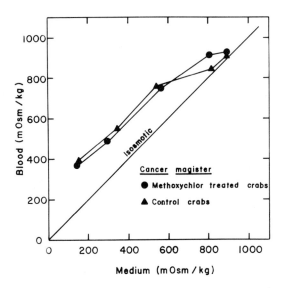

Fig. 1. Blood osmotic concentration of control and methoxychlor treated Cancer magister *as a function of medium osmolarity. Each point represents the mean of analyses on from 7 to 11 crabs.*

hyperosmoregulatory abilities of this species (Dehnel, 1962) although our crabs appeared to be less efficient regulators at the lower salinities than the British Columbia crabs.

Methoxychlor accumulation by the gills and whole body residues in *H. nudus* and *C. magister* following 0.01 mg/l exposures is shown in Table 4. Whole body and gill residues for each species were similar. The slightly higher values, particularly whole body residues, obtained for *C. magister* may be related to the length of exposure which was double that for *H. nudus*. In earlier studies (Armstrong *et al.*, 1974), we determined that methoxychlor uptake by *C. magister* approximates first order kinetics with uptake not yet leveled off in 15 days. Whole body residues of *C. magister* used in the osmotic studies were very similar to levels found by Armstrong *et al.* (1974) in

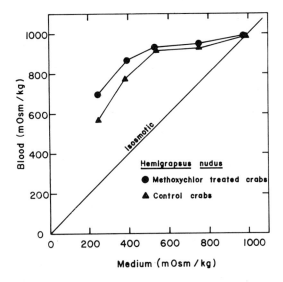

*Fig. 2. Blood osmotic concentra-
tion of control and methoxychlor
treated* Hemigrapsus nudus *as a
function of medium osmolarity.
Each point represents the mean of
analyses on from 7 to 10 crabs.*

crabs that had died as a result of methoxychlor
exposure lending support to the supposition that the
pesticide concentrations employed in the present
study were only slightly sublethal.

Blood Na$^+$ regulation was also unaffected by
methoxychlor treatment in *C. magister* (Fig. 3). The
pattern of regulation of this ion was essentially the
same as that observed for osmotic concentration, consistent with the fact that the Na$^+$ ion is the major
cation found in blood. The pattern of urine Na$^+$ concentrations at salinities less than 30 o/oo was comparable to that of the blood in both groups, confirming the extrarenal regulation of this ion at salinities less than full strength seawater (Alspach, 1972;
Engelhardt and Dehnel, 1973). In full strength seawater salinities and higher, some regulation of blood
Na$^+$ by reabsorption of this ion from the urine has

TABLE 4

Whole Body and Gill Tissue Methoxychlor Residues of H. nudus *and* C. magister *Exposed for 7 and 14 Days, Respectively, to 0.01 mg/liter Methoxychlor*

Species	Methoxychlor Concentration (μg/g wet wt.)	
	Whole Body[a]	Gill[b]
H. nudus	0.31	2.0
C. magister	0.82	2.4, 4.9
	1.1	2.6, 2.5
		2.6

[a]Data for *H. nudus* is for a pooled sample of two crabs and that for *C. magister* represents analyses of two individual crabs.

[b]Data for *H. nudus* is for a pooled sample of gills from eight crabs. For *C. magister*, each number represents an analysis on a sample of gills pooled from two crabs.

been demonstrated (Alspach, 1972; Engelhardt and Dehnel, 1973). Our crabs exposed to 30 o/oo salinity had urine Na^+ levels less than that of the blood, confirming this relationship. Although this response was observed in both the methoxychlor-exposed and control crabs, the differences between blood and urine Na^+ concentrations were greatest in the control animals.

Blood K^+ regulation in *C. magister* (Fig. 4) appeared to be poorer at low salinities than that reported by Engelhardt and Dehnel (1973) but similar to the results reported by Alspach (1972). Our

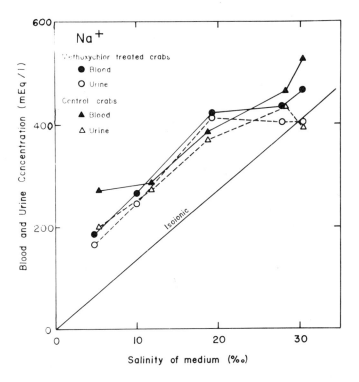

Fig. 3. Blood and urine Na$^+$ concentration of control and methoxychlor treated Cancer magister *as a function of experimental salinity. Each point represents the mean of analyses on from 7 to 11 crabs.*

results with blood K$^+$ levels were highly variable and exhibited certain spurious responses. For example, blood K$^+$ in control crabs was significantly lower (p < 0.05) at 5 o/oo salinity than that found in methoxychlor-treated crabs. However, this response was reversed at about 28 o/oo salinity where methoxychlor-treated crabs had lower blood K$^+$ levels than those found in the control animals. At the higher salinities used, the urine to blood ratio was less than unity supporting the contention of renal K$^+$ regulation in high salinities (Alspach, 1972; Engelhardt and Dehnel, 1973). However, except for

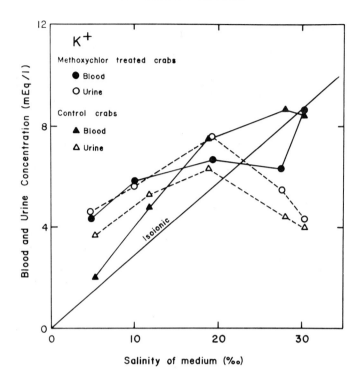

Fig. 4. Blood and urine K+ concentration of control and methoxychlor treated Cancer magister *as a function of experimental salinity. Each point represents the mean of analyses on from 7 to 11 crabs.*

the spurious observations already mentioned, blood and urine K+ levels were not significantly different in comparisons between the control and methoxychlor-treated groups at any of the salinities examined, indicating that the pesticide exposure had no effect on the regulation of this ion.

As previously reported in *C. magister* (Alspach, 1972; Engelhardt and Dehnel, 1973) and in several other crustacean species (Gross, 1964; Gross and Capen, 1966), blood Mg++ was strongly regulated to levels below that of the seawater medium (Fig. 5). The only indication that methoxychlor treatment

affected the blood Mg^{++} concentration was seen at
28 o/oo salinity where the pesticide-treated crabs
had significantly higher ($p < 0.05$) blood Mg^{++} levels
than that seen in the controls. However, this difference was not observed in crabs exposed to 30 o/oo
salinity.

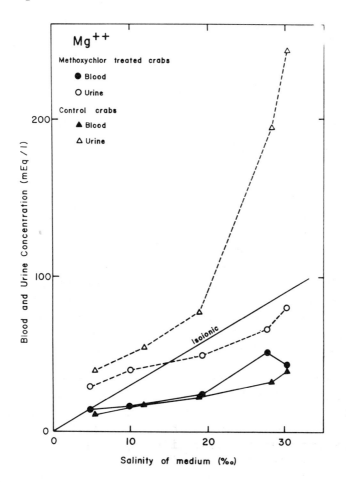

Fig. 5. *Blood and urine Mg^{++} concentration
of control and methoxychlor treated* Cancer
magister *as a function of experimental
salinity. Each point represents the mean
of analyses on from 7 to 11 crabs.*

Control animals exposed to the higher salinities produced urine in which the Mg^{++} concentrations were appreciably elevated over that of the blood and the medium (Fig. 5). However, crabs treated with methoxychlor failed to produce a urine concentrated in Mg^{++}. In view of our inability to demonstrate that methoxychlor affected the regulation of Na^+ and K^+, this observation is of particular interest. The regulation of blood Mg^{++} by crabs has been shown to be a function of the excretory organs, specifically the bladder wall (Gross and Capen, 1966). Although the highest urine Mg^{++} concentrations have been found in crabs immersed in normal and concentrated seawater, Gross and Marshall (1960) demonstrated that, in *Pachygrapsus crassipes*, the total amount of Mg^{++} excreted per day may be highest when crabs are placed in 50% seawater— since the rate of urine production is appreciably increased in dilute media. The rate of Mg^{++} transport across the bladder wall appears to be a direct function of the blood Mg^{++} level (Gross and Capen, 1966), although this property seems to be better developed in *Carcinus maenas* than in *P. crassipes* (Lockwood and Riegel, 1969). Associated with the secretion of Mg^{++}, Na^+ is reabsorbed from the urine, presumably by direct exchange in order to achieve electrochemical balance (Gross and Capen, 1966). We have observed that urine Na^+ concentration is depressed relative to blood Na^+ at a salinity of 30 o/oo (Fig. 3). Although the urine Na^+ concentration of methoxychlor-treated crabs at this salinity is not significantly different than that of the control crabs, the difference between urine and blood concentrations is less, possibly a consequence of the effect of the pesticide on Mg^{++} transport across the bladder wall.

Although the most plausible explanation of the effect of methoxychlor on urine Mg^{++} concentration is that the pesticide inhibits the active transport of the ion across the bladder wall, another explanation can be considered. The rate of urine production in crabs is inversely related to the salinity of the medium because in low salinities excess water is

214

absorbed by the animal. As a result, when crabs are in low salinity seawater, the bladder contents are frequently voided, reducing the possibility of concentrating Mg^{++} in the urine (Gross and Capen, 1966). An alternative explanation, then, for our observations that urine Mg^{++} concentrations were low in crabs exposed to methoxychlor is that the rate of urine production in high salinities was much higher in the pesticide-exposed crabs than in the control crabs. Although we did not measure the rates of urine production at the experimental salinities used, it seems highly unlikely that urine production by pesticide-exposed crabs would be unusually high since we did not see any significant difference between control and treated crabs in the urine Na^+ and K^+ concentrations at high salinities. Also, it is well known that the flow rate of urine in hyperosmoregulating crabs is related to the osmotic gradient between the blood and the medium (Lockwood, 1967), a difference which increases as salinity decreases. It seems most likely, therefore, that our results were caused by a direct inhibition of transbladder Mg^{++} transport by methoxychlor.

Interpretation of the specific effects of methoxychlor on Mg^{++} transport in crab bladder is difficult because little information is available concerning the biochemical nature of this process. There is considerable evidence that different types of transport enzymes exist (Oxender, 1972) so that it is reasonable to expect that the Mg^{++} transport process in crab bladder may be quite different from the more thoroughly studied NaKMg ATPase transport system. Indeed, Gross and Capen (1966) have shown that crabs treated with the Na^+ transport inhibitor, ouabain, were still able to transport Mg^{++} into the urine discounting the possibility that Mg^{++} secretion is enzymatically coupled to active Na^+ uptake from the urine. Similar evidence for the independent nature of divalent cation transport from the NaKMg ATPase transport system was obtained by Schatzmann and Vincenzi (1969). They have shown in their studies that the active transport of Ca^{++} in human red blood cells is also unaffected by ouabain, as well as by oligomycin and Na^+ and K^+ ions.

Although we have obtained evidence that the Mg^{++} transport system in the bladder wall was inhibited by methoxychlor, no evidence was found, in the form of impaired osmotic or Na^+ or K^+ ionic regulation in C. magister or in impaired osmotic regulation in H. nudus, to suggest that the ion transport functions of the gills were affected. Since the gill NaKMg ATPases are presumed to function in the Na^+ transport process, these results suggested that the NaKMg ATPases in crab gills may not have been inhibited by the pesticide. Davis et al. (1972) have shown that the ATPase activities of rainbow trout gill are inhibited by methoxychlor in vitro, but the amount of inhibition produced was considerably less than that found with a number of other organochlorine insecticides. To determine the effect of methoxychlor on crab gill ATPases, gills were excised from adult C. magister which had been held in 0.01 mg/l methoxychlor for a 14-day period. The control group was treated in an identical fashion, but was exposed only to acetone. The gill tissue from two animals was pooled and activity was measured in whole homogenates (Table 5). NaKMg ATPase activity was significantly ($P < 0.05$) inhibited by methoxychlor treatment to the extent of 37%. Inhibition of both the Mg and NaK components of the enzyme appeared to be similar at 36 and 50%, respectively, but the differences were not significant.

The reasons for the apparent contradiction between the effect of methoxychlor on gill ATPase activity and the lack of effect of the pesticide on blood osmotic regulation and on regulation of blood Na^+ and K^+ in the crab are difficult to understand. The activity of the Na^+ transport system in crustacean gills is influenced by both the blood and medium concentrations of this ion (Lockwood, 1962). It is possible that the degree of inhibition of the enzyme was insufficient to significantly affect its function

TABLE 5
ATPase Activity in Gills from Methoxychlor-Exposed
Cancer magister

| Treatment | ATPase Act. (μmoles P_i/mg protein/hr)[a] | | |
	NaKMg	Mg	NaK
Control	2.13 $\pm$ 0.32 (4)	1.83 $\pm$ 0.60 (4)	0.34 $\pm$ 0.35 (4)
Methoxychlor, 0.01 mg/l	1.34 $\pm$ 0.34 (5)	1.17 $\pm$ 0.37 (5)	0.17 $\pm$ 0.25 (5)

[a]Data presented are the means $\pm$ two standard errors of the mean. The numbers in parentheses indicate the number of individual analyses conducted per group.

in osmotic and ionic regulation of the blood if the requirement for activity was less than the maximum capacity of the enzyme in normal crabs. This is plausible since the capacity for enzyme-catalyzed reactions often exceeds the rates required *in vivo* (Chance and Hess, 1959). The correlation between gill ATPase inhibition and gross toxicological effects is equivocal in other instances as well. Davis *et al.* (1972), utilizing literature values of pesticide lethality, reported that the chlorinated hydrocarbon insecticides were more lethal and also more effective inhibitors of gill ATPases than the herbicides, while PCBs were intermediate in effect. However, the same direct relationship was not found within the insecticide group alone, which indicates that the inhibition of gill ATPase is coincidental to a primary lethal lesion. If the primary lesion is associated with ion distributions across axonal membranes, as has been previously suggested (O'Brien, 1967), possibly due to an effect on the membrane NaKMg ATPases of the axon, then a partial but imperfect correlation with gill ATPases is not surprising. Davis *et al.* (1972) also reported that trout given lethal doses of DDT orally accumulated high tissue residue levels in gills, brain and kidney; but that no differences in ATPase activities of these tissues were observed between treated and control groups of fish, further implying that ATPase inhibition is unimportant in lethality.

Our initial assumption in this study was that methoxychlor, as is the case with other organochlorine pesticides and PCBs examined (Davis *et al.*, 1972), would inhibit the NaKMg ATPase activity of gills and consequently impair osmotic and ionic regulation in the crabs, an effect which should ultimately reduce the tolerance of the animals to low salinity exposures. Our results with gill ATPase in *C. magister* confirmed that the pesticide administered to crabs at barely sublethal levels resulted in inhibition of the enzyme (Table 5). However, with the exception of Mg^{++} regulation, we have been unable to show any adverse effect on osmotic and ionic regulation in this species or on osmotic regulation in *H. nudus*. Yet, despite this lack of observed effect on osmotic and ionic regulation, both species of crabs, after exposure to the pesticide, showed a decreased tolerance to low salinity exposures (Tables 2 and 3). Although the role of Mg^{++} regulation is uncertain, these results suggest that impairment of osmotic and ionic regulation in crabs by organochlorine pesticides is not the principal effect leading to decreased survival at low salinities.

Exposure to low salinities, even in normal crabs, is accompanied by a reduction in blood osmotic and ionic concentrations (Alspach, 1972; Dehnel, 1962; Dehnel and Carefoot, 1965; Engelhardt and Dehnel, 1973). It is conceivable, therefore, that the normal reduction of osmotic and ionic content of the blood may be potentiating the known physiological effects of organochlorine pesticides on the nervous system. The familiar symptoms of poisoning by this class of compounds, which were also seen in the present study

in methoxychlor-poisoned crabs, begins with the hyperactive state and at high dosages it is eventually
followed by loss of motion and paralysis. Individual
nerve axons of diverse species poisoned by DDT are
characterized by uncontrolled volleys of impulses
(Yeager and Munson, 1945). It is postulated that the
DDT-like organochlorines exert their effects on nerves
by blocking the membrane Na^+ and K^+ channels and by
inhibiting the nerve ATPases (Matsumura and Narahashi,
1971; Matsumura and Patil, 1969; Narahashi and Hass,
1967). It is reasonable to suppose that normal nerve
function occurs within a finite range of blood ionic
concentrations with optimum conditions near the center
of that range. We postulate that damage to nerve
axons, resulting from sublethal exposure to methoxychlor, serves to reduce the range of blood ionic concentrations compatible with continued nerve function
with the result that survival of crabs at extreme
salinities is reduced.

In summary, the results of this study have shown
that crabs exposed to methoxychlor at concentrations
which are sublethal in acute tests exhibit a decreased
resistance in tolerance to reduced salinities. In
addition, we have shown that these exposures produce
a partial inhibition of the gill NaKMg ATPases in
C. magister. However, we have failed to demonstrate
that the osmotic regulation in *H. nudus* or the osmotic
and ionic regulation in *C. magister* are significantly
impaired by these treatments. The only evidence for
alteration of ionic regulation by methoxychlor involved Mg^{++} transport across the bladder wall membrane. This transport process appeared to be significantly inhibited, but the evidence for alteration of
blood Mg^{++} regulation was equivocal. We suggest,
therefore, that disruption of osmotic or ionic regulation in crabs exposed to methoxychlor is not a principal cause of induced mortalities in brackish water
environments.

ACKNOWLEDGMENTS

The author is indebted to William J. Wilson, L. Fraser Rasmussen, and Stephen G. Lebsack for expert technical assistance; and to Dr. D. R. Buhler for a critical review of the manuscript and the provision of analytical facilities used in the ion analyses. Technical methoxychlor (92% active ingredient) was supplied by E. I. DuPont de Nemours and Company. This research was supported by EPA Contract No. 68-01-0188.

LITERATURE CITED

Alspach, G. S. 1972. Osmotic and ionic regulation in the dungeness crab, *Cancer magister* (Dana). Ph.D. dissertation, Oregon State University, Corvallis, Oregon.

Armstrong, D. A., D. V. Buchanan, M. H. Mallon, R. S. Caldwell, and R. E. Millemann. 1974. Unpublished manuscript.

Burton, R. F. 1973. The significance of ionic concentrations in the internal media of animals. *Biol. Rev.* *48*:195-231.

Chance, B. and B. Hess. 1959. Metabolic control mechanisms. I. Electron transfer in the mammalian cell. *J. Biol. Chem.* *234*:2404-12.

Cutkomp, L. K., H. H. Yap, E. Y. Cheng, and R. B. Koch. 1971. ATPase activity in fish tissue homogenates and inhibitory effects of DDT and related compounds. *Chem.-Biol. Interactions* *3*:439-47.

Davis, P. W., J. M. Friedhoff, and G. A. Wedemeyer. 1972. Organochlorine insecticide, herbicide and polychlorinated biphenyl (PCB) inhibition of NaK-ATPase in rainbow trout. *Bull. Env. Cont. Toxicol.* *8*:69-72.

Dehnel, P. A. 1962. Aspects of osmoregulation in two species of intertidal crabs. *Biol. Bull.* *122*:208-27.

_____ and T. H. Carefoot. 1965. Ion regulation in two species of intertidal crabs. *Comp. Biochem. Physiol. 15*:377-97.

Desaiah, D., L. K. Cutkomp, H. H. Yap, and R. B. Koch. 1972. Inhibition of oligomycin-sensitive and -insensitive magnesium adenosine triphosphatase activity in fish by polychlorinated biphenyls. *Biochem. Pharmacol. 21*:857-65.

Eisler, R. 1967. Tissue changes in puffers exposed to methoxychlor and methyl parathion. Technical Paper No. 17, Bureau of Sport Fisheries and Wildlife. Washington, D. C.: U. S. Department of the Interior.

_____ and P. H. Edmunds. 1966. Effects of endrin on blood and tissue chemistry of a marine fish. *Trans. Am. Fish. Soc. 95*:153-59.

Engelhardt, F. R. and P. A. Dehnel. 1973. Ionic regulation in the Pacific edible crab, *Cancer magister* (Dana). *Can. J. Zool. 51*:735-43.

Fiske, C. H. and Y. SubbaRow. 1925. The colorimetric determination of phosphorus. *J. Biol. Chem. 66*:375-400.

Green, J. W., M. Harsch, L. Barr, and C. L. Prosser. 1959. The regulation of water and salt by the fiddler crabs, *Uca pugnax* and *Uca pugilator*. *Biol. Bull. 112*:34-42.

Gross, W. J. 1957. An analysis of response to osmotic stress in selected decapod crustacea. *Biol. Bull. 112*:43-62.

_____. 1964. Trends in water and salt regulation among aquatic and amphibious crabs. *Biol. Bull. 127*:447-66.

_____ and R. L. Capen. 1966. Some functions of the urinary bladder in a crab. *Biol. Bull. 131*:272-91.

_____ and L. A. Marshall. 1960. The influence of salinity on the magnesium and water fluxes of a crab. *Biol. Bull. 119*:440-53.

221

Janicki, R. H. and W. B. Kinter. 1971. DDT: Disrupted osmoregulatory events in the intestine of the eel *Anguilla rostrata* adapted to seawater. *Science 173*:1146-48.

Kinter, W. B., L. S. Merkens, R. H. Janicki, and A. M. Guarino. 1972. Studies on the mechanism of toxicity of DDT and polychlorinated biphenyls (PCBs): Disruption of osmoregulation in marine fish. *Env. Health Persp. 1*:169-73.

Koch, R. B. 1969/70. Inhibition of animal tissue ATPase activities by chlorinated hydrocarbon pesticides. *Chem.-Biol. Interactions 1*:199-209.

Lockwood, A. P. M. 1962. The osmoregulation of crustacea. *Biol. Rev. 37*:257-305.

_____. 1967. *Aspects of the Physiology of Crustacea*. San Francisco: W. H. Freeman and Company.

_____ and J. A. Riegel. 1969. The excretion of magnesium by *Carcinus maenas*. *J. Exp. Biol. 51*: 575-89.

Lowry, O. H., N. J. Rosebrough, A. L. Farr, and R. J. Randall. 1951. Protein measurement with the Folin phenol reagent. *J. Biol. Chem. 193*: 265-75.

Matsumura, F. and T. Narahashi. 1971. ATPase inhibition and electrophysiological change caused by DDT and related neuroactive agents in lobster nerve. *Biochem. Pharmacol. 20*: 825-37.

_____ and K. C. Patil. 1969. Adenosine triphosphatase sensitive to DDT in synapses of rat brain. *Science 166*:121-22.

Nagel, H. 1934. Die Aufgaben der Exkretionsorgane und der Kieman bei der Osmoregulation von *Carcinus maenas*. *Z. vergl. Physiol. 21*:468-91.

Narahashi, T. and H. G. Hass. 1967. DDT: Interaction with nerve membrane conductance changes. *Science 157*:1438-40.

Nimmo, D. R. and R. R. Blackman. 1972. Effects of DDT on cations in the hepatopancreas of penaeid shrimp. *Trans. Am. Fish. Soc. 101*:547-49.

O'Brien, R. D. 1967. *Insecticides. Action and Metabolism.* New York: Academic Press.

Oxender, D. L. 1972. Membrane transport. *A. Rev. Biochem. 41*:777-809.

Prosser, C. L. 1973. Inorganic ions. In: *Comparative Animal Physiology*, 3rd edition, pp. 79-110, ed. by C. L. Prosser. Philadelphia: W. B. Saunders Company.

Rudy, P. R. Jr. 1967. Water permeability in selected decapod crustacea. *Comp. Biochem. Physiol. 22*:581-89.

Schatzmann, H. J. and F. F. Vincenzi. 1969. Calcium movements across the membrane of human red cells. *J. Physiol., London 201*:369-95.

Shaw, J. 1961. Sodium balance in *Eriocheir sinensis* (M. Edw.). The adaptation of the crustacea to fresh water. *J. Exp. Biol. 38*:153-62.

Whittam, R. and K. M. Wheeler. 1970. Transport across cell membranes. *A. Rev. Physiol. 32*: 21-60.

Yap, H. H., D. Desaiah, L. K. Cutkomp, and R. B. Koch. 1971. Sensitivity of fish ATPases to polychlorinated biphenyls. *Nature 233*:61-62.

Yeager, J. F. and S. C. Munson. 1945. Physiological evidence of a site of action of DDT in an insect. *Science 102*:305-307.

EFFECTS OF ORGANOPHOSPHATE PESTICIDES ON ADULT OYSTERS (*CRASSOSTREA VIRGINICA*)

M. R. TRIPP

Department of Biological Sciences
University of Delaware
Newark, Delaware 19711

Chemicals sprayed to control insect pest may have significant effects on nontarget organisms ranging from outright killing to subtle physiological dysfunctions (Edwards, 1970). The dangers, both short- and long-term, of widespread use of chlorinated hydrocarbons is now recognized (Mrak, 1969) and these substances are being replaced by highly potent, but less persistent compounds, particularly organophosphates. In Delaware, salt marshes have been sprayed for several years with organophosphates to control mosquito populations. Tidal rivers traverse these marshes and contain populations of reproducing oysters, *Crassostrea virginica*. Since oysters are known to accumulate and to be affected by organochlorine materials (Butler, 1966; Lowe *et al.*, 1971), it seemed possible that organophosphate pesticides might also have some deleterious effect on their reproduction. This is a report of experiments on effects of two organophosphate insecticides on survival, gonad maturation, and spawning of adult American oysters *(C. virginica).*

MATERIALS AND METHODS

OYSTERS

Adult oysters were dredged from the Mispillion
River near Slaughter Beach, Delaware, and held in
plastic net bags (50 per bag) suspended from a raft
in the Broadkill River, Lewes, Delaware. Animals were
held there for up to 6 months and were comparable to
freshly dredged oysters in size, growth rate, heart
rate, pumping and feeding. Since a running seawater
system to which pesticides could be continuously added
was not available, exposure to test compounds was
accomplished in the following ways. During the summer
of 1969, oysters were held in indoor tanks to which
freshly prepared pesticide was added every second day
when the water was changed. For this experiment four
groups of about 100 mature oysters each were treated
as follows:

Group 1—"Field Controls"—suspended in net bags
in Broadkill River (salinity 20 to 30%; water temperature 16 to 24°C depending on season, time of day, and
tidal cycle).

Group 2—"Laboratory Controls"—individual oysters in fiberglass-lined plywood tanks containing 40
gallons of Broadkill River water which was replaced
on alternate days. Held in an air-conditioned room
with water temperature 21 ± 2°C from June 20 through
September, 1969 (99 days), at which time surviving
oysters were suspended in Broadkill River.

Group 3—"Abate treated"—held under same conditions as "laboratory controls" (Group 2). On alternate days when water was changed in tanks, fresh Abate
$(O,O,O^1,O^1$-tetramethyl O,O^1-thiodi-p-phenylene phos-
phorothioate; courtesy American Cyanamid Company,
Princeton, New Jersey) was added at a concentration
0.6 ppm (4-E formulation diluted in Broadkill River
water).

Group 4—"Dibrom treated"—held under same conditions as "laboratory controls" (Group 2). Dibrom
(1,2-dibromo-2,2-dichloro-ethyl dimethyl phosphate;

courtesy of California Chemical Company, Ortho Division, Richmond, California), 0.5 ppm, was added on alternate days after water was changed.

These two organophosphate pesticides were chosen for study because they have been used extensively in Delaware, Abate as a larvacide and Dibrom as an adulticide. The manufacturers of Abate state that it is stable at temperatures of 25 through 40°C, even at pH 11, while the manufacturer of Dibrom states that it is unstable in water, particularly under alkaline conditions, and that complete hydrolysis occurs in 48 hrs.

During the summer of 1970, a different experimental regime was used. Adult oysters were dredged in early March, placed in net bags and suspended in the Broadkill River. Beginning in early May they were subjected to twice weekly immersion in test chemicals (Abate or Dibrom, 1 or 10 ppm) for 24 hrs; between treatments they were held in the river. Control oysters were subjected to similar handling but were held in uncontaminated river water. Treatment occurred in plywood tanks 90 × 90 × 15 cm deep into which Broadkill River water was pumped and fresh pesticide solution added. Salinities varied from 24 to 31 o/oo with the tidal cycle; water temperatures were 18 to 25°C depending on ambient air temperature. Waste pesticide was drained into a septic tank buried in sand.

Five oysters were withdrawn from each group weekly and prepared for histologic examination. Other oysters were tested for their ability to spawn. These were placed in white enamel pans and thermally shocked (water temperature 28 to 30°C flowing at 800 ml/min for 60 min in seawater having a salinity of 28 to 32%. If thermal shock did not trigger spawning, oyster sperm were added. Preliminary experiments indicated that physiologically ripe oysters would spawn under these conditions.

HISTOLOGICAL PREPARATIONS

Groups of five or six oysters were removed from each tank at approximately weekly intervals for 15 weeks. The left valve was removed and the oyster placed in Bouins solution for about 6 hrs. Then two transverse sections, each approximately 0.5 cm thick, were cut with a razor blade, one immediately anterior to and the other immediately posterior to the center of the hepatopancreas. These tissues were placed in fresh Bouins for 24 hrs, dehydrated, embedded in Paraplast, sectioned at 8 to 10μ and stained with Lillie's hemalum and eosin. In some experiments tissues were fixed in Davidson's and sections were stained with Harris hematoxylin.

Sections were examined under the scanning (4X) objective for gross pathology and to determine the stage of maturation of the gonad. All oysters were scored on an arbitrary 0 to 4 scale (Table 1).

The "gonad index" was calculated by averaging the scores of all oysters in a sample. The amount of

TABLE 1
Criteria for Rating Oyster Gonad Condition

Rating	Histologic Observation
0	Sexually indifferent; no sign of gonad
1	Isolated pockets of gametocytes
2	Islands of gametocytes enlarged but not confluent
3	Little connective tissue between islands of gametocytes; immature gametes evident
4	Gonads confluent and mature gametes plentiful; "ripe" condition

gonadal tissue in individual oysters was measured by a cross-sectional area technique (Iwantsch, 1970), but since it is not more accurate than the gonad index technique (because of variability of individuals in small samples), it was not used routinely.

RESULTS

MORTALITY

Exposure of adult oysters to two concentrations (1 and 10 ppm) of Abate or Dibrom does not cause dramatic acute mortality, but oysters chronically exposed to 10 ppm of Dibrom under these experimental conditions had considerably higher mortalities than controls (Table 2).

TABLE 2
Mortality of Oysters (C. virginica) *Exposed to Different Concentrations of Two Organophosphate Pesticides*

Response	Control	Abate		Dibrom	
		1 ppm	10 ppm	1 ppm	10 ppm
Dead	2	4	7	1	17
Total	24	62	77	53	67
Mortality (%)	8.3	6.5	9.1	1.9	25.2

HISTOPATHOLOGY

Careful histologic examination of 408 control and 337 experimental (Abate-treated, 184; Dibrom-treated, 153) oysters failed to reveal significant patterns of tissue damage that could be attributed to pesticide

treatment. Gonads were examined particularly carefully, but no cellular abnormality in gametic maturation was detected. Gills, intestine, digestive
diverticula and mantle tissue of experimental animals
could not be distinguished from those of controls.
Most oysters held in laboratory tanks showed some
degree of desquamation of the smaller tubules of the
digestive diverticula when compared to field controls,
presumably because of suboptimal food supply.

Major histopathological reactions were seen in
oysters infected with *Bucephalus* sp. and with *Michinia
nelsoni*. Mature infections with the former often led
to parasitic castration of the host oyster, and such
infected animals were not included in the calculation
of gonad index. Usually less than 5% of the oysters
examined had such mature infections. No special
effort was made to detect *Michinia*-infected oysters
and in only a few cases were heavy infections found.
Major tissue damage in these cases was not in the
gonad.

SPAWNING

Histology

Precise measurement of cross-sectional area of
gonad is possible (Iwantsch, 1970), but it is much
quicker to estimate the stage of sexual maturation
and compute the "gonad index." The relationship
between the two is essentially constant regardless of
sex or size of individual oysters (Fig. 1). The
inverse relationship between gonad index and spawning
is seen in Figure 2. In oysters maintained in estuarine water the gonad index increased steadily from
early June through late July when initial spawning
occurred. The gonads of unspawned oysters continued
to mature until the third week in August when a second spawning occurred. A few animals continued to
spawn during the first few days of September, after
which resorption of unspawned gametes occurred. Control oysters held under laboratory conditions

230

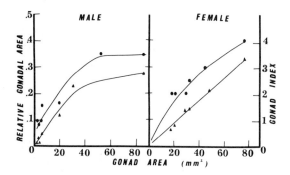

Fig. 1. *Relationship between quantitative measure of gonadal cross-sectional area (■ - ■) and estimation of gonadal index (● - ●) for male and female oysters. All measurements made on control oysters held in laboratory tanks at 20°C for 2 weeks after dredging.*

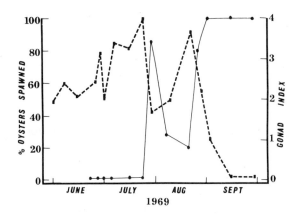

Fig. 2. *Percent oysters spawned (● - ●) and gonad index (■ - ■) of control oysters held in estuarine waters (Broadkill River) after dredging in May 1969. Each point represents average of data from histological examination of six sectioned oysters (males and females). In September sample oysters were given gonad index rating of zero if gametes were being resorbed.*

231

("laboratory controls") showed slower maturation of gonads since maximum development was evident after mid-August and some had not spawned even by mid-September (Fig. 3A).

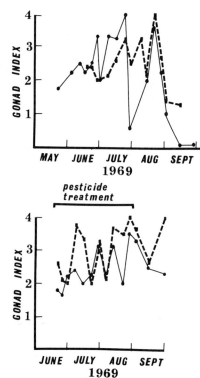

Fig. 3. A. Gonad indices of control oysters held in the Broadkill River (field controls; ● - ●), in laboratory tanks (laboratory controls; ■ - ■) but untreated with pesticides. B. Gonad indices of pesticide-treated oysters; Abate (0.6 ppm; ■ - ■) and Dibrom (0.5 ppm; ● - ●). Members of all groups were dredged May 26, held in the Broadkill River until placed in labora-tory tanks @ 21°C from June 20 until September 2, 1969, at which time they were suspended in the Broadkill River.

By early September control oysters that had been held in the field for 4 months had spawned completely; untreated controls held in laboratory tanks and trans-ferred to the field on September 2 also spawned almost completely (Fig. 3A).

Abate-treated oysters matured through mid-July, some spawned in late July while others matured and spawned in early September (Fig. 3B). By late Septem-ber, when spawning was highly unlikely due to decreas-ing water temperature, many oysters treated with Abate had mature gonads long after controls had spawned. Exposure to pesticide had stopped on September 2, 1969, and the oysters returned to field conditions; thus, it is possible that treatment with Abate under marginal maintenance conditions was responsible for inhibition of spawning. Dibrom-treated oysters matured slowly and did not spawn in appreciable numbers in late July or August. When the experiment was terminated in late September, many had not yet spawned (Fig. 3B) and gamete resorption had begun.

Figure 4 (A and B) shows that under more optimal experimental conditions even greater concentrations (up to 10 ppm) of pesticides do not inhibit severely oyster gonad maturation or spawning. The control

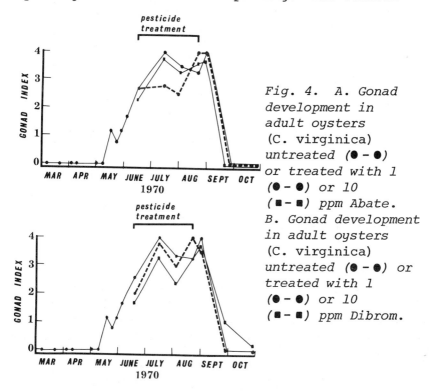

Fig. 4. A. Gonad development in adult oysters (C. virginica) untreated (● - ●) or treated with 1 (● - ●) or 10 (■ - ■) ppm Abate. B. Gonad development in adult oysters (C. virginica) untreated (● - ●) or treated with 1 (● - ●) or 10 (■ - ■) ppm Dibrom.

population showed the characteristic mid-July and late-August spawning peaks. Abate-treated oysters did not show the early spawning peak but did spawn normally in late August (Fig. 4A). The Dibrom

(10 ppm) population did not mature quite as rapidly as the controls and spawned less fully compared to controls (Fig. 4B). Overall there was no dramatic inhibition of gonad maturation or spawning in pesticide-treated oysters.

Live Oysters

Attempts to spawn adult oysters were begun July 1 and ended August 20, 1970. Forty separate attempts were made on 28 different days using about 200 oysters (some more than once). Six males (two each on July 14 and 27 and August 1) released sperm weakly and sporadically. During this time attempts to spawn control oysters (held with treated ones or collected in the field) were uniformly unsuccessful also. Spawning of oysters that had been dredged several months earlier and fed algal cultures to insure gonad maturation was erratic during this period as well. Field observations suggest strongly that several populations of oysters in Delaware Bay and rivers adjacent to the Bay spawned moderately during the last few days of August 1970 (Maurer, personal communication).

DISCUSSION

Spraying of organophosphate pesticides on salt marshes for mosquito control is often intensive and may occur over beds of reproducing oysters. These chemicals might affect oyster populations in several ways: acute toxicity for adults; toxic effects on food organisms; inhibition of gametogenesis (either direct sublethal toxicity or indirectly through dietary imbalance) and release of gametes; inhibition of fertilization; inhibition of larvael development; deleterious effects on veliger larvae; inhibition of growth of juvenile oysters. Any of these effects could have important long-range consequences for reproducing populations.

The experiments reported here were designed to test the possible effects of a larvicide (Abate) and an adulticide (Dibrom), both commonly applied to salt marshes, on oyster reproduction. There is no significant evidence of acute toxicity of either chemical for adult oysters. Even when oysters were held under laboratory conditions with very little food available and exposed to concentrations of pesticides that were at least 300 to 400 times those that they might ordinarily encounter, mortality was not significantly different from that of untreated control animals. Further, histological examination failed to reveal tissue damage in treated oysters that could be attributed to pesticide treatment.

The question of whether or not there is a significant effect on oyster reproduction has not yet been completely resolved. In experiments conducted in 1969, pesticide-treated oysters failed to mature sexually as rapidly as normal, but much of that was probably due to the fact that these oysters had been in laboratory tanks where relatively little food was available. However, when control oysters that had developed gonads under the same conditions were returned to the field in early September, they spawned promptly; pesticide-treated oysters did not. Thus, a combination of stresses might affect reproduction when a single stress would not.

This idea is supported by the 1970 experiments that indicate that well-fed oysters show virtually no detectable effect of pesticide even when chemical concentrations grossly exceed the levels that would be found under most field conditions. One can conclude that under ordinary conditions the two organophosphate pesticides tested here probably would have little effect on oyster reproduction. It is possible, however, that these materials might have an effect under extraordinary conditions (i.e., spraying for mosquito control after very heavy rains) when non-target organisms would already be under stressful conditions.

ACKNOWLEDGMENTS

This research was supported by U.S.P.H.S. Grant UI00793.

Ms. Anna Webers and Mr. Ronald Sasaki gave valuable assistance. The illustrations were prepared by Ms. Patricia A. Cunningham. I thank Dr. Donald Maurer, Shellfish Laboratory, and Dr. Kent Price, Bayside Laboratory, Lewes, and their staffs for use of facilities and valuable cooperation.

LITERATURE CITED

Butler, P. A. 1966. The problem of pesticides in estuaries. *Amer. Fish. Soc. Spec. Publ. 3*: 110-15.

Edwards, C. A. 1970. *Persistent Pesticides in the Environment*. Cleveland, Ohio: CRC Press.

Iwantsch, G. F. 1970. The effect of high temperature on the gonadal area of representative cross-sections of *Mytilus edulis*, L. M. S. thesis, University of Delaware, Newark, Delaware.

Lowe, J. I., P. D. Wilson, A. J. Rick, and A. J. Wilson, Jr. 1971. Chronic exposure of oysters to DDT, toxaphene and Parathion. *Proc. Natl. Shellfisheries Assoc. 61*:71-79.

Mrak, E. M. 1969. *Report on the Secretary's Commission on Pesticides and Their Relationship to Environmental Health*. Washington, D. C.: U. S. Department of Health, Education and Welfare.

EFFECTS OF EXPOSURE TO A SUBACUTE CONCENTRATION OF PARATHION ON THE INTERACTION BETWEEN CHEMORECEPTION AND WATER FLOW IN FISH

H. KLEEREKOPER

Department of Biology
Texas A & M University
College Station, Texas 77843

Much attention has been and continues to be given to acute physiological effects on fish of pollutants of various kinds, particularly as they relate to lethality. Although there are numerous situations in the field in which acutely deleterious concentrations of pollutants are present, early dilution processes commonly lower the acute concentrations of the pollutant to subacute levels, frequently at fast rates. In this manner large volumes of water may become affected, exposing the fauna chronically to subacute concentrations of the pollutant in large areas of the hydrographic basin. Insufficient information is available on the physiological effects of chronic exposure to pollutants generally, whereas the behavioral toxicology of such exposure has hardly been studied. The few data that have been obtained on fish point to subtle effects of potentially great import for the viability of the species affected (Davy *et al.*, 1972; Hatfield and Anderson, 1972;

Welsh and Hanselka, 1972; Jackson *et al.*, 1970;
Ogilvie and Anderson, 1965; Saunders, 1969; Warner
et al., 1966). With respect to commonly used organic
pesticides this is not surprising since the physiological action of both chlorinated hydrocarbons and
organophosphates is known to affect neural functions
at various levels (e.g., Aubin and Johansen, 1969;
Mazur and Bodansky, 1946; Weiss, 1958, 1959, 1961;
Kayser *et al.*, 1962; Anderson, 1968; Anderson and
Peterson, 1969; Anderson and Prins, 1970; Desi *et al.*,
1966; O'Brien and Matsumura, 1964; Welsh and Gordon,
1947).

Aspects of behavioral toxicology of some pesticides and heavy metals related to orientation mechanisms are being examined in this laboratory (Davy
et al., 1972, 1973; Kleerekoper *et al.*, 1973; Timms
et al., 1972; Westlake *et al.*, 1974; Kleerekoper,
1973; Rand *et al.*, 1974) parallel with a continuing
study of locomotor behavior of sharks and teleosts.
The aim of these studies is to elucidate the basic
mechanisms of orientation control in these animals.
The rationale for the experimental approach assumes
that the orientation problem consists of at least two
parts. The first, "spontaneous" locomotor behavior,
occurs in the absence of overt directional cues from
the environment. The second is modification of this
behavior in response to environmental stimuli which
reach the animal through sensory systems; this information may involve one or more modalities and in some
instances conveys directional information. By a comparison of locomotor behavior in the absence and
presence of a particular environmental cue it should
be possible not only to assess the latter's significance as a modifier of locomotor behavior but also to
describe the resulting modification and infer which
locomotor and control mechanisms are implicated.

Pattern and direction of locomotion are the net
outcome of the interplay of a number of distinct
locomotor variables. The most obvious of these variables are the magnitude, direction, frequency and
ratios of turns, the step length and velocity between

turns, and the correlations between these variables
in space and time. Once the values of these variables and their interdependencies are known, it
becomes possible to develop mathematical models of
"spontaneous" locomotion for normal animals of a
certain species in a "constant" environment. Even
modifications of variables which may escape direct
observation, such as a locomotor response to an environmental variable or exposure to pollutants, would
be detectable quantitatively through the model. The
above approach depends on the availability of monitoring techniques capable of acquiring the information
needed to statistically describe the locomotor variables mentioned under controlled conditions and during periods of sufficient length to account for
variability of locomotor behavior over time. Such
techniques were developed (Kleerekoper, 1967, 1969)
and involve the tracking of locomotion by means of
arrays of photocells in two monitor systems. One of
these employs photo sensors in a 44 × 44 matrix with
the cells on 10 cm centers in the floor of a tank
providing a free area of almost 500 × 500 cm for the
fish's movements. As the fish moves about and inter-
cepts light from an overhead collimated light source,
the "addresses" of the photocells and the elapsed
time between events are computed by an electronic
interface and computer. These data form the basis
for the subsequent computation of the locomotor variables mentioned above and for the analysis of their
relationships in time and space. The second system
offers the fish the choice of 16 freely accessible
compartments monitored by photocell gates. Entries
and time spent in each compartment are computed; from
these data the pathways of the fish between compartments can be reconstructed. In either system environmental conditions are controlled and a variety of
chemical and physical conditions can be varied in the
entirety or in defined parts of the monitors.
  Statistical application of the data obtained by
means of these techniques to various aspects of the
above problem has resulted in the recognition and

description of the interdependence of a number of
locomotor variables (Kleerekoper *et al.*, 1969, 1970;
Matis *et al.*, 1973, 1974) and it has become possible
to detect and quantify subtle orientation responses
to environmental cues such as food odor and water
flow conditions (Kleerekoper, 1974; Kleerekoper
et al., 1974). Furthermore, behavioral effects of
exposure to subacute concentrations of pesticides
were detected which would have escaped notice by more
conventional methods of observation (Davy *et al.*,
1972, 1973). New information on the latter problem
seems relevant to the topic of this symposium.

Recent monitor data obtained in this laboratory
by A. Fokakis and G. Rand indicate that mullet *(Mugil
cephalus)* which normally display positive rheotaxis
become negatively rheotactic after short-term exposure to .016 ppm parathion. The acetone carrier had
no effect. The consequences of such a change in
rheotaxis for the capability to localize food may be
important as was demonstrated in a study of locomotor
orientation in response to flow and food odor in the
goldfish (Rand *et al.*, 1974). In a free choice situation food odor was significantly attractive to normal
fish in the absence of a differential of water flow.
However, when the odor was combined with a mean increased rate of flow of about 2.5 percent, significant avoidance occurred. Factorial analysis revealed
an interaction between the increased flow rate and
the food odor.

The above responses changed drastically after
a short-term exposure (24 hrs) to a subacute concentration of parathion (.33 ppm). Food odor in the
absence of increased flow rate, attractive to the
normal fish, now was significantly avoided whereas
the response to the "food odor-increased flow" situation was not so affected. It was established that
this result of the parathion exposure was not caused
by a change in response to the flow conditions *per se*,
but that it was caused by a reversal of the interaction between the food odor and the increased rate
of flow. Recently we described an interaction

phenomenon between copper ions and a slight temperature increase affecting the locomotor orientation of
fish (Kleerekoper *et al.*, 1973).

There is strong evidence, therefore, that behavioral responses to environmental variables can, and
probably do normally, result from the interaction
between two or more of these variables. If this interpretation is correct, it implies that ecologically
meaningful experimentation on the significance of an
individual environmental variable must not only consider the effects of the variable in question but
also the results of its interaction with other common
variables. Such experimentation becomes complex,
technically as well as statistically, the more so
since many behavioral responses are subtle and subject to variation over time.

Interactions must be expected to be biologically
adaptive and may well be the mechanism which enables
the organism to respond to variations in the environment as functions of the whole rather than as isolated
phenomena. The fact that exposure to a subacute,
"safe" concentration of perception and water flow in
the experiments cited above strongly suggests that
the biological effects of certain pollutants may be
vastly more complex and possibly more serious than
hitherto suspected.

ACKNOWLEDGMENT

This research was supported in part by National
Institute of Health Grant No. 5 R01 ES00427-04 to
H. Kleerekoper on behavioral toxicology induced by
insecticides.

LITERATURE CITED

Anderson, J. M. 1968. Effect of sublethal DDT on
the lateral line of brook trout *(Salvelinus
fontinalis)*. *J. Fish. Res. Bd. Canada* 25:2677-82.

_____ and M. R. Peterson. 1969. DDT: Sublethal effects on brook trout nervous system. *Science 164*:440-41.

_____ and H. B. Prins. 1970. Effects of sublethal DDT on a simple reflex in brook trout. *J. Fish. Res. Bd. Canada 27*:331-34.

Aubin, A. E. and P. H. Johansen. 1969. The effects of an acute DDT exposure on the spontaneous electrical activity of goldfish cerebellum. *Can. J. Zool. 47*:163-66.

Davy, F. B., H. Kleerekoper, and P. Gensler. 1972. Effects of exposure to sublethal DDT on the locomotor behavior of the goldfish *(Carassius auratus)*. *J. Fish. Res. Bd. Canada 29*:1333-36.

_____, _____, and J. H. Matis. 1973. Effects of exposure to sublethal DDT on the exploratory behavior of goldfish *(Carassius auratus)*. *Water Resources Research 9*:900-05.

Desi, I., I. Farkas, and T. Kemeny. 1966. Changes of central nervous function in response to DDT administration. *Acta Physiol. Sci. Hung. Tomus 30*:275-82.

Hatfield, C. T. and J. M. Anderson. 1972. Effects of two insecticides on the vulnerability of Atlantic salmon *(Salmo salar)* parr to brook trout *(Salvelinus fontinalis)* predation. *J. Fish. Res. Bd. Canada 29*:27-29.

Jackson, D. A., J. M. Anderson, and D. R. Gardner. 1970. Further investigations of the effect of DDT on learning in fish. *Can. J. Zool. 48*:577-80.

Kayser, H., A. Luedemann, and H. Neumann. 1962. Nerve cell injuries caused by insecticides poisoning in fishes and land crabs. *Z. angew Zool. 49*:135-48.

Kleerekoper, H. 1967. Some effects of olfactory stimulation on locomotor patterns in fish. In: *Int. Symp. on Olfaction and Taste, Tokyo, 1965*, pp. 625-45. Pergamon Press.

_____. 1969. *Olfaction in Fishes*. Bloomington, Ind.: Indiana University Press.

_____. 1973. Effects of copper on the locomotor orientation of fish. EPA-R3-73-045, Ecological Research Series. Washington, D. C.: U. S. Environmental Protection Agency.

_____. 1974. Locomotor responses to environmental variables by sharks and teleosts. In: *Physiological Adaptation to the Environment*. Proceedings of AIBS Symposium, Amherst, Mass. 1973. New York: Intext Educational Publishers. In press.

_____, D. Gruber, J. Matis, P. Maynard, and P. Gensler. 1974. Orientation of the nurse shark *(Ginglymostoma cirratum)* in response to chemical stimulation in stagnant and flowing water. Submitted.

_____, A. Timms, G. Westlake, F. B. Davy, T. Malar, and V. M. Anderson. 1969. Inertial guidance system in the orientation of the goldfish *(Carassius auratus)*. *Nature 223*:501-02.

_____, _____, _____, _____, _____, and _____. 1970. An analysis of locomotor behaviour of the goldfish *(Carassius auratus)*. *Anim. Behav. 18*:317-30.

_____, J. B. Waxman, and J. Matis. 1973. Interaction of temperature and copper ions as orienting stimuli in the locomotor behavior of the goldfish *(Carassius auratus)*. *J. Fish. Res. Bd. Canada 30*:725-28.

_____, G. F. Westlake, J. H. Matis, and P. J. Gensler. 1972. Orientation of goldfish *(Carassius auratus)* in response to a shallow gradient of a sublethal concentration of copper in an open field. *J. Fish. Res. Bd. Canada 29*:45-54.

Matis, J. H., D. R. Childers, and H. Kleerekoper. 1974. A stochastic locomotor control model for the goldfish *(Carassius auratus)*. *Acta Biotheoretica*. In press.

_____, H. Kleerekoper, and P. Gensler. 1973. A time series analysis of some aspects of loco-motor behavior of goldfish *(Carassius auratus)*. *J. Interdiscipl. Cycle Res.* 4:145-58.

Mazur, A. and O. Bodansky. 1946. The mechanism of *in vitro* and *in vivo* inhibition of cholinesterase activity by diisopropyl fluorophosphate. *J. Biol. Chem. 163*:261-76.

O'Brien, R. D. and F. Matsumura. 1964. DDT: a new hypothesis of its mode of action. *Science 146*: 657-58.

Ogilvie, D. M. and J. M. Anderson. 1965. Effects of DDT on temperature selection by young Atlantic salmon, *Salmo salar*. *J. Fish. Res. Bd. Canada 22*:503-12.

Rand, G., H. Kleerekoper, and J. Matis. 1974. Interaction of odor and flow perception and the effects of parathion in the locomotor orienta-tion of the goldfish. Submitted.

Saunders, J. W. 1969. Mass mortalities and behavior of brook trout and juvenile Atlantic salmon in a stream polluted by agricultural pesticides. *J. Fish. Res. Bd. Canada 26*:695-99.

Timms, A. M., H. Kleerekoper, and J. Matis. 1972. Locomotor response of goldfish, channel catfish, and largemouth bass to a "copper-polluted" mass of water in an open field. *Water Resources Research 8*:1574-80.

Warner, R. E., K. K. Peterson, and L. Borgman. 1966. Behavioural pathology in fish: a quantitative study of sublethal pesticide toxication. In: *Pesticides in the Environment and Their Effects on Wildlife*. *J. Appl. Ecol.* (Suppl.) 3:223-47.

Weiss, C. M. 1958. The determination of cholines-terase in the brain tissue of three species of fresh water fish and its inactivation *in vivo*. *Ecology 39*:194.

_____. 1959. Response of fish to sub-lethal exposures of organic phosphorus insecticides. *Sewage and Industr. Wastes 31*:580-93.

_____. 1961. Physiological effect of organic phosphorus insecticides on several species of fish. *Trans. Amer. Fish. Soc. 90*:143-52.

Welsh, J. H. and H. T. Gordon. 1947. The mode of action of certain insecticides on the arthropod nerve axon. *J. Cell. Comp. Physiol. 30*:147.

Welsh, M. J. and C. W. Hanselka. 1972. Toxicity and sublethal effects of methyl parathion on behavior of Saimese fighting fish *(Betta splendens)*. *Texas J. Sci. 23*:519-29.

Westlake, G. F., H. Kleerekoper, and J. Matis. 1974. The locomotor response of goldfish to a steep gradient of copper ions. *Water Resources Research 10*:103-05.

THE SOURCES, FATES AND EFFECTS
OF OIL IN THE SEAS

D. G. AHEARN

Department of Biology
Georgia State University
Atlanta, Georgia   30303

Hydrocarbons found in the oceans are from a variety of sources, including decaying phyto- and zooplankton, routine tanker and shipping operations, terrestrial runoff, atmospheric fallout, natural seepage, and shipping and offshore well disasters. The quantities from each of these sources are imprecisely known and probably vary widely by category and in their total sum from year to year. The relative magnitudes of sources of oceanic hydrocarbons for the early 1970s are estimated in Table 1. Decaying plants and animals probably constitute the greatest source, whereas crude oil constitutes only a minor portion of the seas' hydrocarbon content. The total petroleum input into the oceans for recent years has been estimated at close to 6 million metric tons per year (Butler, Morris, and Sass, 1973). Blumer (1973) estimated a total annual oil influx to the oceans of between 5 and 10 million tons. Man's activities account for the majority of crude oil in the oceans. Data collated from various reports (Study of Critical Environment Problems, 1970; Luten, 1971; Ahearn and Meyers, 1973; Butler, Morris, and

Sass, 1973; Blumer, 1973; Wilson *et al.*, 1974) im-
plicate shipping operations as the major source of
oceanic crude (Table 2). The analysis is imprecise
due to the scarcity of data. Assuming a 0.1% loss
of oil transported overseas (Button, 1971), it can
be estimated that for the past few years, between

TABLE 1
Sources of Hydrocarbons Found in Oceans (1970-1973)

Source	% of Total[a]
Decaying animals and plants	50
Shipping operations	18
Terrestrial runoff	17
Atmospheric fallout	8
Natural seepage	4
Shipping and production accidents	3

[a]Estimated on the basis of an annual input of
6 to 12 million metric tons.

TABLE 2
*Estimated Inputs of Crude Oil into the Oceans
(1970-1973)*

Source	Metric Tons $\times 10^6$
Shipping operations	1.15 - 4
Offshore oil operations	0.09 - 0.1
Shipping and well disasters	0.3
Natural seeps	0.3 - 0.6
Total crude	1 - 5

1 and 2 million metric tons were spilled into the
oceans each year just from routine tanker operations.
A load-on-top process which conserves oil wastes in
bunker wash waters is rapidly being instituted in
shipping practices. Consequently, bilge losses and
the problem of tar lumps, which are considered to be
mainly of tanker origin (Butler *et al.*, 1973), should
gradually decrease in magnitude. In contrast to the
open seas, more harmful chronic and catastrophic pollution of inshore areas will be an increasing threat.
Common use of supertankers (200,000 to 500,000 dead-
weight tons), development of more offshore oil fields
and construction of deep-water offshore superports
will increase the probabilities of chronic leakage
and cataclysmic oil spills. New regions will now
become potential sites of these disasters. The North
Sea most likely will be subjected to crude oil pollution of at least 30,000 tons per annum from well production alone by the mid-1970s. Superports along the
Gulf and Atlantic coasts of the United States will be
sites of potential disasters by the 1980s. In general, a steady increase in development of offshore
oil production and processing can be expected until
at least 2000.

The longevity and ultimate effects of oils on
the marine ecosystems vary with the oceanic and atmospheric conditions, the type of hydrocarbon and
mode of the spill. Physical and biological processes
affecting the fate of the oil include:

slick formation	oil in water emulsions
dissolution	photooxidation
evaporation	microbial attack
polymerization	sedimentation
emulsification	plankton ingestion
water in oil emulsions	tar lump formation

The immediate processes of thin surface slick
formation, evaporation of volatile components and
dissolution of hydrocarbons may reduce the volume of
spilled oil by as much as 50% during the first few
days. Refined products, such as gasoline or kerosene, may almost completely disappear during this

time, whereas viscous crudes such as Kuwait oil may
undergo less than a 25% reduction. The degree of
wave action and the type of crude governs emulsifica-
tion which in turn affects biodegradation. Water in
oil emulsions may be formed, particularly by the
sudden release of heavy asphaltic oils into calm
seas. Such emulsions usually are slowly degraded by
bacteria because of the reduced oil-water surface
ratio. The types of microorganisms and nutrients
present can be critical. Hydrocarbonoclastic bacteria and fungi, particularly species with emulsify-
ing properties, are most numerous in regions of
chronic oil pollution, but often in low numbers in
the open seas. The growth of these microorganisms
appears to be dependent upon available nitrogen and
phosphorus (Atlas and Bartha, 1973).

The direct lethal effects of oil pollution on
fauna and flora have been amply documented for
coastal areas. Possible subtle effects, such as food
chain alteration and reduced resistence to environmental stress have received only cursory study. The
classic investigations of Blumer and his colleagues
(for a complete list of references, see Blumer et al.,
1973) have indicated that certain hydrocarbon fractions are remarkably stable in the marine environment,
and that hydrocarbons may detrimentally affect feeding responses of lobsters. Takahashi and Kittredge
(1973) found that water-soluble extracts of oil
interferred with both the mating and feeding responses
of an intertidal crab. Beneficial effects, such as
an overall increase in productivity, may result from
low level oil input into the seas. Further research
into all facets of hydrocarbon interactions in the
oceans are warranted.

LITERATURE CITED

Ahearn, D. G. and S. P. Meyers. 1973. *The Microbial
Degradation of Oil Pollutants*. LSU-SG-73-01.
Baton Rouge, La.: Louisiana State University.

Atlas, R. M. and R. Bartha. 1973. Stimulated bio-degradation of oil slicks using oleophilic fertilizers. *Envr. Sci. Tech.* 7:538-41.

Blumer, M. 1973. Oil contamination and the living resources of the sea. In: *Marine Pollution and Sea Life*, pp. 476-81, ed. by M. Ruivo. London: Fishing News (Books) Ltd.

_____, J. M. Hunt, J. Atema, and L. Stein. 1973. *Interaction Between Marine Organisms and Oil Pollution.* EPA-R3-73-042. Washington, D. C.: Ecological Research Series.

Butler, J. N., B. F. Morris, and J. Sass. 1973. *Pelagic Tar from Bermuda and the Sargasso Sea.* New York: Bermuda Biological Station for Research.

Button, D. K. 1971. Biological effects of petroleum in the marine environment. In: *Impingement of Man on the Oceans*, pp. 421-29, ed. by D. W. Hood. New York: Wiley.

Luten, D. B. 1971. The economic geography of energy. *Sci. Amer.* 225:164-75.

Study of Critical Environment Problems. 1970. *Man's Impact on the Global Environment: Assessment and Recommendation for Action.* Cambridge, Mass.: MIT Press.

Takahashi, F. T. and J. S. Kittredge. 1973. Sub-lethal effects of the water soluble component of oil: chemical communication in the marine environment. In: *The Microbial Degradation of Oil Pollutants.* LSU-SG-73-01. Baton Rouge, La.: Louisiana State University.

Wilson, R. D., P. H. Monaghan, A. Osanik, L. C. Price, and M. A. Rogers. 1974. Natural marine oil seepage. *Science.184*:857-65.

EFFECTS OF BENZENE (A WATER-SOLUBLE COMPONENT OF CRUDE OIL) ON EGGS AND LARVAE OF PACIFIC HERRING AND NORTHERN ANCHOVY

JEANNETTE W. STRUHSAKER, MAXWELL B. ELDRIDGE,
and TINA ECHEVERRIA

National Marine Fisheries Service
Southwest Fisheries Center
Tiburon Laboratory
Tiburon, California 94920

The effects of oil spills on the marine environment have received increasing attention over the past decade. A major portion of the research on these effects has followed a holistic approach of subjecting organisms to crude oil. Although these studies have contributed considerably to an understanding of the problem, oil is a complex amalgam of several components, varying in toxicity. It is difficult to ascertain specific effects on the survival and physiology of marine organisms using whole crude oil. While some studies have focused on the effect of tar components in oil (Bargmann, 1971; Moulder and Varley, 1971), little attention has been given to those compounds in crude oil thought most toxic to aquatic organisms, i.e., the low boiling point, water-soluble, aromatic hydrocarbons.

253

Most research to the present time at Tiburon Laboratory has concentrated on the effects of the aromatic component, benzene, on juvenile fishes (Brocksen and Bailey, 1973; Bailey and Brocksen, 1974; Benville and Korn, 1974; Korn and Macedo, 1974). Benzene is among the most abundant of the aromatic components, comprising at least 20% of the total aromatic hydrocarbons in crude oil. Benzene is relatively soluble in water (up to 2000 ppm in fresh water; Benville and Korn, 1974), and is among the most toxic of all oil components. Experiments are also in progress to test the effects of other toxic aromatics (toluene, xylene, 1, 2, 4-trimethylbenzene) on fishes, including synergistic and/or antagonistic effects of their combinations.

Since eggs and larvae are generally the life history stages most sensitive to environmental stress, and because few studies of the effects of oil components have been performed on these stages, experiments have been extended to test the effects of benzene on eggs and larvae of Pacific herring and northern anchovy.

The Pacific herring *(Clupea pallasi)* is an important commercial species in the San Francisco Bay area and along the coasts of Canada and Alaska (Eldridge and Kaill, 1973). In addition to a fishery for adults, there is a commercial fishery for the eggs which are principally exported to Japan. The eggs are demersal, being deposited on algae and substrate in bays from about October to March. Herring eggs and larvae occur inshore in bays where oil is likely to accumulate or be spilled.

The northern anchovy *(Engraulis mordax)* is the most abundant species with immediate harvest potential in the California Current System (Frey, 1971). It is also numerous in west coast bays where it has been found to be the most abundant fish, both numerically and in terms of biomass (Aplin, 1967). The anchovy spawns not only in San Francisco Bay but extensively throughout the California Current System of California (Ahlstrom, 1966). In offshore waters spawning occurs in every month of the year, usually peaking in late winter and early spring with another minor peak in early fall. In San Francisco Bay, however, spawning has been delimited from about May through October. Pelagic anchovy eggs are abundant in surface layers of San Francisco Bay and, like herring, they may possibly be exposed to oil.

This paper presents the results of preliminary experiments testing lethal and sublethal concentrations of benzene on eggs and larvae of the Pacific herring and northern anchovy. Developing embryos were exposed at two stages to contrast their sensitivity to benzene: (1) eggs a few hours after spawning and fertilization and (2) larvae a few hours before or after completion of yolk absorption. Parameters measured include percent mortality, percent abnormal larvae, types of abnormalities, length of larvae and growth, yolk utilization, feeding and respiration.

METHODS

Herring eggs were collected inshore on the Tiburon Peninsula a few hours after spawning occurred. Eggs, attached to intertidal algae (several species of *Fucus*, *Laminaria*, and *Gracilaria*) were in early cleavage stages when collected. Anchovy eggs were collected from Raccoon Straits between Tiburon Peninsula and Angel Island with a 0.5 meter plankton net (333μ mesh) towed on the surface. Although varying somewhat in developmental stage, most anchovy eggs were in gastrula stage when collected. Experiments performed on herring and anchovy eggs and larvae are summarized in Table 1.

After collection, eggs were sorted and placed into green polyethylene containers containing 8 liters of filtered seawater. Seawater was filtered through a series of two 5μ cellulose filters and one charcoal filter, removing most organisms and debris. To increase hatching, erythromycin gluceptate was added to a final concentration of 5 to 10 ppm. Rearing methods

TABLE 1

Summary of Conditions in Experiments Testing Effects of Benzene on Eggs and Larvae of Pacific Herring and Northern Anchovy. NM = Not Measured

Exp. No.	Date Begun	Day and Stage First Exposed	Mean Benzene Concentrations (ppm)	Total Exposure (Hrs)	Range and Mean Temp. (°C)	Sal. (ppt)	O_2 (ppm)	pH	NH_3 (ppm)	Food Spec. (No./ml)	Algae (No./ml)
Herring											
1	2-05-73	Egg Day 0 Early Cleavage	0,3.8,16.5,39.1	24,48,96	14.0-17.5 (15.2)	(24.0)	NM	NM	NM	None	None
2	3-01-73	Egg Day 0 Early Cleavage	0,4.8,17.7,45.0	24,48,96	10.0-17.0 (13.2)	(24.0)	NM	NM	NM	5-10 Artemia nauplii	None
3	11-28-73	Larval Day 2	0,6.7,12.1	48	11.5-14.0 (12.9)	(22.0)	4.7-7.9 (7.0)	7.4-8.1 (7.9)	.011-.910 (.128)	9-20 (13) Brachionus	(10^5)
4	1-22-74	Larval Day 2	0,13.0,31.9	24,48	12.2-13.5 (12.9)	(10.0)	6.1-8.0 (7.3)	7.4-7.8 (7.6)	NM	4-9 (6) Brachionus	(10^5)
Anchovy											
1	7-06-73	Day 0	0,0.4,2.7,7.5	24	17.5-19.2 (18.5)	(29.5)	7.1-8.0 (7.5)	7.7-7.9 (7.8)	.007-.030 (.014)	None	None
2	7-31-73	Day 0	0,4.7,10.5,24.0	24,48	16.5-18.0 (17.5)	(28.0)	7.0-8.0 (7.5)	7.6-7.8 (7.7)	.005-.040	None	None
3	8-14-73	Day 0	0,40.0	24,48	17.0-18.2 (17.5)	(28.0)	7.2-7.5 (7.4)	7.4-7.9 (7.6)	.010-.030	None	None
4	8-22-73	Day 0	0,13.5,19.5,53.5	24,48	16.8-18.4 (17.6)	(30.0)	(6.9-7.6) (7.3)	7.5-8.1 (7.9)	.010-.050	None	None

generally approximate those of Lasker *et al.* (1970) and Struhsaker *et al.* (1973).

All experiments were designed to permit multiway analysis of variance with two or more variables and three to five replicates per treatment combination. Replicate containers were placed randomly in water tables. In most experiments, 50 eggs were placed in each container. After sorting eggs into containers, various concentrations of benzene were added except for controls. On following days during development, benzene was added to give varying total exposure times of 0, 24, 48 and 96 hrs. Mean concentrations of benzene and exposure times are summarized in Table 1. After initial exposures, benzene concentrations declined over the 24-hr period due to volatization. For testing effects of 48- and 96-hr exposure times, additional benzene was added daily to achieve the initial concentration specified. The decline of three different initial benzene concentrations was measured every 2 hrs for 24 hrs under experimental conditions comparable to those in larval experiments. The decline in concentration is shown in Figure 1. The equation for the three calculated regression lines is $\hat{Y} = ae^{-bX}$. The regression coefficients for each concentration (Fig. 1) were not significantly different ($P < 0.01$) (Sokol and Rohlf, 1969). After 24 hrs, 25% to 30% of the initial concentration remained for all three concentrations. All benzene concentrations were measured with a Model MT-220 Tracor gas chromatograph to the nearest 0.05 ppm (Benville and Korn, 1974).

Ranges and means of physical variables, algae and food concentrations are also given in Table 1. During initial experiments with herring (1 and 2), the controlled seawater system was not completed and a greater range of conditions prevailed. In all other experiments these variables were more closely controlled.

Herring eggs or larvae were subsampled daily from each treatment combination and examined for indications of abnormal rates and modes of development.

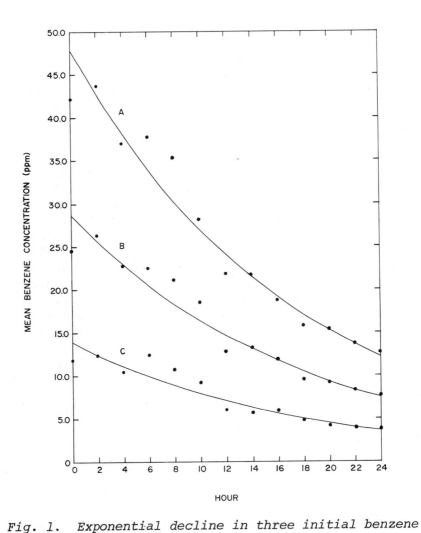

Fig. 1. Exponential decline in three initial benzene concentrations over a 24-hr period. Each point is the mean of three replicate containers. Range of temperature: 11.2°C-13.5°C. Equation for the calculated curve is: $\hat{Y} = ae^{-bX}$
A-a = 47.8, b = .05697; B-a = 28.7, b = .05608; C-a = 13.9, b = .05648. All three regression coefficients were significant (P < 0.01). There was no significant difference between the three regression coefficients, (b) (Sokal and Rohlf, 1969).

In some experiments, measurements of the standard
length of larvae (most anterior point of snout to end
of notochord) were made to the nearest 0.1 mm.

Herring larvae hatched in the laboratory after
9 to 11 days of incubation, while anchovy larvae
hatched after 1 to 2 days of incubation. For herring,
developmental days were designated as egg day or
larval day depending upon experiment being discussed.
Egg day 0 corresponds to the first 24 hrs after fertilization and larval day 0 to the day that larvae
hatch. For anchovy, day 0 corresponds to the day of
collection (gastrula) and egg and larval days are not
separated since larvae hatch only 1 to 2 days after
collection. At hatching, the following counts were
made: number of actively swimming normal larvae,
number of dead eggs remaining and number of abnormal
larvae. Percent hatching and percent abnormal larvae
among those hatching were calculated for each treatment combination.

In herring experiments 1 and 2 an attempt was
made to feed larvae *Artemia salina* nauplii. Few
larvae were able to feed (about 5%) because the
nauplii were too large. In herring experiments 3 and
4, larvae were given rotifers *(Brachionus plicatilus)*,
on which they fed readily. Culture methods for
rotifers are given by Theilacker and McMaster (1971).
Nephroselmis sp., isolated from San Francisco Bay,
was added to cultures in the latter two experiments.
Concentrations of food organisms and *Nephroselmis* are
in Table 1. Counts of rotifers and *Nephroselmis* were
made with a Model ZBI Coulter counter. In experiments
exposing herring larvae, benzene was added on larval
day 2, just after disappearance of the yolk sac.
Daily counts from hatching to larval day 7 were made
to determine percent survival.

Daily counts were not made in experiments with
anchovy, because eggs and larvae were too small to be
seen readily. Counts were made at the termination of
the experiment and percent survival and abnormal
larvae calculated. The standard length and size of
yolk sac of surviving larvae were also measured at

259

the end of the experiment. At the time of these experiments, rotifers had not been cultured and anchovy larvae were not fed. The experiments were ended prior to occurrence of starvation mortality (counted on days 3 or 6).

Larvae were observed for indications of abnormal swimming and feeding behavior. Also the percent of subsamples with food in guts was determined. All samples were preserved for more detailed studies of the effects of benzene on various morphometric characters.

Respiration of larvae was measured using a Gilson Respirometer. Ten larvae and 10 ml of sea-water were placed in each 25 ml flask. Correction was applied from blanks containing only seawater.

Data were analyzed using the UCLA BMD computer programs 01V and 02V (Dixon, 1970).

RESULTS

HERRING

Exposed Eggs; Effect on Early Development and Hatching

After exposure of eggs to benzene, the developmental sequence appeared normal except at the higher concentrations (35-45 ppm range). In embryos treated with an initial mean concentration of 45.0 ppm, developmental rate was obviously delayed over that in controls. The heartbeat rate of normal embryos ranged from about 70 to 90 beats per minute. In embryos at 45.0 ppm the heartbeat rate was often irregular, sometimes increasing to 110 beats a minute, while at other times a few beats were missed altogether.

Hatching occurred from 9 to 11 days in the laboratory. At hatching, approximately 20% to 25% of the larvae appeared abnormal in all treatment combinations, including controls. Eggs from the same

spawn collected in the field just prior to hatching
also exhibited 20% to 25% abnormal larvae at hatching.
Of these abnormal embryos, many simply did not develop and died before hatching. Others appeared to
be retarded at an earlier stage of development,
characterized by incomplete development of the body,
fins and jaw. Yolk was not absorbed to the degree
observed in normal larvae and the embryos remained
dorsoventrally coiled over the yolk sac instead of
straightening as in normal larvae at time of hatching.
The high percent of abnormal, inviable larvae was observed among those hatching in all experiments with
herring eggs collected from San Francisco Bay. In
one experiment, using eggs from Tomales Bay, fewer
abnormal larvae were observed. The possible reasons
for this high percent of abnormal larvae at hatching
are discussed below.

For brevity, and because data from the two
experiments exposing eggs are comparable, only
results from the second experiment (Table 1) are
presented to illustrate the effects of benzene on
eggs. The percent larvae hatching and percent of
abnormal larvae at hatching are shown in Figure 2.
Analysis of variance of the data showed a significant
difference (P < 0.01) between concentrations, but not
between exposure times. Significantly fewer larvae
survived and significantly more larvae were abnormal
at 45.0 ppm. Although the effects of different times
of exposure were not significantly different, there
is an indication that more larvae were abnormal at
48 hrs than at 24 and 96 hrs. At higher concentrations the time of exposure may have been more influen-
tial. The reason for the unusually high percent
abnormal larvae at 17.7 ppm, 48 hrs is unclear. Most
of these abnormal larvae occurred in only one replicate of the treatment combination and may be due to
errors in application of benzene (too high concentration) added to the containers at the beginning of egg
day 2. Other replicates at 17.7 ppm and 48 hrs were
comparable to those at 24 and 96 hrs. Generally, it
appears that herring eggs are relatively resistant to

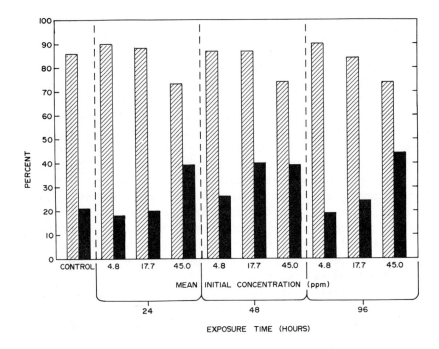

Fig. 2. Percent survival and percent abnormal among herring larvae at time of hatching after exposure of eggs to benzene. Herring experiment 2. Cross-hatched bars = percent survival, solid bars = percent abnormal.

benzene exposure. It should be remembered, however, that the benzene concentration declined over each 24-hr exposure period (Fig. 1). An initial concentration of 45 ppm is only 12 ppm or less after 24 hrs.

From the results of exposing herring eggs in experiment 1, it was noted that the 50% mortality level among larvae at time of hatching occurred at an initial mean concentration of approximately 40 to 45 ppm benzene when exposed for 96 hrs.

Among larvae exposed to a mean initial concentration of 45 ppm, all three exposure times, and larvae at 17.7 ppm, 96-hr exposure, abnormalities were observed that did not occur in controls and

lower concentrations. Most common was some form of
flexure along the body. Many larvae exhibited severe
lateral or ventral bends just posterior to the head.
Several were laterally bent to varying degrees of
severity in the body and tail region; some both in
the nape and tail region forming an s-shaped larva.
A few had only one eye. Development of the lower jaw
appeared incomplete. Of the abnormal larvae hatching,
those most severely retarded or bent died shortly
after hatching. A few of those less severely bent
survived for several days, but were unable to swim or
feed normally and eventually died.

Exposed Eggs; Survival of Larvae after Hatching

In herring experiment 2 (Table 1), an attempt
was made to rear larvae past the period of yolk
absorption by feeding larvae *Artemia salina* nauplii
as reported by Talbot and Johnson (1972). The percent
of larvae feeding and surviving daily for 33 days
after hatching was recorded. It was observed that
only a few larvae were feeding on the nauplii (5-10%),
although these began feeding on larval day 1. The
inability of most larvae to feed was probably due to
the nauplii being too large. Yolk was absorbed by
larval days 2 to 3 in all treatment combinations
except at 45 ppm (approximately 1 day later). From
larval days 1 to 7, survival of larvae at 4.8 ppm at
all three exposure times was comparable to controls.
In larvae at 17.7 ppm, 96 hrs, and 45 ppm, 48 and
96 hrs, survival was less than controls for these
periods. This was due to the death of larvae with
less obvious abnormalities. Starvation mortality
began on about larval day 6 in all treatment combinations. Survival of feeding larvae was not significantly different between treatment combinations for
the remainder of the experiment (33 days).

The standard length of surviving larvae (33 days
from spawning) was measured and larvae of similar
size from the field examined. Most major developmental

characteristics were comparable. The standard length
of laboratory larvae ranged from 12.4-17.0 mm; mean,
14.7 mm; there were two modal groups of 12.4-13.0 mm
and 14.8-16.1 mm; most larvae were from 14.8-16.1 mm.
There was no significant difference in standard length
between larvae from different treatment combinations.

Larvae surviving for 33 days after higher benzene
exposures of eggs appeared normal in all respects.
At least some larvae are able to recover from all
effects of the early exposure. However, at least 30%
to 50% more larvae at 35 to 45 ppm benzene suffered
irreversible effects leading ultimately to their
death than did control larvae. The effects of benzene on exposed eggs are summarized in Table 2. Subsamples and surviving larvae were preserved and a
more complete study of their morphometrics and abnormalities is underway and will be published
separately.

Exposed Larvae; Effects on Feeding, Behavior, Development and Survival

Eggs were incubated for 9 to 11 days to obtain
larvae for benzene exposure. Only normal larvae were
used in experiments. Because brine shrimp nauplii
were obviously inadequate food, larvae were given
rotifers *(Brachionus plicatilus)* immediately after
hatching. Larvae were observed feeding on rotifers
on larval day 1. *Nephroselmis* were also added to
provide food for rotifers. After completion of yolk
absorption on larval day 2, benzene was added to
containers (concentrations in Table 1). We chose to
expose larvae at this stage in the assumption that
they would be most sensitive to pollutant stress.
Larvae reacted immediately to the benzene, swimming
erratically and contracting violently. During the
7-day experimental period, the most obvious effect on
larvae was on their swimming and feeding behavior.
Many exposed larvae were lethargic, often lying on
the bottom with reduced swimming and feeding movements, while controls were actively swimming or

TABLE 2

Summary of Effects of Benzene on Eggs and Larvae of Pacific Herring. Percents Are Approximate Estimates of the Significant Differences from Control Levels (P < 0.05). NS = Not Significant, LD = Larval Day

Stage Exposed	Approximate Concentration Range (ppm)	Exposure Time (Hrs)	Survival Hatching LD 0	Survival Larvae LD 3	Abnormal Larvae Hatching	Mean % Feeding LD 1-7	Respiration Rate	Stand. Length LD 1-7	Developmental Rate
Egg Day 0	3- 5	24	NS	NS	NS	---	---	---	NS
	15-20		NS	NS	NS	---	---	---	NS
	35-45		-15%	-10%	+20%	---	---	---	Delayed
Egg Day 0,1	3- 5	48	NS	NS	NS	---	---	---	NS
	15-20		NS	NS	NS	---	---	---	NS
	35-45		-15%	-10%	+20%	---	---	---	Delayed
Egg Day 0,1,2,3	3- 5	96	NS	NS	NS	---	---	---	NS
	15-20		NS	-10%	NS	---	---	---	Delayed
	35-45		-15%	-10%	+25%	---	---	---	Delayed
Larval Day 2	5-10	24	Not exp. until LD 2	LD 7 NS		---	---	NS	NS
	10-15			-10%		-25%	+25%	NS	Delayed
	30-35			-25%		-30%	+45%	NS	Delayed
Larval Day 2,3	5-10	48		NS		-25%	---	Smaller	Accelerated
	10-15			NS		-25%	---	Smaller	Delayed
	30-35			-70%		-50%	---	Smaller	Delayed

suspended in the water column and feeding normally.
This lethargic behavior was most notable during the
first 1 to 2 days after exposure. Subsequently, surviving larvae recovered, exhibiting normal swimming
and feeding behavior. Their ability to recover and
feed is shown by the percent of larvae with food
observed in the gut (Table 3) over the 7-day period.

TABLE 3
Percent of Herring Larvae Subsample with Food
(Rotifers) in Gut. Total Benzene Exposure of 48 Hrs,
LD 0 and 1. Experiment 3

Concentration (ppm)	0	1	2	3	4	5	6	7	Mean (%)
0 - control		100	100	100	100	100	100	100	100
6.7		45	40	100	85	85	85	85	75
12.1		35	50	65	65	100	100	100	74

Unlike larvae hatching from exposed eggs, exposed
larvae showed few obvious morphological abnormalities.
Most notable was the delay in development. The developmental sequence appeared unaltered, but the treated
larvae were approximately 1 day behind control larvae
in rate of development. While most exposed larvae
regained ability to feed, some appeared to feed rarely
and subsequently the foregut and hindgut collapsed.
After this occurred, no further feeding took place
and larvae died. This was particularly noticeable in
larvae exposed to higher concentrations of benzene
(above 10 ppm). All larvae were preserved for further
study of their morphometrics and abnormalities.
Figure 3 shows percent survival of larvae on
day 7, 48-hr exposure of 6.7 and 12.1 ppm benzene

(herring experiment 3). Analysis of variance of the data shows a significant difference between concentrations (P < 0.05). Significantly fewer larvae survived in 12.1 ppm than in controls. Apparently, larvae exposed to 6.7 ppm for 48 hrs were able to recover sufficiently to survive through the critical period of yolk absorption. Most remaining larvae from all treatments survived beyond larval day 7 without further significant mortality for 21 days.

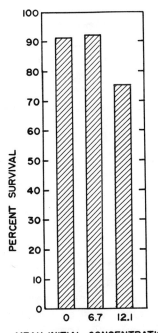

Fig. 3. Percent survival of herring larvae on larval day 7 after 48-hr exposure on larval day 2 and 3 to different concentrations of benzene. Herring experiment 3.

Figure 4 shows percent survival of larvae on the larval day preceding starvation mortality (herring experiment 4). The density of rotifers was inadequate to sustain normal growth and survival in this experiment (see Table 1). Significant differences in survival between treatments were found (P < 0.05). There was a significant difference between both concentrations and times of exposure. Fewer larvae survived at either concentration of benzene than in controls. Significantly fewer larvae survived at 48 hrs than at 24 hrs.

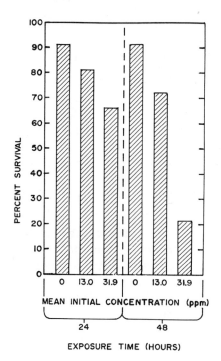

Fig. 4. Percent survival of herring larvae on larval day preceding starvation mortality after exposure to benzene on larval day 2 and/or 3. Herring experiment 4.

The 50% mortality level on larval day 7 is approximately a mean initial concentration of benzene of 20 to 25 ppm for 48 hrs exposure. Larvae seem to be more sensitive to lower concentrations of benzene than are eggs. Larvae, however, have the capacity to recover to some degree from the exposure while the affected embryos in eggs undergo irreversible effects.

Exposed Larvae; Effect on Growth and Respiration

Although many larvae treated with higher concentrations showed reduced feeding and delayed development, many were able to recover. Table 4 shows the mean standard length of larvae sampled on larval day 8 at the end of the experiment. Exposed larvae were significantly smaller than control larvae (P < 0.05). A difference of approximately 1.0 mm suggests that larval growth was retarded by benzene. In the future, measurements will be taken for a longer period to determine if exposed larvae are able to recover and attain size of control larvae.

Larvae exposed to benzene were placed in respiration flasks in the Gilson respirometer on larval day 3, 24 hrs after initial exposure. The total amount of oxygen consumed over a 24-hr period was recorded (Table 5). The results are preliminary, but show a significant increase (P < 0.05) in respiration exposed larvae over that of controls. Additional experiments are being performed on eggs and larvae of different developmental stages. The oxygen consumption is being measured at shorter 3-hr intervals for a total period of 24 hrs.

A summary of results of exposure of herring larvae is in Table 2.

TABLE 4
Mean Standard Length of Surviving Larvae (Day 8).
Total Exposure of 48 Hrs. Experiment 3

Concentration (ppm)	Exposure Time (hrs)	Mean Standard Length (mm)	N
0 - control	---	10.3	80
6.7	48	9.4	80
12.1	48	9.2	80

TABLE 5

Summary of Preliminary Experiment Measuring Respiration in Herring Larvae (Larval Day 3) Exposed to Benzene for 24 Hrs. Larvae Acclimated for 30 Min Before Test. Test Run for 24 Hrs

Mean Concentration Benzene	Flask Number	Number Larvae	Oxygen Consumed in 24 Hrs Microliters/Larva	
0 - control	1	10	7.5	
	2	10	6.9	
	3	10	7.7	
	4	Malfunction	____	
Mean				7.4
13.0 ppm	1	10	8.0	
	2	10	9.7	
	3	10	12.3	
	4	9	10.4	
Mean				10.1
39.1 ppm	1	10	13.2	
	2	10	13.6	
	3	10	15.0	
	4	10	10.5	
Mean				13.1

A significant difference exists between all three treatments ($P < 0.05$).

ANCHOVY

Exposed Eggs and Larvae; Effect on Development and Survival

The effect of benzene on anchovy eggs and larvae was tested by exposing both stages in each experiment. Four experiments were done (Table 1). The first experiment exposed eggs to a single 24-hr exposure; the latter three studies included both 24- and 48-hr exposures. At the time of the experiments, rotifers had not been cultured and larvae were not fed. Survival of larvae on days 3 and 6 was determined. Because eggs at time of collection varied in developmental stage, the results of anchovy experiments are more variable than those for herring.

Figure 5 shows percent survival of anchovy larvae on days 3 and 6 for different mean initial

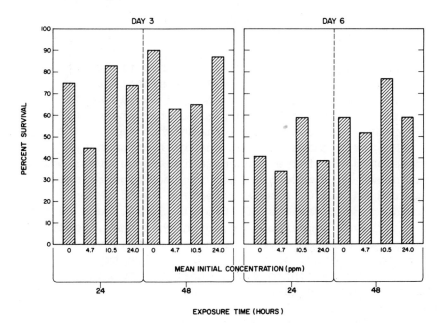

Fig. 5. Percent survival of anchovy larvae on days 3 and 6 after exposure of eggs and larvae to benzene. Anchovy experiment 2.

concentrations and exposure times. Refer to Figure 1 for actual concentrations over the 24-hr exposure period. On day 3, 24-hr exposure, survival of larvae was significantly lower (P < 0.05) at 4.7 ppm than in other concentrations and controls. Development at this concentration was accelerated over controls, whereas development at the higher concentrations was delayed (Table 6). On day 3, 48-hr exposure, an indication of lower survival is shown for larvae at 4.7 and 10.5 ppm. The difference, however, is not significant. On day 6, both 24- and 48-hr exposure, the pattern of larval survival was similar to that for day 3, 24 hrs, but differences were not significant. Except for larvae at 10.5 ppm, 48 hrs, on day 3, the same survival pattern persists throughout. There appears to be an acceleration of development and decreased survival at 4.7 ppm and delayed development and increased survival at 10.5 and 24.0 ppm. Larvae died earlier at 4.7 ppm, while at higher concentrations death of larvae was probably delayed due to narcotization. The slightly lower survival of larvae at 24.0 ppm was due to increased mortality of grossly abnormal larvae not occurring in other treatments. For both days, larval survival was significantly higher at 48 hrs because of delayed development. Additional experiments are planned to determine effects later in larval development when larvae are fed.

Figure 6 shows percent survival of anchovy larvae on day 6 for a concentration range overlapping that shown in Figure 5. When the two figures are compared on day 6, similarities are seen. Mean initial concentrations of 10.5, 13.5, 19.5 and 24 ppm show survival comparable to, or higher than, controls because of delayed development and mortality. At 53.5 ppm, however, survival is significantly less (P < 0.05) than controls and other treatments. Many abnormal larvae died at this concentration and, among those surviving (34%), approximately 30% were still abnormal in some way. Development of the surviving larvae (in 53.5 ppm) was greatly delayed over that of controls and they were also inactive. Even when fed

272

TABLE 6

Summary of Effects of Benzene on Eggs and Larvae of Northern Anchovy. Approximate Estimates of the Differences from Control Levels (P < 0.05) Derived from the Combination of All Experiments. NS = Not Significant

Stage Exposed	Approximate Concentration Range (ppm)	Exposure Time (hrs)	Larval Survival Day 3	Larval Survival Day 6	Abnormal Larvae Day 3	Abnormal Larvae Day 6	Yolk Utilization Day 3	Standard Length Day 3	Standard Length Day 6	Developmental Rate Day 3	Developmental Rate Day 6
Day 0 Egg	4- 5	24	-10%	NS	+25%	+20%	Accelerated	NS	NS	Accelera.	Accelera.
	5-10		NS	NS	+20%	+20%	Accelerated	NS	NS	Accelera.	Accelera.
	10-15		NS	NS	NS	+25%	Accelerated	NS	NS	Accelera.	Accelera.
	20-25		NS	NS	NS	+20%	Delayed	NS	NS	Delayed	Delayed
	40-55		NS	NS	+20%	+20%	Delayed	Smaller	Smaller	Delayed	Delayed
Day 0 Day 1	4- 5	48	-10%	NS	+20%	NS*	Accelerated	NS	Larger	Accelera.	Delayed
	5-10		-10%	NS	+20%	NS*	Delayed	NS	Larger	Delayed	Delayed
	10-15		-10%	NS	+25%	NS*	Delayed	NS	Larger	Delayed	Delayed
	20-25		NS	NS	+50%	NS*	Delayed	NS	Larger	Delayed	Delayed
Egg and Larva	40-55		NS	-15%	+50%	+30%*	Delayed	Smaller	Smaller	Delayed	Delayed

*Most abnormal larvae died by day 6, thus no significant difference from control as in day 3.

273

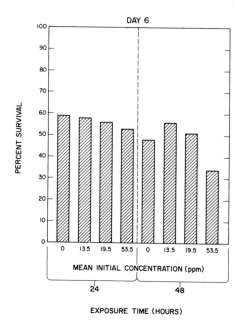

Fig. 6. Percent survival of anchovy larvae on day 6 after exposure of eggs and larvae to benzene. Anchovy experiment 4.

they would die a few days later. Assuming these abnormal larvae will die, the lowest mean initial concentration range at which 50% mortality occurs is approximately 20 to 25 ppm.

Figure 7 shows percent abnormal larvae among those surviving on days 3 and 6 for the same experiment as in Figure 5. On day 3, 24- and 48-hr exposure, there was a significant difference ($P < 0.05$) between controls and exposed larvae. On day 3, 48 hrs, there were significantly ($P < 0.05$) more abnormal larvae at 24 ppm than controls and other treatments. On day 6, there were fewer abnormal larvae among survivors than on day 3, because most grossly abnormal larvae had already died. The only significant difference in percent abnormal larvae on day 6 was between controls and larvae exposed to 24 ppm, 48 hrs ($P < 0.05$).

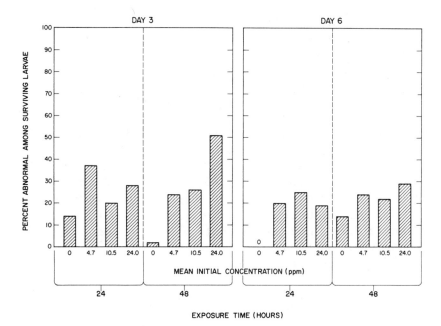

Fig. 7. Percent of abnormal anchovy larvae among larvae surviving to days 3 and 6 after exposure of eggs and larvae to benzene. Anchovy experiment 2.

The development of embryos before hatching was not observed in these experiments. When the experiment was ended, control larvae were examined and compared to larvae at approximately the same stage collected from the bay. No obvious differences were observed. Anchovy larvae treated with benzene exhibited abnormalities similar to those described earlier for herring larvae, the most obvious being bending of the notochord and somatic musculature ("bent larvae"). They were preserved for study of their morphometrics and abnormalities.

Unlike herring, few anchovy larvae were abnormal in the controls. The maximum percent of abnormal larvae observed in any experiment was from 10% to 15%. These abnormalities were not severe and usually occurred either when eggs were not treated with antibiotics before hatching or in eggs spawned later in the spawning season.

Benzene affected the developmental rates of the exposed eggs and larvae. Figure 8 shows the percent of larvae with yolk remaining on day 3 after 24-hr exposure of the egg. This illustrates the accelerated utilization of yolk at lower concentrations. Up to 10 ppm the differences from controls are significant (P < 0.05). Utilization of yolk at 24 ppm and, in other experiments, at 40 to 55 ppm is delayed. A difference in rate of yolk utilization occurs in larvae at concentrations of 4.7 and 10.5 ppm depending upon the length of exposure. When exposed for only 24 hrs, yolk is utilized more rapidly in both these concentrations than in controls. When exposed for 48 hrs, yolk utilization is delayed at both concentrations (see Table 6).

This pattern of yolk absorption is also reflected in the mean standard length of larvae on day 3 (Fig. 9). Differences in mean standard length are

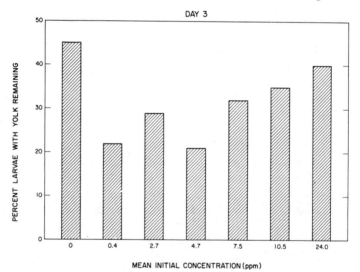

Fig. 8. *Percent of larval anchovies with yolk remaining on day 3 after exposure of eggs for 24 hrs to different concentrations of benzene. Anchovy experiments 1 and 2.*

276

not significant between concentrations or exposure
times on day 3. On day 6, however, there are significant differences (P < 0.01) in the size of larvae
correlated with developmental rate and yolk absorption. Control larvae decreased in size from day 3
to day 6. They were active, but not fed, and thus
resorbed tissue. When eggs were exposed for 24 hrs
to 4.7 ppm, larvae on day 6 were larger than in controls, but smaller than those in 10.5 ppm. They
utilized yolk more rapidly, but size remained about
the same as day 3. Even though more active than
larvae at higher concentrations, they were not as
active as control larvae and resorbed less tissue.
When eggs were exposed for 24 hrs to 10.5 ppm, larvae

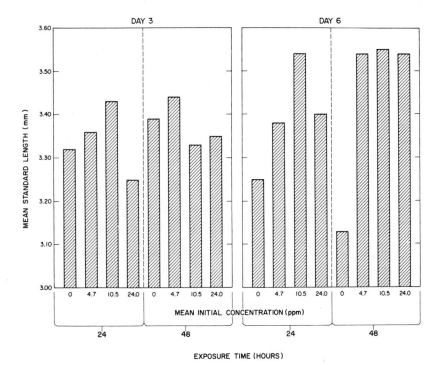

Fig. 9. Mean standard length of anchovy
larvae on days 3 and 6 after exposure of
eggs and larvae to benzene. Anchovy
experiment 2.

on day 6 were larger than in other treatments. They were probably narcotized and absorbed yolk more slowly. Because the larvae were less active, energy derived from yolk was channeled into growth. Resorption of tissue would probably occur later. When eggs were exposed for 24 hrs to 24 ppm, larvae on day 6 were larger than controls because of slower yolk absorption and development. Larvae at 24 ppm were less in mean standard length than those at 10.5 ppm because smaller, abnormal larvae were included in the measurements.

When eggs and larvae were exposed for 48 hrs, by day 6 all exposed larvae were larger than controls. Only normal larvae in all treatments were measured. At all concentrations, development was delayed and larvae were relatively inactive. Most energy derived from yolk absorption was expended in growth and tissue was not resorbed through activity as in control larvae. Whether the exposed larvae are able to recover later in development as did herring larvae is not known. Experiments are planned for next spawning season in which larvae will be fed and the delayed effects examined.

DISCUSSION

Benzene induces considerable physiological stress on eggs and larvae of herring and anchovy over much of the initial concentration range tested (approximately 5 to 50 ppm). Both lethal and sublethal effects were noted in measurements of survival and other parameters indicative of metabolic rate. The lethal effects of benzene are summarized in Table 7 as 50% mortality levels at certain stages. Because benzene volatilizes, calculated concentrations are given for various times. Maintenance of a constant concentration of benzene in an open system, such as that used in tests with juveniles and adults (Benville and Korn, 1974), was not possible. When a constant concentration is maintained, a fungus-like organism, apparently

TABLE 7

Summary of Concentrations and Exposure Times of Benzene Causing 50% Mortality in Herring and Anchovy Eggs and Larvae

Species	Stage Exposed	50% Mortality	Approximate Calculated Concentrations Over 24 Hrs*					Initial Conc. Range (ppm)	Exposure Time (Hrs)
			Initial	6	12	18	24		
Herring	Eggs	At Hatching	40	29	20	15	10	40-45	96
			45	32	23	16	12		
Herring	Larval Day 2	Larval Day 7	20	14	10	7	5	20-25	48
			25	18	13	9	7		
Anchovy	Egg and Larval Day 1	Larval Day 3	20	14	10	7	5	20-25	48
			25	18	13	9	7		

*Calculated from the equation $\hat{Y} = ae^{-.0565X}$ (refer to Fig. 1).

using benzene as an energy source, grows over the chorionic surface and the eggs die. We are trying to identify this organism and attempting to define its growth requirements as well as means of eliminating it from laboratory cultures.

A single exposure of benzene may approximate what occurs in the field. The initial benzene put into solution in the field undoubtedly volatilizes and the concentration declines over time. We plan to conduct spatial and temporal sampling in San Francisco Bay in order to determine actual concentrations of benzene and other aromatic components in the water column, both in areas of chronic oil pollution and in the event of a spill. Benzene may interact synergistically with other water-soluble aromatic components to produce greater effects than that of benzene alone. Data on concentrations of aromatics in either estuaries or oceans, however, are incomplete.

Sublethal effects of benzene on yolk absorption, growth and respiration show that benzene influences metabolic rate and energy utilization of embryos and larvae. At lower concentrations, metabolic rate is accelerated with a resultant energy cost in the alleviation of stress. At higher concentrations, metabolic rate is delayed, probably because larvae are narcotized. We plan to do additional measurements of respiration and growth to determine effects of the aromatic components on the energy budget of the larvae.

Reference to Table 7 shows that a greater exposure of benzene is required to affect mortality in exposed eggs than in exposed larvae. The exposure of eggs, however, induces abnormalities in the embryos. The effect is permanent, larvae eventually dying. It is interesting that many exposed herring and anchovy larvae are able to recover even though they undergo temporary effects. Those that survive 3 days after the last exposure often recover all feeding and swimming activity. There may be, however, delayed effects in development and growth not obvious until later in larval development. Herring and anchovy larvae showed

50% mortality at approximately the same concentration range of 20 to 25 ppm (Table 7).

Brocksen and Bailey (1973) found similar effects of benzene on the metabolic rate of juvenile striped bass and chinook salmon, evaluated by measuring their respiration rate at different concentrations and exposure times. They found an increase in respiration at lower concentrations and shorter exposure times. In contrast, at higher concentrations and longer exposures, narcotization occurred and respiration was reduced. They also found that fish recovered from exposures. Brocksen and Bailey (1973) provide an explanation of the possible biochemical pathways involved in benzene metabolism which may account for these effects on respiration.

Most other studies of the effect of oil on larval fish have been with whole crude oil or crude oil fractions. Kühnhold (1969, 1970) has tested the effect of several crude oils and their water-soluble fractions on Atlantic herring *(Clupea harengus)* larvae. At the National Marine Fisheries Service laboratory at Auke Bay, Alaska, the effect of water-soluble fractions of Prudhoe Bay crude oil was tested on Pacific herring *(C. pallasi)* and pink salmon fry *(Oncorhynchus gorbuscha)* (Stanley Rice, personal communication). Since we are studying the isolated physiological effects of benzene, our results cannot be legitimately compared.

It is probable that oil, acting synergistically with other pollutants such as pesticides deleteriously affects the young stages of fish and invertebrates in polluted estuaries. Many of these fish and invertebrates are important sport and commercial fisheries resources. Although the decline in their populations cannot be attributed solely to the effects of oil, oil and other pollutants may act together in significantly reducing larval recruitment in estuarine-dependent species.

The higher percent of abnormal larvae observed in larval herring hatching from eggs spawned in

San Francisco Bay than in the relatively unpolluted waters of Tomales Bay farther north may indicate that San Francisco Bay water is already polluted sufficiently to stress eggs after spawning. Although some genetic variation in viability could be expected in species producing such large numbers of eggs, it is unlikely that 20% to 25% would be genetically inviable. Abnormal eggs may also result from deleterious effects of pollutants on the development of eggs in gonads of adults. Since aromatics are highly soluble in lipid tissues, such as eggs, we plan to study this problem further.

The applied benefit of this research is obvious in view of the increasing possibility of oil spills, blowouts, effluents, and ballast dumping in the future. Recommendations on allowable levels of oil must involve knowledge of specific effects at sublethal concentrations as well as immediate lethal effects.

ACKNOWLEDGMENTS

We would like to thank Dr. Robert Brocksen and Norman Abramson for their suggestions and advice. We also appreciate the assistance of Pete Benville, Sid Korn, Lloyd Richards, and Jeff O'Neill in conducting these experiments. The identification of phytoplankton was made by E. Peter Scrivani and Andrea Alpine, University of California at Berkeley.

LITERATURE CITED

Ahlstrom, E. H. 1966. Distribution and abundance of sardine and anchovy larvae in the California Current Region off California. 1951-62: a summary. *U. S. Fish and Wildl. Serv., Spec. Sci. Rept.—Fish. 534:1-71.*

Aplin, J. W. 1967. Biological survey of San Francisco Bay, 1963-1966. *Mar. Res. Operations Branch, Ref. No. 67-4, Calif. Dept. Fish. and Game.*

Bailey, H. C. and R. W. Brocksen. 1974. Growth of juvenile Chinook salmon exposed to benzene, a water-soluble component of crude oil. *J. Fish. Res. Bd. Canada.* In press.

Bargmann, G. 1971. Bibliography of effects of oil pollution on aquatic organisms. Publ. College Fisheries, University of Washington, Seattle.

Benville, P. E. and S. Korn. 1974. A simple apparatus for metering volatile liquids into water. *J. Fish. Res. Bd. Canada 31*:367-68.

Brocksen, R. W. and H. C. Bailey. 1973. The respiratory response of juvenile Chinook salmon and striped bass exposed to benzene, a water-soluble component of crude oil. In: *Proceedings of the 1973 Conference on Prevention and Control of Oil Spills*, pp. 783-91. Washington, D. C.

Dixon, W. J. (Ed.). 1970. *BMD Biomedical Computer Programs.* Berkeley, Calif.: University of California Press.

Eldridge, M. E. and W. M. Kaill. 1973. San Francisco Bay area's resource—a colorful past and a controversial future. *Mar. Fish. Rev. 35*:25-31.

Frey, H. E. (Ed.). 1971. *California's Living Marine Resources and Their Utilization.* Calif. Dept. Fish and Game, Resources Agency.

Lasker, R. E., H. M. Feder, G. H. Theilacker, and R. C. May. 1970. Feeding growth, and survival of *Engraulis mordax* larvae reared in the laboratory. *Mar. Biol. 5*:345-53.

Korn, S. and D. Macedo. 1974. Determination of fat content in fish with a nontoxic, noninflammable solvent. *J. Fish. Res. Bd. Canada 30*:1880-81.

Kühnhold, W. W. 1969. Der einfluss wasserlöslicher bestandteile von rohölen und rohölfractionen auf die entwicklung von heringsbrut. *Ber. Dt. Wiss. Komm. Meeresforsch. 20*:165-71.

_____. 1970. The influence of crude oils on fish fry. In: *FAO Tech. Conf. on Marine Pollution and Its Effect on Living Resources and Fish.* Rome, Italy.

Moulder, D. S. and A. Varley. 1971. *A Bibliography on Marine and Estuarine Oil Pollution.* Plymouth: Publ. Laboratory Marine Biology Assoc., U. K.

Sokal, R. R. and F. J. Rohlf. 1969. *Biometry: The Principles and Practice of Statistics in Biological Research.* San Francisco, Calif.: W. H. Freeman and Co.

Struhsaker, J. W., D. Y. Hashimoto, S. M. Girard, F. T. Prior, and T. D. Cooney. 1973. Effect of antibiotics on survival of carangid fish larvae *(Caranx mate)*, reared in the laboratory. *Aquaculture 2*:53-88.

Talbot, G. B. and S. I. Johnson. 1972. Rearing Pacific herring in the laboratory. *Prog. Fish. Cult. 34*:2-7.

Theilacker, G. H. and M. F. McMaster. 1971. Mass culture of the rotifer *Brachionus plicatilis* and its evaluation as food for larval anchovies. *Mar. Biol. 10*:183-88.

THE EFFECTS OF OIL ON ESTUARINE ANIMALS: TOXICITY, UPTAKE AND DEPURATION, RESPIRATION

J. W. ANDERSON, J. M. NEFF, B. A. COX,
H. E. TATEM, and G. M. HIGHTOWER

*Department of Biology
Texas A & M University
College Station, Texas   77843*

Petroleum contamination of the marine environment can occur from a variety of sources. A significant portion of the petroleum hydrocarbons present in the world's oceans may have been transported there through the atmosphere. However, chronic low level contamination of estuaries and coastal waters originates primarily from the discharge of oil from near shore ship operations and via urban and industrial sewage effluents. On a more intermittent basis, discharge of ballast waters by tankers and other vessels and accidental spills from offshore wells and shipping mishaps may heavily contaminate localized regions of the coastal zone (Nelson-Smith, 1970; Blumer, 1971). If the fauna and flora of the affected region has been thoroughly studied previous to a spill and very careful ecological and chemical analyses of organisms is conducted after the spill, much meaningful information may be gained regarding the impact of oil on the marine and estuarine environment. Ideally, field and laboratory investigations should be combined to determine population and organismic responses to pollutants. Insight into the impact of a pollutant on a marine or estuarine species may be gained from four general types of biological and chemical investigation. These include:

1. Short-term toxicity studies—to determine the range of tolerance to the pollutant, to set exposure levels for sublethal studies, and to evaluate the significance of environmental concentrations.
2. Organism-environment transfer studies—to determine rates of accumulation and release of the pollutant by the organism.
3. Physiological studies—to determine the extent and nature of modification of metabolic parameters in response to exposure to sublethal concentrations of the pollutant.
4. Field studies—to compare responses of marine animals to natural and perhaps chronic exposures to pollutants with responses observed in the laboratory, and to determine the effects of pollutants on the metabolism and structure of marine communities.

It should be emphasized that these phases of research need not and should not be separated. During short bioassay and physiological studies, important data may be collected on environment-organism transfer of the pollutant under investigation. Following exposure to the pollutant, the survivors can be returned to clean seawater and the rates of pollutant release from their tissues determined. In laboratory and field studies alike, considerable stress must be placed on the accurate determination of the concentration of pollutant to which the animal is actually exposed and of the resulting pollutant concentration in the tissues of the animals. Our research to date has been concerned with the first three topics listed above and research on the fourth topic is just getting under way. This research was supported by a contract from the American Petroleum Institute. Fortunately, additional funds were

provided to Dr. Scott Warner of Battelle Memorial
Laboratories, Columbus, Ohio, for detailed gas
chromatographic/mass spectrophotometric analyses of
the hydrocarbon composition and concentration in oil,
water and tissues. During the early phases of this
work, an ultraviolet spectrophotometetric technique
for the quantitative determination of naphthalenes
in water and tissues was developed in our laboratory
(Neff and Anderson, unpublished data) and has been
used extensively during this research.

RESULTS AND DISCUSSION

CHARACTERISTICS OF THE TEST OILS

    The four test oils used in our research were
furnished by the American Petroleum Institute.
Reasonable quantities of these materials will be
supplied upon request to other investigators to
facilitate comparisons of biological effects studies.
Two crude oils (South Louisiana and Kuwait) and two
refined oils (a high aromatic #2 fuel oil and
Venezuelan bunker C residual oil) were used. Test
organisms were exposed to oil-water mixtures of two
types, oil-in-water dispersions (OWD) and water-
soluble fractions of oil (WSF). OWDs were prepared
by adding a known volume of oil to 500 or 750 ml of
synthetic seawater (Instant Ocean) of the appropriate
salinity in 1-liter bottles and shaking the mixture
for 5 min at approximately 200 cycles per minute on
a shaker platform. The dispersions were allowed to
equilibrate for 1 hr before the aqueous phase was
sampled for hydrocarbon analysis or animals were
added for exposure. The WSFs were prepared by adding
1.5 liters of oil to 13.5 liters synthetic seawater
in a 5-gallon pyrex bottle and mixing (vortex approximately one-fourth the depth) with a magnetic stirrer
for 20 hrs. The water phase was siphoned off for use
within 8 hrs and this stock WSF was diluted to specific concentrations with synthetic seawater.

The results of gas chromatographic analyses of the four test oils and the stock WSFs prepared from them are shown in Table 1. The petroleum hydrocarbons analyzed represent only 4.6 to 16.56% by weight of the oils and only 2 to 23.6% of the total petroleum hydrocarbons present in the WSFs. Anderson *et al.* (1974) have shown that the majority of the remaining hydrocarbons in the WSFs are the light hydrocarbons ethane through trimethylbenzenes. Among the hydrocarbons analyzed, the normal paraffins present at

TABLE 1

The Concentrations of C_{12}-C_{24} n-Paraffins and Di-tri-aromatic Hydrocarbons in the Test Oils and in Water-Soluble Fractions (WSFs) Prepared from Them. Hydrocarbon Concentrations in the Oils Are Given As Percent (g/100 ml) and in the WSFs As Parts Per Billion (μg/L) in 20 o/oo Seawater

Name of Compound	S. Louisiana Whole Oil (%)	WSF (ppb)	Kuwait Whole Oil (%)	WSF (ppb)	#2 Fuel Oil Whole Oil (%)	WSF (ppb)	Bunker C Whole Oil (%)	WSF (ppb)
n-paraffins								
C_{14}	0.44	10.0	0.46	<0.5	0.82	5.0	0.11	0.8
C_{15}	0.48	10.0	0.41	<0.5	1.06	7.0	0.11	0.9
C_{16}	0.54	12.0	0.43	0.6	1.20	8.0	0.15	1.2
C_{17}	0.41	9.0	0.42	0.8	0.98	6.0	0.12	1.9
C_{18}	0.30	7.0	0.28	0.5	0.60	4.0	0.10	1.0
Total C_{12}-C_{24} n-paraffins	3.98	89.0	4.00	2.9	7.38	47	1.26	12
aromatics								
naphthalene	0.04	120	0.04	20	0.40	840	0.10	210
1-methylnaphthalene	0.08	60	0.05	20	0.82	340	0.28	190
2-methylnaphthaelne	0.09	50	0.07	8.0	1.89	480	0.47	200
dimethylnaphthalenes	0.36	60	0.20	20	3.11	240	1.23	200
trimethylnaphthalenes	0.27	8.0	0.19	3.0	1.84	30	0.88	100
biphenyls	<0.01	2.0	<0.01	1.0	0.16	28	<0.01	1.0
fluorenes	0.02	<1.0	<0.01	<1.0	0.36	20	0.24	11.0
phenanthrenes	0.06	4.0	0.04	3.0	0.53	20	1.11	23.0
dibenzothiophene	0.02	1.0	0.01	<1.0	0.07	4.0	<0.01	<1.0
Total aromatics	0.94	305	0.60	75	9.18	2,002	4.31	935
Total hydrocarbons measured	4.92	394	4.60	78	16.56	2,049	5.57	947
Total hydrocarbons present (IR anlaysis)	19,800		10,400		8,700		6,300	

highest concentration in the oils and their WSFs are
those with chain lengths of 14 to 17 carbons. The
n-alkanes as a group are present at very low concentration in the WSFs, reflecting their low water solubility (McAuliffe, 1966). The dimethylnaphthalenes
are the hydrocarbons present at highest concentration
in the oils (0.20 to 3.11%). However, presumably due
to its greater water solubility, naphthalene represents the most prominant diaromatic hydrocarbon in
the WSFs. The WSFs of the two refined oils contain
significantly higher levels of diaromatic hydrocarbons
than do those of either of the crude oils. Biphenyls,
fluorenes, phenanthrenes and the sulfur-containing
compound dibenzothiophene are present in the WSFs at
very low concentrations, reflecting their extremely
low water solubility. Anderson *et al.* (1974) have
shown that OWDs have a hydrocarbon composition very
closely resembling that of the parent oil. This is
to be expected since most of the oil in the aqueous
phase of these preparations is present in the form of
dispersed droplets of whole oil.

Aeration of oil-water mixtures during bioassays
and other exposures can be expected to cause a rapid
decline in the concentration of oil hydrocarbons in
the aqueous phase. Infrared (API, 1958) and gas
chromatographic analyses of the aqueous phase of OWDs
following gentle aeration for different lengths of
time reveal that as much as 90% of the oil hydrocarbons originally present are lost in 24 hrs
(Anderson *et al.*, 1974). The n-alkanes appear to be
lost from the oil-water mixtures more rapidly than
the aromatic hydrocarbons.

Because of the difficulties inherent in the
measurement of total or specific oil hydrocarbons in
water, most investigators in the past have assumed
that their test organisms were being exposed to the
concentration of oil added to the water. The possibility of alteration in the concentration of oil
hydrocarbons in the water with time was not usually
taken into consideration. It is important to note
that the aqueous phase of OWDs usually contains one

to two orders of magnitude less oil than the amount originally added during the preparation of the dispersion. The remaining oil is present as a surface oil slick. Even when the initial concentration and composition of the test mixture is known, aeration and other factors tend to rapidly change the composition and lower the concentration of aqueous hydrocarbons during the time course of exposure. The test organisms are therefore exposed to high hydrocarbon concentrations for only a brief period and the nature and concentration of the materials remaining in the aqueous phase over longer time periods will depend upon the specific characteristics and composition of the parent oil. Thus, the composition, relative solubilities and dispersion-forming capabilities of the test oils must be considered in assessing their relative toxicities to marine organisms.

RELATIVE TOXICITY OF OILS

At the outset of this investigation, standard bioassays were conducted to determine the relative toxicities of the test oils to our experimental organisms. It was felt that such information would be useful in establishing the sensitivities of the test organisms to the different oils and in setting concentration limits within which subsequent sublethal studies should be conducted.

One to five organisms, depending on size, were placed in each bioassay bottle containing 500 to 2000 ml of an oil-water mixture. A minimum of ten animals were exposed to each concentration of oil. Air was gently bubbled through each bottle during the exposure period to provide aeration. Periodically the bottles were examined and the number of dead animals recorded. Dead animals were removed from the bottles as soon as discovered. The results of these studies utilizing three species of estuarine animals are summarized in Table 2. Concentrations of oil-water dispersions (OWDs) are listed in terms of the amount of oil added to the water and not the amount

290

TABLE 2

LC50 Values for Three Species of Estuarine Animals and Statistical Analysis of the Bioassay Results by the Method of Litchfield and Wilcoxon (1949). Ninety-five Percent Confidence Intervals (95% C.I.) and Slope Functions (S.F.) Are Also Given

Hydrocarbon Solution	Time (hrs.)	Cyprinodon variegatus		Palaemonetes pugio			Penaeus aztecus (postlarvae)		
		24	96	24	48	96	24	48	96
S. La. crude OWD	LC50 / 95% C.I. / S.F.	80,000[a]	29,000[a]	1,700 / 567-5,100 / 5.48	1,650 / 589-4,620 / 5.48	200 / 133-302 / 2.00	[b]1000[a]	[b]1000[a]	[b]1000[a]
WSF	LC50 / 95% C.I. / S.F.	[b]19.8[a]	[b]19.8[a]	[b]19.8[a]	[b]19.8[a]	[b]19.8[a]	[b]19.8[a]	[b]19.8[a]	[b]19.8[a]
Kuwait OWD	LC50 / 95% C.I. / S.F.	[b]80,000[a]	[b]80,000[a]	13,500 / 6,750-27,000 / 3.09	9,000 / 3,462-23,400 / 6.38	6,000 / 2,400-15,000 / 4.31	--	--	--
WSF	LC50 / 95% C.I. / S.F.	--	--	[b]10.2[a]	[b]10.2[a]	[b]10.2[a]	--	--	--
#2 fuel oil	LC50 / 95% C.I. / S.F.	250 / 156-400 / 2.55	93 / 66-130 / 1.78	3.8 / 3.0-4.9 / 1.51	3.4 / 2.8-4.2 / 1.42	3.0 / 2.7-3.3 / 1.18	9.4 / 7.6-11.6 / 1.26	9.4 / 7.6-11.6 / 1.26	9.4 / 7.6-11.6 / 1.26
WSF	LC50 / 95% C.I. / S.F.	[b]6.9[a]	[b]6.9[a]	4.4 / 3.3-5.6 / 1.48	4.1 / 2.7-5.5 / 1.77	3.5 / 2.4-4.9 / 1.92	5.0 / 4.6-5.5 / 1.12	5.0 / 4.6-5.5 / 1.12	4.9 / 4.6-5.5 / 1.06
bunker C WSF	LC50 / 95% C.I. / S.F.	4.7[a]	3.1[a]	3.2 / 2.4-4.3 / 1.40	2.8 / 2.1-3.8 / 1.40	2.6 / 2.0-3.3 / 1.40	3.8 / 3.3-4.5 / 1.31	3.5 / 2.9-4.1	1.9 / 1.0-3.5 / 1.94

[a]Denotes data points which could not be analyzed by the above method.

[b]Denotes 'greater than.'

actually present in the aqueous phase. Concentrations of water-soluble fractions (WSFs) are listed as ppm total hydrocarbons actually present in the aqueous phase. Before dilution to various concentrations, the hydrocarbon content of each stock water-soluble fraction was determined by infrared analysis (API, 1958).

Whenever possible, LC_{50} values, their 95% confidence intervals and slope functions were calculated by the graphical method of Litchfield and Wilcoxon (1949). In those cases where the nature or distribution of data points precluded computation by this method, LC_{50} values were computed on semilogarithmic paper by standard APHA method (APHA, 1971).

The confidence limits indicated must be interpreted with caution. They are merely indicators of what might be expected if the same stock of animals were immediately retested under identical conditions. It can be expected that bioassays with similar animals at different times of year will give somewhat different LC_{50} values. The slope function has been included with the LC_{50} values in the table and represents the factor by which a dose must be multiplied or divided to produce a standard deviation change in response. Thus, large "s" values are indicative of large standard deviations in the test results and produce rather broad 95% confidence limits.

Inspection of Table 2 reveals that there was a relatively uniform and consistent pattern of toxicity of the OWDs and WSFs to the test organisms. The dispersions and soluble fractions of the two refined petroleum products, #2 fuel oil and bunker C residual oil, were in all cases considerably more toxic to the organisms tested than were those of the two crude oils, South Louisiana and Kuwait. It should be noted that the hydrocarbon concentration in the aqueous phase of OWDs produced by the addition of 1,000 to 10,000 ppm oil to water only varied from a low of 27 ppm for Kuwait (1,000 ppm added) to a high of 75 ppm for South Louisiana crude oil (10,000 ppm added) (Anderson *et al.*, 1974). It would appear that the water-soluble fraction of bunker C was slightly more toxic than the #2 fuel oil WSF. Little information is available on the relative toxicity of bunker C dispersions, due to the difficulties encountered in handling this heavy, viscous oil. It should be noted that the full strength water-soluble fraction of the crude oils (19.8 ppm for S. Louisiana and 10.2 ppm for Kuwait) failed to produce 50% mortality among the test animals during the 96-hr bioassay period.

The bioassays with the WSFs of bunker C and #2 fuel oil and the OWD of the latter show that the post-larvae of *Penaeus aztecus* are intermediate in toler-ance between *Palaemonetes* and *Cyprinodon*. Cox and Anderson (1973) have shown that the sensitivity of *Penaeus aztecus* to these oil-water mixtures increases as the shrimp mature. This is not too surprising since the postlarvae of this species undergo consider-able natural environmental stress during their migra-tions into the estuaries. Additional data (Tatem, unpublished data) indicate that *Palaemonetes* post-larvae are only slightly less (0.5 to 6 ppm differ-ence) tolerant than adults to the oil-water mixtures.

Because the WSFs of the refined oils were found to be considerably more toxic than those of the crude oils to the test organisms, the relative toxicities of those hydrocarbons shown to be most prominent in the refined oil WSFs were determined in additional bioassays. Table 3 summarizes the results of bio-assays with naphthalene and different alkylnaphtha-lenes. The different naphthalenes were toxic to these estuarine animals at levels between 0.08 and 5.1 ppm. For both species of shrimp, the dimethyl-naphthalenes were shown to be the most toxic, with 24-hr LC_{50} values of 80 and 700 ppb. Other LC_{50} values were in the general range of 1 to 3 ppm.

The relatively low concentrations of naphthalenes required to produce 50% mortality in 24 hrs indicate that they may contribute significantly to the toxicity of the WSFs, particularly of the refined oils. It should be recalled that the total concentration of these compounds, including a small amount of

TABLE 3

*The Tolerance of Three Species of Estuarine Animals
to Specific Petroleum Hydrocarbons (Naphthalenes)*

Hydrocarbon	24-Hr LC_{50} Values (ppm)		
	Cyprinodon variegatus	*Palaemonetes pugio*	*Penaeus aztecus*
Naphthalene	2.4	2.6	2.5
1-methylnaphthalene	3.4	---	---
2-methylnaphthalene	2.0	1.7	0.7
Dimethylnaphthalenes	5.1	0.7	0.08

trimethylnaphthalenes, was 0.90 ppm for the WSF of
bunker C and 1.93 ppm for the WSF of #2 fuel oil.

UPTAKE AND DEPURATION OF PETROLEUM HYDROCARBONS

In earlier investigations, we exposed oysters,
Crassostrea virginica, and estuarine clams, *Rangia
cuneata*, to OWDs of the four test oils for different
lengths of time and then determined the concentration
of a wide variety of petroleum hydrocarbons in the
mollusc tissues by gas chromatography (J. W. Anderson,
1973; R. D. Anderson, 1973). We showed that these
bivalves were able to accumulate a wide variety of
different hydrocarbons in their tissues. However, in
all cases, the different naphthalenes were the hydrocarbons concentrated to the greatest extent. When
the clams and oysters were returned to oil-free seawater, they rapidly released the accumulated hydrocarbons from their tissues, complete depuration
requiring from 10 to 52 days. The alkylnaphthalenes

were always the last petroleum hydrocarbons to reach
undetectable levels in the mollusc tissues. These
results prompted us to develop and evaluate a simple
rapid ultraviolet spectrophotometric technique for
the quantitative determination of naphthalenes in
animal tissues (Neff and Anderson, 1974). This UV
technique has been used to study the accumulation and
release of naphthalenes by *Cyprinodon*, *Penaeus*, and
Palaemonetes.

Sheepshead minnows, *C. variegatus*, were placed
in seawater containing either naphthalene or
1-methylnaphthalene in solution at a concentration
of 1 ppm (Fig. 1). During the 4-hr exposure period,
the fish accumulated approximately 205 ppm 1-methyl-
naphthalene or 60 ppm naphthalene in their tissues.
When the remaining fish were placed in clean seawater,
they rapidly released the accumulated hydrocarbons
from their tissues. After 29 hrs in clean seawater,
the tissue concentrations of naphthalene and
1-methylnaphthalene had dropped to approximately 10
and 30 ppm, respectively. These results demonstrate
the rapidity with which the diaromatic hydrocarbons
are exchanged between the environment and the tissues
of the fish.

A group of brown shrimp, *Penaeus aztecus*, was
exposed to a 30% WSF of #2 fuel oil (1.95 ppm total
hydrocarbons in solution) for 30 min and then returned
to oil-free seawater. At the end of the exposure
period and at different time periods thereafter, small
groups of shrimp were sacrificed and the concentrations of naphthalene, methylnaphthalenes, and dimethyl-
naphthalenes in their tissues were determined (Fig. 2).
The methylnaphthalenes were accumulated by the shrimp
tissues to a considerably greater extent than were
either naphthalene or dimethylnaphthalenes. When the
shrimp were returned to oil-free seawater, the tissue
concentrations of these aromatics gradually decreased
to undetectable levels (< 20 ppb) in about 10 hrs
(600 min). Similar experiments have been conducted
with the grass shrimp *Palaemonetes pugio*, and a
similar pattern of rapid accumulation followed by

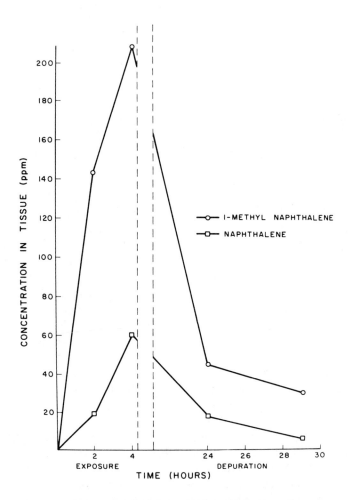

Fig. 1. *Accumulation and release of naphthalene and 1-methylnaphthalene by* Cyprinodon variegatus. *Groups of fish were exposed for 4 hrs to 1 ppm solutions of each hydrocarbon in seawater and then transferred to clean seawater. Each point represents the average of the analyses of three fish.*

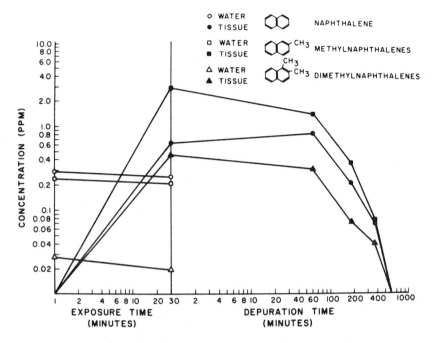

Fig. 2. Uptake and release of naphthalenes from a 30% WSF of #2 fuel oil by the brown shrimp Penaeus aztecus. *The exposure water initially contained approximately 1.95 ppm total oil hydrocarbons.*

complete depuration was demonstrated. In the numerous exchange experiments conducted thus far with several species of estuarine animals, we find that the naphthalenes are accumulated rapidly by the animal tissues, but in periods between a few hours to a few weeks after the animals have been returned to hydrocarbon-free seawater the compounds have been released back to the environment.

RESPIRATION

There is a considerable body of published information dealing with the respiratory responses of marine invertebrates and fish to the physiological stresses induced by changes in environmental salinity, temperature and dissolved oxygen (Fry, 1971; Newell, 1970, 1973). Chronic or acute exposure to sublethal concentrations of petroleum may also constitute a physiological stress to marine organisms. If this is so, marine organisms might be expected to show a respiratory response to exposure to sublethal concentrations of oil-water mixtures. To test this hypothesis we have measured the respiratory rates of several species of estuarine fish and invertebrates before, during, and after exposure to different oil-water mixtures.

Nine groups of six sheepshead minnows, *Cyprinodon variegatus*, were exposed to different concentrations of the WSFs of the four test oils for 24 hrs. A tenth group of 36 fish was maintained in oil-free seawater during the exposure period and served as a control. After exposure, each fish was placed in a separate 500 ml sealed glass respiratory chamber containing clean 15 o/oo seawater at 22°C, and its respiratory rate was measured with a Clark-type oxygen electrode and O_2 monitor (Yellow Springs Instruments). With one exception, no mortality occurred during the exposure period. Mortality was observed among the fish exposed to the 100% WSF of South Louisiana crude oil, necessitating termination of the exposure to this WSF after 2 hrs. The results of this experiment are summarized in Figure 3. The concentration of WSF which elicited a significant respiratory response was dependent on the test oil used. Significant departures from control respiratory rates occurred at lower concentrations of the WSFs of #2 fuel oil and bunker C than of South Louisiana crude and Kuwait crude, correlating well with the toxicity data. However, insufficient data are available to draw any diffinitive conclusions about the respiratory response of *Cyprinodon* to low concentrations of any of the WSFs.

Differences in the respiratory response of the fish to the WSFs of #2 fuel oil and bunker C residual oil may be related to the differences in their hydrocarbon compositions. As indicated above, the dominant

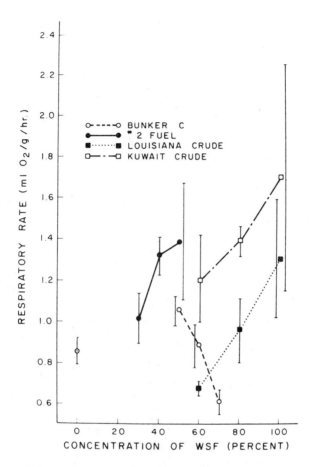

Fig. 3. Respiratory rates of
Cyprinodon variegatus *following*
exposure for 24 hrs to different
concentrations of the WSF of the
four test oils in comparison to
control (0 concentration) rates.
The control respiratory rate is the
mean of measurements for 36 fish
and other data points are the mean
respiratory rate of six individuals.
Vertical lines represent the
standard errors.

hydrocarbons in the WSF of #2 fuel oil are the naphthalenes. Exposure to the WSFs of this oil caused a
stimulation of the respiratory rate of the fish. The
bunker C WSFs, on the other hand, contain significant
concentrations of alkylphenols (Warner, personal
communication) in addition to relatively high concentrations of naphthalenes. Exposure to WSFs of this
oil caused a depression of oxygen consumption by the
fish. In general, the standard errors increased as
the exposure concentrations of the WSFs increased.
This is not surprising since the effects of natural
individual variations in the tolerance of the fish to
oil can be expected to become more intense as the
exposure concentration is increased.

The effects of exposure to the WSFs of oil on
the respiratory rate of postlarval brown shrimp,
Penaeus aztecus, was also investigated. In these
experiments, respiratory rates of the shrimp were
determined during exposure of the shrimp to the WSFs
of either #2 fuel oil or South Louisiana crude oil.
Shrimp were placed in sealed glass respiratory chambers containing a WSF of the appropriate concentration and the amount of oxygen consumed by the shrimp
during exposure was determined by the micro-Winkler
titration technique. The results of three experiments
are summarized in Figure 4. A different size class
of postlarvae was used in each experiment, and it
sould be noted that control respiratory rates vary as
expected from a log-log linear regression of weight-
specific oxygen consumption versus weight. Those
postlarvae of the smallest size class (2.25 mg mean
dry weight) consumed significantly greater amounts of
oxygen per unit weight under control conditions than
did the larger shrimp. In each experiment, animals
at the higher exposure concentrations exhibited increased activity and a spiral swimming pattern while
some were on their backs with only the pleopods beat-
ing at the termination of the exposure.

The pattern of respiratory response was remarkably similar in the three experiments. *Penaeus* postlarvae responded in general by lowering their oxygen

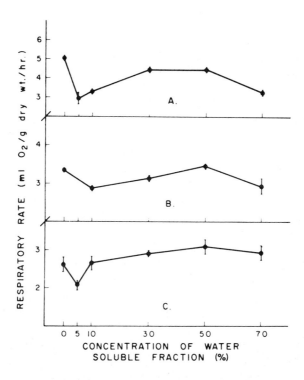

Fig. 4. Respiratory rates of different size classes of postlarval Penaeus aztecus *during exposure to WSFs of oil Each point represents the mean of the respiratory rate of five groups of two or three postlarvae. Vertical bars represent standard errors.*
A. Four-hr exposure to WSFs of #2 fuel oil. Mean dry weight/shrimp = 2.25 mg.
B. Five-hr exposure to WSFs of South Louisiana crude oil. Mean dry weight/shrimp = 3.46 mg.
C. Four-hr exposure to WSFs of South Louisiana crude oil. Mean dry weight/shrimp = 12.08 mg.

consumption, particularly at the lower exposure concentrations. At the intermediate WSF concentrations, the shrimp generally had respiratory rates near those of the controls. The respiratory response of the shrimp exposed to the WSFs of #2 fuel oil (Fig. 4A)

was considerably greater than that of the shrimp exposed to the WSFs of South Louisiana crude oil (Figs.
4B and 4C). At all exposure concentrations of the
#2 fuel oil WSF, mean respiratory rate was found to
be significantly different from the control rate by
the Student's T-test for unequal varience with α set
at 0.025. On the other hand, the respiratory rates
of shrimp exposed to the WSFs of South Louisiana
crude oil were not significantly different from
control rates. This difference may be due either to
the smaller size of the #2 fuel oil-exposed shrimp
(2.25 mg mean dry weight versus 3.46 and 12.08 mg
mean dry weight for the South Louisiana crude-exposed
shrimp) or to the differences in the hydrocarbon composition of the two WSFs. There are interesting
differences in the respiratory responses of small and
large postlarvae to the South Louisiana crude oil
WSFs. The smaller shrimp had a mean respiratory rate
lower than that of the controls at the lowest exposure concentration, but mean respiratory rates were
nearly the same as those of the controls at all
higher exposure concentrations (Fig. 4B). Likewise,
the mean respiratory rate of the larger shrimp was
lower than that of the controls at the lowest exposure concentration. However, at exposure concentrations between 30 and 70% WSF, mean respiratory rates
of the larger shrimp were higher than those of the
controls (Fig. 4C). These differences may be related
to the observation of Cox and Anderson (1973) that
the sensitivity of *Penaeus* to oil-water mixtures
increases as their size increases.

   Additional respiratory studies were conducted
utilizing the estuarine grass shrimp *Palaemonetes
pugio*. Four groups of 10 to 12 individuals were
exposed to each of three concentrations of the WSF of
Kuwait crude oil. The respiratory rate of each group
was determined immediately before exposure, after
24 hrs of exposure, and 3 days after exposure. Each
group of shrimp was maintained and exposed separately
and therefore served as its own control. All respiratory rate determinations were conducted by the oxygen

electrode method in respiratory chambers containing oil-free seawater. Between the second (exposed) and final (recovered) respiratory measurements, each group of shrimp was maintained separately in filtered seawater and was fed Tetramin fish food.

The results of this experiment are summarized in Figure 5. At all three exposure concentrations, the shrimp had respiratory rates immediately after

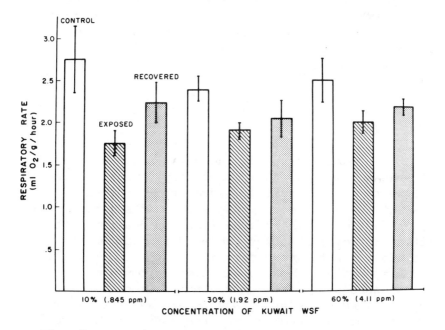

Fig. 5. Respiratory rates of the grass shrimp Palaemonetes pugio before, immediately after, and 3 days after exposure to WSFs of Kuwait crude oil. All respiratory measurements were made in clean synthetic seawater at 15 o/oo salinity and 22°C. The heights of the columns represent the mean of the rates of four groups of 10 to 12 animals and the vertical bars represent the standard deviations. At each exposure concentration, the control, exposed and recovered columns are the rates for the same groups of individuals.

exposure which were significantly lower than the preexposure control rates. The greatest respiratory
depression appeared to occur following exposure to
the lowest concentration of Kuwait WSF (10% WSF containing 0.845 ppm total hydrocarbons in solution).
At all exposure concentrations, the respiration rates
determined 3 days after exposure had increased to
levels not significantly different from preexposure
control rates. It would therefore appear that exposure to Kuwait WSFs caused a temporary but significant depression of oxygen consumption by the grass
shrimp, but upon their return to oil-free seawater,
their respiratory rates rapidly returned to near-
normal levels.

CONCLUSIONS

When marine or estuarine animals are exposed to
oil-water mixtures, they ordinarily come in contact
only with those petroleum hydrocarbons actually
present in the aqueous phase either in solution or in
dispersed form. Therefore, research dealing with the
effects of oil on marine and estuarine organisms must
include a careful consideration of the concentration
and composition of petroleum hydrocarbons present in
the aqueous phase during exposure. Crude oils and
even refined petroleum products are extremely complex
mixtures of hundreds or even thousands of different
hydrocarbons and related compounds. Crude and refined
oils from different sources may differ substantially
in their hydrocarbon composition and in the relative
concentrations of different hydrocarbons. Such
differences have a profound effect on the solubility
and dispersion-forming characteristics of the oils.
It is therefore necessary for the investigator to
empirically characterize the oil-water mixtures used
in his biological investigations.

In the present investigation we showed that WSFs
of the two refined oils were in all cases considerably
more toxic to our test animals than were those of the

two crude oils. The observation that the stock WSFs
of the refined oils contained considerably higher
concentrations of naphthalene and alkylnaphthalenes
than did those of the two crude oils led us to suspect
that the greater toxicity of the former might be due
to their high naphthalenes concentrations. Subsequent
bioassays with pure naphthalenes and alkylnaphthalenes
confirmed that the majority of the toxicity of the
#2 fuel oil and bunker C WSFs could be attributed to
the naphthalenes in them.

In the bioassays reported here, as well as in
bioassays, we reported earlier (Anderson *et al.*,
1974), we observed that in most cases the majority of
the mortalities recorded occurred during the first
12 to 24 hrs of the bioassay. This is demonstrated
by the minor changes in the LC_{50} values between 24
and 96 hrs for the refined oil bioassays in Table 2.
These observations suggest that the toxicity of the
oil-water mixtures decreased rapidly during the exposure period. Infrared and gas chromatographic analyses of the aqueous phase of our OWDs before and after
24 hrs gentle aeration indicated that there was as
much as 90% decrease in the total aqueous hydrocarbon
concentration during the first 24 hrs of aeration.
The concentration of n-paraffins dropped more rapidly
than that of the aromatic hydrocarbons (Anderson
et al., 1974).

Blumer *et al.* (1970) reported that when oysters
Crassostrea virginica are exposed to oil-water mixtures, they nonselectively accumulate in their tissues
a wide variety of petroleum hydrocarbons in direct
proportion to their concentrations in the exposure
water. They suggested that once accumulated in the
oyster tissues, these petroleum hydrocarbons, and
particularly the aromatic hydrocarbons, are retained
indefinitely. More recently, Lee *et al.* (1972a)
exposed the mussel *Mytilus edulis* to isotopically
labeled petroleum-derived alkanes and aromatic hydrocarbons and showed that the molluscs released more
than 90% of the accumulated hydrocarbons within
2 weeks of return to isotope-free seawater. In

addition, several species of marine fish were shown
to first accumulate and then metabolize and excrete
naphthalene and benzo[a]pyrene (Lee *et al.*, 1972b).
Recently Corner *et al.* (1973) demonstrated that the
spider crab *Maia squinodo* was also able to rapidly
detoxify and excrete naphthalene. Stegeman and Teal
(1973) exposed *Crassostrea virginica* to low concentrations of a #2 fuel oil for 49 days in a flowing
seawater system. The oysters accumulated in their
tissues a maximum of 334 ppm total petroleum hydrocarbons. When returned to a "clean" seawater system
(11 ppb total hydrocarbons), the oysters released
nearly 90% of the accumulated hydrocarbons in 2 weeks.
The remaining hydrocarbons were released much more
slowly reaching a level of 34 ppm after 4 weeks.

In the present and prior publications (Anderson,
1973; Neff and Anderson, 1974), we showed that
molluscs, crustaceans and fish, when exposed to OWDs
and WSFs of oil, rapidly accumulated a wide variety
of petroleum hydrocarbons in their tissues. Although
alkanes and aromatic hydrocarbons were both accumulated, present evidence strongly suggests that the
aromatic hydrocarbons were accumulated more rapidly
to higher concentrations and were retained in the
tissues for longer periods of time than were the
alkanes. Furthermore, at least with respect to the
naphthalenes, different hydrocarbons were not accumulated nonselectively in direct proportion to their
concentration in the exposure water. In all the
estuarine animals we have investigated, the alkylnaphthalenes were accumulated more rapidly and to a
greater extent than was naphthalene. When animals
were returned to hydrocarbon-free seawater, naphthalene was usually released from the tissues more
rapidly than were the alkylnaphthalenes. Because of
this, it can be expected that the petroleum hydrocarbon composition in the tissues of oil-exposed
marine animals will be substantially different from
that of the oil to which they were exposed.

We have also observed substantial species-
specific differences in the rates and patterns of

petroleum hydrocarbon uptake, retention and release. Generally, the crustaceans we have so far investigated very rapidly accumulated petroleum hydrocarbons, but retained them only for a brief period of time, releasing them very rapidly (Cox and Anderson, 1973; Tatem and Anderson, 1973). In most cases tissue hydrocarbon concentrations dropped to undetectable levels (0.02 to 0.10 ppm) in the crustacean tissues in from 1 to 7 days following exposure. On the other hand, molluscs took up petroleum hydrocarbons somewhat more slowly, but accumulated them in their tissues to considerably higher concentrations and released them more slowly than did crustaceans (Anderson, 1973; Neff and Anderson, 1974). Complete depuration by oysters may require as long as 2 months.

The physiological responses of marine organisms to the stress induced by exposure to sublethal concentrations of pollutants are poorly understood. Since respiratory rate may reflect the overall functional well-being of an animal, we have studied the respiratory response of several species of estuarine animals to exposure to sublethal concentrations of oil-water mixtures. The nature and magnitude of the respiratory response was species-dependent. The sheepshead minnow, *Cyprinodon variegatus*, showed a respiratory response which increased somewhat in intensity as the exposure concentration was increased. The response of the fish to WSFs of the four oils varied with respect to (1) the concentration required to elicit a significant change in respiratory rate, (2) the direction of change (stimulation or suppression), and (3) the magnitude of the respiratory response. The crustaceans studied (*Penaeus* and *Palaemonetes*) showed slight depressions in respiratory rate during or immediately after exposure to WSFs of oil. Surprisingly, in most cases, the magnitude of the respiratory response was greatest at the lowest exposure concentration. When *Palaemonetes* were returned to oil-free seawater after the exposure period, their respiratory rates rose again, approaching the preexposure control rates in 3 days. The respiratory rates of fish also returned to normal within a short time after exposure to sublethal concentrations of oil or specific petroleum hydrocarbons (Anderson *et al.*, unpublished data; Brocksen and Bailey, 1973).

These preliminary results indicate that the nature and magnitude of the respiratory response of marine animals to exposure to sublethal concentrations of oil is species-dependent and also apparently dependent on the hydrocarbon composition and concentration of the exposure water. The respiratory response appears to be transitory, being most significant during or immediately after exposure and rapidly returning to normal when the animals are returned to clean seawater. This suggests that an important contributing factor to the respiratory response is the concentration of petroleum hydrocarbons in the tissues of the test animals. We have some evidence that, following exposure, the respiratory rates of our test animals return to control values when the tissue hydrocarbon concentration returns to a relatively low level. We are at present determining the distribution of hydrocarbons in the tissues and organs of oil-exposed marine animals and attempting to relate this distribution to a variety of physiological responses of the test animals to oil exposure.

LITERATURE CITED

American Petroleum Institute. 1958. Determination of volatile and non-volatile oil material. Infrared spectrometric method, No. 733-58.

American Public Health Association. 1971. *Standard Methods for the Examination of Water and Waste-Water*. Washington, D. C.: A.P.H.A.

Anderson, J. W. 1973. Uptake and depuration of specific hydrocarbons from oil by the bivalves *Rangia cuneata* and *Crassostrea virginica*. Ocean Affairs Board of the National Academy of Sciences, National Research Council.

_____, J. M. Neff, B. A. Cox, H. E. Tatem, and G. M. Hightower. 1974. Characteristics of dispersions and water-soluble extracts of crude and refined oils and their toxicity to estuarine crustaceans and fish. *Mar. Biol.* In press.

Anderson, R. D. 1973. Effects of petroleum hydrocarbons on the physiology of the American oyster *Crassostrea virginica* Gmelin. Ph.D. dissertation, Texas A & M University.

Blumer, M. 1971. Scientific aspects of the oil spill problem. *Environ. Affairs 1*:54-73.

_____, S. Souza, and J. Sass. 1970. Hydrocarbon pollution of edible shellfish by an oil spill. *Mar. Biol. 5*:195-202.

Brocksen, R. W. and H. T. Bailey. 1973. Respiratory response of juvenile chinook salmon and striped bass exposed to benzene, a water-soluble component of crude oil. *Proc. Joint. Conf. on Prevention and Control of Oil Spills*, pp. 783-92.

Corner, E. D. S., C. C. Kilvington, and S. C. M. O'Hara. 1973. Qualitative studies on the metabolism of naphthalene in *Maia squinado* (Herbst). *J. Mar. Biol. Assn., U. K. 53*:819-32.

Cox, B. A. and J. W. Anderson. 1973. Some effects of #2 fuel oil on the brown shrimp *Penaeus aztecus. Amer. Zool. 13*:262.

Fry, F. E. J. 1971. The effect of environmental factors on the physiology of fish. In: *Fish Physiology*, Vol. 6, pp. 1-99, ed. by W. S. Hoar and D. J. Randall. New York: Academic Press.

Lee, R. F., R. Sauerheber, and A. A. Benson. 1972. Petroleum hydrocarbons: uptake and discharge by the marine mussel *Mytilus edulis. Science 177*: 344-46.

_____, _____, and G. H. Dobbs. 1972. Uptake, metabolism and discharge of polycyclic aromatic hydrocarbons by marine fish. *Mar. Biol. 17*:201-208.

Litchfield, J. T. and F. Wilcoxon. 1949. A simplified method for evaluating dose effect experiments. *J. Pharmac. Exp. Therap. 96*:99-113.

McAuliffe, C. 1966. Solubility in water of paraffin, cycloparaffin, olefin, acetylene, cycloolefin, and aromatic hydrocarbons. *J. Phys. Chem. 70*: 1267-75.

Neff, J. M. and J. W. Anderson. 1974. Uptake and depuration of petroleum hydrocarbons by the estuarine clam, *Rangia cuneata*. *Proc. Nat. Shellfish. Assoc.* In press.

Nelson-Smith, A. 1970. The problem of oil pollution in the sea. *Adv. Mar. Biol. 8*:215-306.

Newell, R. C. 1970. *Biology of Intertidal Animals*. New York: American Elsevier Publishing Company.

_____. 1973. Factors affecting the respiration of intertidal invertebrates. *Amer. Zool. 13*: 513-28.

Stegeman, J. J. and J. M. Teal. 1973. Accumulation, release and retention of petroleum hydrocarbons by the oyster *Crassostrea virginica*. *Mar. Biol. 22*:37-44.

Tatem, H. E. and J. W. Anderson. 1973. The toxicity of four oils to *Palaemonetes pugio* (Holthuis) in relation to uptake and retention of specific petroleum hydrocarbons. *Amer. Zool. 13*:261.

THE ACUTE EFFECTS OF EMPIRE MIX
CRUDE OIL ON ENZYMES
IN OYSTERS, SHRIMP AND MULLET

JAMES R. HEITZ, LANCELOT LEWIS,
JANICE CHAMBERS, and JAMES D. YARBROUGH

*Department of Biochemistry
and Department of Zoology
Mississippi State University
Mississippi State, Mississippi   39762*

Since oil represents an obvious biological
hazard in the estuarine environment, it is essential
that its impact on the organisms of this ecosystem be
ascertained. The usual measure of the environmental
effect of any pollutant on animals is mortality;
however, other effects which are more subtle and
indicative of physiological change may ultimately be
as detrimental to a population's survival. Although
acute effects of oil pollution may be physical, such
as the coating of fish gill tissue which results in
suffocation, long-term effects would be due primarily
to physiological alterations. Of the many parameters
which may be used to investigate these subtle deleterious effects from an environmental challenge, one of
the most fundamental would be changes in enzyme
levels, since these cellular catalysts control formation of biochemical intermediates essential to all
normal physiological functions. If a physiological
effect were to occur as the result of acute oil

exposure, it is likely that a change in enzyme activity would be seen. Changes in enzyme activity have been observed in the clinical diagnosis of various physiological dysfunctions such as liver or heart diseases. Therefore, by assaying a broad spectrum of enzymes, we should be able to determine the subtle effects of oil in biological systems. The method involves monitoring enzyme activity changes in tissues that may occur due to *in vivo* oil exposure. Comparisons of enzyme levels between oil-treated and control mullet *(Mugil cephalus)*, shrimp *(Penaeus* sp.*)*, and oyster *(Crassostrea virginica)* have been made.

A list of enzymes monitored in this study is shown in Table 1. These enzymes have all been involved in the diagnosis of various physiological states in vertebrates and standard methodology for their detection has been developed (Bergmeyer, 1965). Although these enzymes have been assayed primarily in the serum of mammals, in this study the enzymes have been assayed in tissue fractions due to the lack of sufficient quantities of serum from the experimental animals used.

METHODS AND RESULTS

All animals used in this study were collected on the Mississippi Gulf Coast. Animals were held in 128 l all-glass aquaria at 28°C and were not fed during the tests. The test aquaria were equipped with a siphon system which circulated aerated water into the aquarium below the surface to insure that the surface was not disturbed. Aquaria were filled with artificial seawater (Rila Marine Mix) at 15 o/oo. The animals were introduced into the aquaria 1 day prior to the addition of emulsified Empire Mix crude oil. Control aquaria received no oil. During the course of the experiments the oil formed a layer on the surface and settled on the sides of the aquaria. The actual oil concentration in the water column dropped dramatically during the first 24 hrs, reaching

TABLE 1
Enzymes Studied

	Enzyme	Referenced Method
LDH	Lactate dehydrogenase (L-lactate:NAD oxidoreductase; 1.1.1.27)	Wroblewski et al., 1955
MDH	Malate dehydrogenase (L-malate:NAD oxidoreductase; 1.1.1.37)	Bergmeyer and Bernt, 1965a
CYT RED	NADPH$_2$ cytochrome $\underline{c}$ reductase (NADPH$_2$:cytochrome $\underline{c}$ oxidoreductase; 1.6.2.3)	Masters et al., 1967
CYT OX	Cytochrome oxidase (cytochrome $\underline{c}$:O$_2$ oxidoreductase; 1.9.3.1)	Wharton and Tzagoloff, 1967
GOT	Glutamic-oxaloacetic transaminase (L-aspartate:2-oxoglutarate aminotransferase; 2.6.1.1)	Bergmeyer and Bernt, 1965b
GPT	Glutamic-pyruvic transaminase (L-alanine:2-oxoglutarate aminotransferase; 2.6.1.2)	Bergmeyer and Bernt, 1965c
ACHE	Acetylcholinesterase (acetylcholine acetylhydrolase; 3.1.1.7)	Ellman et al., 1961
ALP	Alkaline phosphatase (orthophosphoric monoester phosphohydrolase; 3.1.3.1)	Bessey et al., 1946
ACP	Acid phosphatase (orthophosphoric monoester phosphohydrolase; 3.1.3.2)	Andersch and Szczpinski, 1947
B-GLU	β-glucuronidase (β-D-glucuronide glucuronohydrolase; 3.2.1.31)	Fishman et al., 1967
LAP	Leucine aminopeptidase (3.4.1.1)	Bernt and Bergmeyer, 1965

a relatively constant level after 72 hrs for the
duration of the exposure of less than 4 ppm as
determined by infrared analysis.

All enzymes were assayed spectrophotometrically
at 30°C, with the exception of B-GLU, which was
assayed at 56°C. Specific activities are expressed
as milliunits/mg protein in which 1 unit is defined
as 1 micromole of product formed/min. Protein was
determined by the method of Lowry *et al.* (1951). For
ease of sampling and consistency, most enzyme assay
conditions were not optimized.

MULLET STUDIES

Animals were sacrificed after 4 days exposure
to oil; and gill, liver, brain, and muscle tissues of
four to six specimens were pooled and homogenized in
0.10 M sodium phosphate buffer, pH 7.6, in 0.25 M
sucrose. Three fractions were prepared from the
whole homogenate: a homogenate fraction (H), which
is the supernatant from an 800 g, 10-min centrifugation of the whole homogenate; a mitochondrial fraction (M), which is that portion of the 800 g fraction
sedimented at 8000 g for 15 min; and a supernatant
fraction (S), which is the supernatant from the 8000 g
centrifugation.

With one possible exception the enzymes investigated in the mullet do not appear to be affected by
acute 4-day oil exposures. The tissues expected to
be least affected by acute oil exposures would be
muscle and brain; however, these are the two tissues
in which there is a change in enzyme level. There is
a statistically significant decrease in mitochondrial
B-GLU activity from muscle (Table 2), which represents
about an 80% decrease relative to the control value.
In brain tissue, the only statistically significant
change in enzyme activity was again the B-GLU of the
mitochondrial preparation (Table 3). In this case,
the enzyme level increased approximately 50%. There
were no differences observed for any enzymes studied
in either liver (Table 4) or gill tissue (Table 5).

314

TABLE 2

Enzyme Levels in Muscle Tissue from Mullet Exposed to 75 ppm Empire Mix Crude Oil for 4 Days

| Enzyme | Fraction[b] | Specific Activity[a] | |
		Control	Treated
ACHE	H	348.82 ± 68.10 (5)	544.43 ± 65.39 (7)
ACP	H	4.83 ± 1.71 (4)	3.14 ± 1.46 (8)
ALP	H	2.64 ± 0.40 (4)	1.87 ± 0.34 (8)
CYT OX	M	38.71 ± 12.74 (3)	42.03 ± 9.14 (8)
CYT RED	S	1.00 ± 0.62 (2)	1.25 ± 0.36 (6)
B-GLU	M	506.10 ± 73.40 (2)	73.30 ± 42.38 (6)+
B-GLU	S	141.43 ± 33.99 (2)	23.01 ± 26.15 (7)
GOT	M	328.60 ± 65.67 (4)	293.57 ± 47.88 (8)
GOT	S	216.33 ± 40.74 (4)	227.59 ± 29.07 (8)
GPT	M	24.22 ± 9.37 (4)	31.42 ± 6.83 (8)
GPT	S	26.01 ± 7.31 (4)	20.51 ± 5.33 (8)
LDH	M	2579.42 ± 585.64 (3)	2523.43 ± 500.57 (4)
LDH	S	6504.14 ± 778.93 (3)	4813.70 ± 665.77 (4)

[a]Specific activity is expressed as millimicromoles of product formed/min/mg protein, mean ± SE. Number in parentheses is number of replications.

[b]H fraction is defined as the supernatant from an 800 g/10-min centrifugation.
M fraction is defined as that portion of an 800 g supernatant sedimented at 8000 g for 15 min.
S fraction is defined as the supernatant from an 8000 g centrifugation.
+ Treated significantly different from control ($\underline{P}$ < 0.05).

TABLE 3

Enzyme Levels in Brain Tissue from Mullet Exposed to 75 ppm Empire Mix Crude Oil for 4 Days

Enzyme	Fraction[b]	Specific Activity[a]	
		Control	Treated
ACHE	H	191.42 ± 14.46 (5)	158.33 ± 13.24 (8)
ACP	H	8.76 ± 1.31 (4)	5.60 ± 1.12 (8)
ALP	H	13.85 ± 5.30 (4)	12.87 ± 4.51 (8)
CYT OX	M	95.06 ± 12.49 (3)	86.93 ± 7.82 (9)
CYT RED	S	3.64 ± 0.77 (2)	2.68 ± 0.49 (5)
B–GLU	M	53.60 ± 28.45 (2)	120.31 ± 21.88 (7)+
B–GLU	S	65.38 ± 12.53 (2)	71.02 ± 9.64 (7)
GOT	M	308.27 ± 50.78 (4)	173.47 ± 37.02 (8)
GOT	S	360.51 ± 45.11 (4)	349.70 ± 32.89 (8)
GPT	M	40.91 ± 13.62 (4)	20.45 ± 9.93 (8)
GPT	S	32.25 ± 8.23 (4)	33.71 ± 6.00 (8)
LDH	M	598.50 ± 64.71 (3)	614.08 ± 55.31 (4)
LDH	S	982.34 ± 310.54 (3)	1150.43 ± 265.42 (4)

[a]Specific activity is expressed as millimicromoles of product formed/min/mg protein, mean ± SE. Number in parentheses is number of replications.

[b]H fraction is defined as the supernatant from an 800 g/10-min centrifugation.
 M fraction is defined as that portion of an 800 g supernatant sedimented at 8000 g for 15 min.
 S fraction is defined as the supernatant from an 8000 g centrifugation.
 + Treated significantly different from control ($\underline{P} < 0.05$).

TABLE 4
*Enzyme Levels in Liver Tissue from Mullet
Exposed to 75 ppm Empire Mix Crude Oil for
4 Days*

		Specific Activity[a]	
Enzyme	Fraction[b]	Control	Treated
ACP	H	47.91 ± 4.41 (5)	44.13 ± 3.65 (11)
ALP	H	28.91 ± 3.37 (5)	20.82 ± 2.79 (11)
CYT OX	M	118.78 ± 31.10 (3)	125.71 ± 19.47 (9)
CYT RED	S	5.20 ± 1.05 (2)	5.22 ± 0.56 (7)
B–GLU	M	4144.73 ± 1439.56 (2)	3295.85 ± 1107.47 (7)
B–GLU	S	1669.00 ± 363.40 (2)	1622.58 ± 279.57 (7)
GOT	M	331.37 ± 80.50 (4)	472.96 ± 58.69 (8)
GOT	S	454.26 ± 51.50 (4)	304.48 ± 37.55 (8)
GPT	M	204.73 ± 64.13 (4)	305.53 ± 46.76 (8)
GPT	S	258.33 ± 24.82 (5)	239.11 ± 18.10 (8)
LDH	M	216.35 ± 107.63 (3)	109.30 ± 92.00 (4)
LDH	S	124.75 ± 23.47 (3)	130.88 ± 20.06 (4)

[a]Specific activity is expressed as millimicromoles of product
 formed/min/mg protein, mean ± SE. Number in parentheses is
 number of replications.

[b]H fraction is defined as the supernatant from an 800 g/10-min
 centrifugation.
 M fraction is defined as that portion of an 800 g supernatant
 sedimented at 8000 g for 15 min.
 S fraction is defined as the supernatant from an 8000 g
 centrifugation.

TABLE 5

*Enzyme Levels in Gill Tissue from Mullet
Exposed to 75 ppm Empire Mix Crude Oil for
4 Days*

Enzyme	Fraction[b]	Specific Activity[a]	
		Control	Treated
ACP	H	33.79 ± 4.58 (4)	25.13 ± 3.90 (8)
ALP	H	43.13 ± 13.60 (4)	38.34 ± 11.59 (8)
CYT OX	M	79.48 ± 19.71 (3)	77.44 ± 12.34 (9)
CYT RED	S	1.91 ± 0.80 (2)	2.15 ± 0.43 (7)
B-GLU	M	591.95 ± 106.16 (2)	644.08 ± 81.67 (7)
B-GLU	S	295.75 ± 42.90 (2)	224.84 ± 33.01 (7)
GOT	M	512.75 ± 60.99 (4)	366.57 ± 44.47 (8)
GOT	S	180.34 ± 47.78 (4)	221.01 ± 34.84 (8)
GPT	M	67.97 ± 25.41 (4)	77.89 ± 18.52 (8)
GPT	S	18.01 ± 13.15 (4)	26.60 ± 9.59 (8)
LDH	M	479.81 ± 61.20 (3)	302.43 ± 52.31 (4)
LDH	S	980.80 ± 75.12 (3)	1011.65 ± 64.20 (4)

[a]Specific activity is expressed as millimicromoles of product
formed/min/mg protein, mean ± SE. Number in parentheses is
number of replications.

[b]H fraction is defined as the supernatant from an 800 g/10-
min centrifugation.
M fraction is defined as that portion of an 800 g supernatant
sedimented at 8000 g for 15 min.
S fraction is defined as the supernatant from an 8000 g
centrifugation.

It would have been expected that gill and liver would be the tissues most likely to be affected since these tissues are involved with initial contact (gill) and detoxification (liver).

SHRIMP STUDIES

Shrimp (46-109 mm) were exposed to 8 ppm emulsified Empire Mix crude oil for 12-hr periods. The 12-hr exposure period was necessary because of excessive mortality of the oil-treated shrimp in longer exposure periods. Stomach-hepatopancreas homogenates from four to six pooled specimens were made and the enzyme levels were assayed only in the mitochondrial (M) and supernatant (S) fractions as previously described. The two exceptions were CYT OX, measured only in the mitochondrial fraction, and CYT RED, measured only in the supernatant fraction. There were no detectable effects from oil exposure on the enzymes assayed (Table 6).

OYSTER STUDIES

Oysters (5 to 44 g) were exposed to 75 ppm emulsified Empire Mix crude oil for up to 7 days. Enzyme levels from single oysters were determined in the mitochondrial (M) and supernatant (S) fractions described earlier, with the addition of a fraction designated as the sonified soluble (SS) prepared as follows: The total oyster homogenate was sonified at 35 w for 0.5 min, and the resulting solution was centrifuged at 110,000 g for 30 min and the supernatant was retained for analysis. In this study more enzyme determinations have been made for oysters than for either shrimp or mullet; therefore, a better basis for evaluation of the effects of oil on this organism is possible.

The results of acute oil exposure over 2-, 4-, and 7-day periods show no statistically significant effects related to acute oil exposure for those enzymes presented in Table 7. Those enzyme levels

TABLE 6
Shrimp Stomach-Hepatopancreas Enzymes after a 12-Hr Exposure to 8 ppm Empire Mix Crude Oil

Enzyme	Enzyme Specific Activity[a]			
	Mitochondrial[b]		Supernatant[c]	
	Control	Treated	Control	Treated
ACP	4.97 ± 1.22 (9)	5.20 ± 0.83 (12)	8.60 ± 1.25 (9)	9.78 ± 0.92 (12)
MDH	593.02 ± 125.00 (7)	609.14 ± 96.77 (12)	1.52 ± 0.24 (6)	1.41 ± 0.32 (11)
GOT	98.49 ± 17.21 (6)	86.44 ± 12.71 (11)	2.52 ± 0.79 (6)	5.84 ± 3.78 (11)
GPT	85.58 ± 18.12 (6)	76.07 ± 12.39 (11)	0.47 ± 0.09 (6)	1.34 ± 0.69 (11)
B-GLU	96.25 ± 33.29 (9)	159.02 ± 29.27 (11)	258.27 ± 71.68 (9)	325.60 ± 63.85 (12)
CYT OX	121.90 ± 12.21 (9)	123.61 ± 12.14 (12)		
CYT RED			3.13 ± 0.55 (5)	2.84 ± 0.58 (8)

[a] Specific activity is expressed as millimicromoles of product formed/min/mg protein, mean ± SE. Number in parentheses is number of replications.

[b] Mitochondrial fraction is defined as that portion of an 800 g/10-min supernatant sedimented at 8000 g for 15 min.

[c] Supernatant fraction is defined as the supernatant of an 8000 g/15 min centrifugation.

TABLE 7
Enzyme Levels in the Oyster Treated with 75 ppm Empire Mix Crude Oil

Enzyme	Fraction[b]	Specific Activity[a]			
		Control	2 Day	4 Day	7 Day
ACHE	SS	6.74 ± 0.36 (35)	7.40 ± 0.44 (22)	7.67 ± 0.64 (18)	6.19 ± 0.40 (25)
ACP	SS	17.48 ± 1.17 (36)	20.27 ± 1.44 (22)	16.00 ± 1.90 (19)	18.53 ± 1.32 (25)
ALP	SS	1.26 ± 0.12 (35)	1.00 ± 0.15 (22)	0.92 ± 0.19 (19)	1.09 ± 0.14 (24)
CYT OX	M	6.82 ± 1.96 (19)		6.94 ± 1.71 (15)	6.95 ± 1.21 (15)
GOT	M	68.80 ± 18.01 (21)	61.54 ± 18.94 (19)	52.22 ± 36.79 (6)	124.94 ± 22.06 (4)
GOT	S	70.90 ± 4.42 (18)	78.73 ± 3.97 (19)	63.73 ± 7.76 (6)	77.61 ± 4.90 (13)
LAP	M	13.89 ± 0.93 (17)	11.49 ± 0.82 (6)	14.22 ± 0.89 (15)	13.01 ± 0.90 (11)
MDH	S	1665.57 ± 357.28 (23)	1334.56 ± 329.40 (15)	2377.73 ± 411.21 (15)	2077.11 ± 366.50 (17)

[a] Specific activity is expressed as millimicromoles of product formed/min/mg protein, mean ± SE. Number in parentheses is number of replications.

[b] M fraction is defined as that portion of an 800 g/10-min supernatant sedimented at 8000 g for 15 min.
S fraction is defined as the supernatant from an 8000 g/15-min centrifugation.
SS fraction is defined as the supernatant of a 110,000 g/30-min centrifugation of a homogenate which had been sonified at 35 w for 0.5 min.

found to be statistically different during acute oil exposure are presented in Table 8. Four of the enzymes investigated show statistically significant differences. No clear trend as to stimulation or inhibition is observed. It is unlikely that the patterns we are seeing are due to normal enzyme fluctuations because of the large sample sizes and the small standard errors.

At 7 days, the B-GLU level was significantly different from the 4-day level. The enzyme level increased gradually through the 2nd and 4th days, with a subsequent decrease to control levels. Both the mitochondrial and soluble GPT levels show statistically significant differences. The soluble enzyme level peaked at day 2 and this level differs from day 4 and day 7 levels, which are similar to the control level. The mitochondrial enzyme level is significantly higher at day 7. The soluble LAP level is highest on day 2 of treatment and is essentially the same as the control level by day 4. There were no changes in the mitochondrial LAP levels during the treatment period. Although the soluble MDH level showed no changes during this period, the mitochondrial enzyme levels showed significant changes at 2, 4, and 7 days. The MDH level dropped on day 2, but by day 7 it was in the range of the control level. This pattern was the reverse of what we have seen for the other enzymes.

The enzyme patterns observed may be the result of the normal response of an organism to an environmental challenge. Another possible explanation may be related to the very rapid decrease in the available concentration of oil in the water column. In the first 24 hrs, there is at least a 50% decrease in the available oil concentration as determined by infrared analysis. This concentration continues to diminish rapidly until about 72 hrs, at which time the concentration is stabilized below 4 ppm. If it were possible to maintain a constant oil level during the 7-day exposure period, the data suggest that the enzyme levels might continue to increase.

TABLE 8
Enzyme Levels in Oyster Treated with 75 ppm Empire Mix Crude Oil

Enzyme	Fraction[b]	Specific Activity[a]			
		Control	2 Day	4 Day	7 Day
B-GLU	SS	958.05 ± 82.99 (23)	1275.21 ± 114.25 (11)	1370.97 ± 112.35 (15)	848.18 ± 148.10 (17)[c]
GPT	M	34.31 ± 3.09 (27)	39.00 ± 3.68 (19)	26.45 ± 7.74 (13)	58.94 ± 7.26 (21)[c]
GPT'	S	39.58 ± 2.51 (26)	51.20 ± 2.36 (19)	37.74 ± 3.17 (15)	40.91 ± 2.58 (20)[d]
LAP	S	421.92 ± 74.14 (21)	160.37 ± 69.95 (14)	501.79 ± 76.52 (15)	312.84 ± 71.39 (17)[e]
MDH	M	440.00 ± 36.54 (22)	326.05 ± 30.38 (15)	341.11 ± 38.13 (15)	468.21 ± 34.03 (17)[f]

[a] Specific activity is expressed as millimicromoles of product formed/min/mg protein, mean ± SE. Number in parentheses is number of replications.

[b] M fraction is defined as that portion of an 800 g/10-min supernatant sedimented at 8000 g for 15 min.
S fraction is defined as the supernatant from an 8000 g/15-min centrifugation.
SS fraction is defined as the supernatant of a 110,000 g/30-min centrifugation of a homogenate which had been sonified at 35 w for 0.5 min.

[c] 4-day treatment mean significantly different from 7-day treatment mean ($p < 0.05$).

[d] 2-day treatment mean significantly different from 4- and 7-day treatment means ($p < 0.05$).

[e] 2-day treatment mean significantly different from 4-day treatment mean ($p < 0.05$).

[f] Control and 7-day treatment means significantly different from 2- and 4-day treatment means ($p < 0.05$).

DISCUSSION

The relative lack of effects by acute oil exposures on the animals examined is not surprising. It is obvious that oil does not kill immediately by poisoning enzyme systems. The subtle cellular effects, such as membrane permeability changes, fatty infiltration of the liver or modification of subcellular organelles, that might be indicative of damage would require longer exposure periods. The immediate effects of oil are physical, probably involving disruption of the activity of gaseous exchange at the gill level and is the result of oil coating the gills. In the oil-exposed animals the only deleterious symptom exhibited was a spiraling by shrimp. Although this spiraling is difficult to explain, it might represent a normal stress reaction to oil, an attempt at purging the animal of oil, or a reaction to lack of oxygen.

We are dealing with representatives from diverse animal groups. Of the three test animals, shrimp appeared most sensitive to crude oil. Various studies have shown that the level of oil toxicity to shrimp is related to season and/or salinity. In the fall (September to December) the 96-hr TL_m value was about 35 ppm emulsified oil, while in the winter (January to March) the TL_m value dropped to 15 ppm. During this period the salinity of the Mississippi Gulf dropped from 15-22 o/oo to a level of from 1-4 o/oo (Lytle, 1973). For oyster and mullet no satisfactory TL_m values have been determined. In oil exposures of 75 ppm for 7 days no mortality occurred in oyster. Mullet have been exposed to 400 ppm oil with little mortality.

The lack of oil effects based on the enzyme determinations might indicate that oil is not entering the test animals during acute exposures. However, qualitative residue analysis indicates that certain fractions of oil are entering the tissues (Miles, 1973). The uptake of oil by oysters has been documented in other studies (Cahnmann and Kuratsune, 1957;

Ehrhardt, 1972). Furthermore, oil in the tissues is indicated by the fact that certain fatty acid ratios are altered during acute oil treatments (Wesley, 1973). Therefore, oil is entering the test animals, even though the available oil in the water column rapidly decreases.

Of the enzymes that were affected by acute oil exposure, most are related to carbohydrate metabolism. It is interesting to note that there is an apparent reversal of a portion of the tricarboxylic acid (TCA) cycle in oyster under anaerobic conditions (Hammen, 1969). MDH, a component of the TCA cycle, is involved in this reversal and it may be hypothesized that under conditions of acute oil exposure the oyster closes and functions anaerobically. The decreased activity at 2 and 4 days of exposure with the return to normal at 7 days probably represents a normal compensation by the oyster to oil stress.

Carbohydrate (tricarboxylic acid cycle) and amino acid metabolism are enzymatically connected by GOT and GPT. B-GLU is also involved in carbohydrate metabolism. Furthermore, the presence of increased levels of B-GLU suggest that increased detoxification may be taking place. It is difficult to explain the increased levels of LAP in the oyster unless under anaerobic conditions there is mobilization of tissue proteins for energy.

This study indicates that acute oil exposures to aquatic animals have few effects on the enzymes studied. The enzyme changes seen in the oyster may represent a compensation response which often occurs when animals are exposed to toxicants. Based on this study, other enzymes might serve as more sensitive indicators of oil exposure, such as those involved in fatty acid metabolism and/or detoxification. With information from an expanded study a more complete understanding of the impact of oil on marine organisms will be possible.

ACKNOWLEDGMENTS

This research was supported by Environmental Protection Agency Contract 68-01-0745. This is Contribution No. 2705, Mississippi Agriculture and Forestry Experiment Station.

LITERATURE CITED

Andersch, M. A. and A. J. Szczypinski. 1947. Use of p-nitrophenylphosphate as the substrate in determination of serum acid phosphatase. *Amer. J. Clin. Path. 17*:571-74.

Bergmeyer, H.-U. 1965. *Methods of Enzymatic Analysis*. New York: Academic Press.

_____ and E. Bernt. 1965a. Malic dehydrogenase. In: *Methods of Enzymatic Analysis*, pp. 757-60, ed. by H.-U. Bergmeyer. New York: Academic Press.

_____ and _____. 1965b. Glutamate-oxaloacetate transaminase. In: *Methods of Enzymatic Analysis*, pp. 837-45, ed. by H.-U. Bergmeyer. New York: Academic Press.

_____ and _____. 1965c. Glutamate-pyruvate transaminase. In: *Methods of Enzymatic Analysis*, pp. 846-53, ed. by H.-U. Bergmeyer. New York: Academic Press.

Bernt, E. and H.-U. Bergmeyer. 1965. Colorimetric determination of leucine aminopeptidase in serum. In: *Methods of Enzymatic Analysis*, pp. 833-36, ed. by H.-U. Bergmeyer. New York: Academic Press.

Bessey, O. A., O. H. Lowry, and M. J. Brock. 1946. A method for the rapid determination of alkaline phosphatase with five cubic millimeters of serum. *J. Biol. Chem. 164*:321-29.

Cahnmann, H. J. and M. Kuratsune. 1957. Determination of polycyclic aromatic hydrocarbons in oysters collected in polluted water. *Anal. Chem. 29*:1312-17.

Ehrhardt, M. 1972. Petroleum hydrocarbons in oysters from Galveston Bay. *Environ. Pollut.* *3*:257-71.

Ellman, G. L., K. D. Courtney, V. Andres, Jr., and R. M. Featherstone. 1961. A new and rapid colorimetric determination of acetylcholinesterase activity. *Biochem. Pharmacol. 7*: 88-95.

Fishman, W. H. 1965. β-Glucuronidase. In: *Methods of Enzymatic Analysis*, pp. 869-74, ed. by H.-U. Bergmeyer. New York: Academic Press.

Hammen, C. S. 1969. Lactate and succinate oxidoreductases in marine invertebrates. *Mar. Bio. 4*:233-38.

Lowry, O. H., N. J. Rosebrough, A. L. Farr, and R. J. Randall. 1951. Protein measurement with the Folin-phenol reagent. *J. Biol. Chem. 193*: 265-75.

Lytle, J. 1973. Fate and effect of oil in the aquatic environment—Gulf Coast Region (Fourth Quarterly Progress Report to Environmental Protection Agency), pp. 20-21. Mississippi State, Miss.: Mississippi State University.

Masters, B. S. S., C. H. Williams, Jr., and H. Kamin. 1967. The preparation and properties of microsomal TPNH-cytochrome c reductase from pig liver. In: *Methods in Enzymology*, Vol. 10, pp. 565-73, ed. by R. W. Estabrook and M. E. Pullman. New York: Academic Press.

Miles, H. 1973. Fate and effect of oil in the aquatic environment—Gulf Coast Region (Fourth Quarterly Progress Report to Environmental Protection Agency), p. 3. Mississippi State, Miss.: Mississippi State University.

Wesley, D. 1973. Fate and effect of oil in the aquatic environment—Gulf Coast Region (Fourth Quarterly Progress Report to Environmental Protection Agency), p. 12. Mississippi State, Miss.: Mississippi State University.

Wharton, D. C. and A. Tzagoloff. 1967. Cytochrome oxidase from beef heart mitochondria. In: *Methods in Enzymology*, Vol. 10, pp. 245-50, ed. by R. W. Estabrook and M. E. Pullman. New York: Academic Press.

Wroblewski, F. and J. S. LaDue. 1965. Lactic dehydrogenase activity in blood. *Proc. Soc. Exp. Biol. Med. 90*:210-13.

HYDROCARBONS IN SHELLFISH CHRONICALLY EXPOSED TO LOW LEVELS OF FUEL OIL

JOHN J. STEGEMAN

Department of Biology
Woods Hole Oceanographic Institution
Woods Hole, Massachusetts   02543

Hydrocarbons are pervasive components of marine waters, including estuaries. Marine hydrocarbons may be recently biosynthesized, for instance by algae (Clark and Blumer, 1967), and they may also be petroleum in origin in many areas, resulting from natural seeps, oil fields, refinery wastes, normal shipping operations and atmospheric transport. From the composition as described in part by Barbier *et al.* (1973), Johnson and Calder (1973), Keizer and Gordon (1973) and Stegeman and Teal (1973), both petroleum and biogenic origins are in fact indicated for hydrocarbons in many coastal regions and concentrations reported in the water have generally been less than 100 ppb (μg hydrocarbon per liter). Whether biogenic or petroleum in origin, these hydrocarbons can occur in solution and micelles, adsorbed to sediments, detritus and particulate matter, and may be lost to as well as contributed by the atmosphere (Boehm and Quinn, 1973; Meyers and Quinn, 1973; Mackay and Matsugu, 1973).

The gross effects on estuarine biota resulting from extensive oil spills, and the use of oil

disperants, have been substantially documented (North
et al., 1964; Sanders *et al.*, 1972). Yet the type
and extent of biological or physiological effect resulting from low level petroleum hydrocarbon contamination of estuarine and coastal waters has not been
adequately described. Mackin and Hopkins (1961) have
postulated that in some coastal areas the continuous
input of petroleum has little or no demonstrable
adverse effect on populations or productivity, while
other investigators have suggested that biological
processes in estuarine organisms can be affected by
low levels of petroleum (Gilfillan, 1973; Jacobson
and Boylan, 1973; LaRoche, 1973).

In general, the biological or biochemical activity exerted by foreign compounds is related to their
tissue concentration, their biological half-life, and
the nature of the chemical groups on the compounds
(Pelkonen and Kärki, 1973). Accordingly, these are
among important parameters which must be taken into
consideration in experimentally evaluating the effects
of petroleum hydrocarbons on marine organisms. Re-
stated they are: (a) the levels to which petroleum
hydrocarbons are accumulated; (b) the duration of
hydrocarbon residence in the organism; and (c) the
composition of the hydrocarbon mixture in the water
and, subsequently, in the organism. With aquatic
organisms it is necessary to consider how the above
parameters are modified by biological factors such as
lipid content, hydrocarbon uptake efficiency, route
of entry and disposal.

Although recently biosynthesized hydrocarbons
and petroleum-derived compounds may both be present,
it is unlikely that the former should have any adverse effect on estuarine organisms. Consequently,
my concern here is to clarify aspects of uptake level,
residence duration, and particularly the composition
of the mixture for petroleum hydrocarbons in shellfish exposed to very low rather than high levels of
contaminant. This will be done with oysters in the
context of a flow-through aquaculture system, and the
results will be considered in terms of potential

330

biological effects, particularly in reference to effects on individuals. A thorough discussion of the effects of chronic petroleum-derived hydrocarbon contamination in estuaries should include effects at varied stages of organisms' life and reproductive cycles, and possible synergism with other environmental stresses. These concerns fall beyond the scope of the present contribution, although examples of such investigations are contained elsewhere in these proceedings.

EXPERIMENTAL

The materials and methods employed are essentially those of Stegeman and Teal (1973). Groups of oysters, *Crassostrea virginica*, were obtained from stocks (Long Island Oyster Farms) kept in a seawater facility (Group 1) or in a local harbor (Group 2). Animals averaged 5 cm in length. Those in Group 1 contained 1.63% pentane extractable lipid and in Group 2 an average of 0.93%. Oysters from each group were marked and placed together in a fiberglass tray through which flowed seawater filtered to 1 micron and regulated at 20°C. A nutrient grown mixed phytoplankton culture was introduced into the tank at a constant rate (Tenore and Dunstan, 1973). The oysters were acclimated to this system for 1 week prior to addition of hydrocarbons.

A locally purchased No. 2 fuel oil with an n-alkane range from n-decane to n-tricosane, and an n-heptadecane/pristane ratio of 2.36, was dissolved in aqueous ethanol and added to the inflowing seawater by means of an infusion pump (Harvard Apparatus) at a concentration of 106 μg/liter as determined by analysis of the input. A second "clean" tray was maintained without added hydrocarbons, and oysters from Group 1 were transferred to this tray after 50 days exposure in the hydrocarbon contaminated system. No hydrocarbons were added to this tank, yet analysis showed that the water flowing into it

contained 11 μg hydrocarbons/liter in a range from
n-tetradecane to at least n-tetratriacontane, with
the most prominent alkanes occurring between n-
eicosane and n-triacontane.

Oyster samples were extracted with pentane and
the nonsaponifiable fraction chromatographed on a
column on alumina and partially deactivated silica,
using three column volumes of pentane (fraction A),
one of 10% benzene (fraction B), and one of 20%
benzene (fraction C) to elute the hydrocarbons. The
hydrocarbon fractions were analyzed gravimetrically
for hydrocarbon content, and subjected to TLC and UV
absorbance and fluorescence analysis. Aliquots of
each sample were gas chromatographed on a 10' stainless steel column of 3% Apiezon L on Chromosorb W in
a Hewlett-Packard Model 700 gas chromatograph equipped
with a flame-ionization detector. The column oven
was temperature programmed from 80° to 280°C.
N-alkanes were identified by comparison with internal
standards, isoprenoids by their retention indices.

RESULTS AND DISCUSSION

Oysters exposed for extended periods to low
levels of petroleum hydrocarbons accumulate these
hydrocarbons in their tissues at levels determined by
a variety of factors. The apparent equilibrium in
hydrocarbon uptake upon prolonged exposure is shown
in Figure 1. The extent of accumulation exhibited in
Figure 1 appears to be linked to the neutral lipid
content of the organisms, although this effect becomes
evident only as the tissue concentration exceeds
50 ppm. Below this level correlation with wet weight
rather than lipid weight is evident. This relationship between lipid content and hydrocarbon accumulation by oysters has been discussed by Stegeman and
Teal (1973) as being indicative of some degree of
equilibration between hydrocarbons in the water and
hydrocarbons which become associated with oyster
lipid. The involvement of lipid in hydrocarbon

accumulation is further emphasized by the efficiency data discussed below.

It must be cautioned that the observed experimental correlation between lipid content and hydrocarbon accumulation in no way argues that such correlation should be generally observable in the environment. In the present experiment the organisms possessed very low lipid levels, were the same species and age, and were examined under the same environmental conditions and exposure history. Variation in species, age, thermal environment, diet and reproductive phase all affect animal lipid qualitatively and quantitatively (Love, 1970), resulting in lipids of presumably differing capacities to retain hydrocarbons removed from the water. This has been discussed by Harvey *et al.* (1974) with regard to chlorinated hydrocarbons. Given similarities in the above parameters it would presumably still be necessary for organisms to contain a sufficiently high ratio of hydrocarbon concentration to lipid content for lipid to limit the uptake of hydrocarbons.

Entry of hydrocarbons into marine organisms both by ingestion of hydrocarbon contained in food (Blumer *et al.*, 1970) and by direct adsorption of hydrocarbons from the water are known to occur (Lee *et al.*, 1972), the latter presumably via unmediated transport (Kotyk, 1973). Although both routes can contribute to the oysters, direct absorption is considered as a major source of hydrocarbon accumulation in the present system (Stegeman and Teal, 1973). There is, however, no information on the relative contribution of hydrocarbon by ingested phytoplankton.

It is apparent (Fig. 1) that oysters discharge or dispose of some accumulated hydrocarbons upon lowering the exposure level, yet they retain a portion of accumulated hydrocarbons for a period of weeks even when none are added to the system. Potential means by which shellfish can reduce tissue levels of hydrocarbons include unmediated transport out, sloughing of hydrocarbon-containing cells and perhaps metabolism of hydrocarbons. Although it is clear that

disposal must occur whether the organisms are in
contaminated or clean water, a reduction in the body
burden will obviously occur only in the latter.

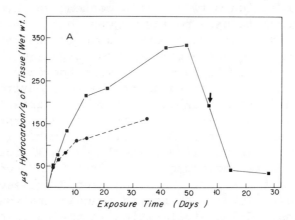

*Fig. 1. Uptake and release of total fuel oil
hydrocarbons by* Crassostrea virginica.
*Hydrocarbon concentration is expressed on a
wet weight basis. The hydroconcentration in
the water was 106 μg/liter. (■) Oysters
with 1.63% pentane extractable lipid.
(●) Oysters with 0.93% lipid. Prior to
exposure the oysters contained 1 μg hydrocarbon per g wet weight. At day 50 the
higher lipid content oysters were transferred
to an uncontaminated system. Arrow: Gas
chromatograms of this sample are shown in
Figure 3. (After Stegeman and Teal, 1973.)*

Whether a total discharge of contaminants, observed in some experimental systems (Anderson, 1973)
is a normal occurrence in nature is still a matter of
conjecture. This is emphasized by the partial retention of hydrocarbons observed here and elsewhere
(Clark and Finley, 1974). It is unlikely the retention of petroleum hydrocarbons in the present system
would be permanent, yet it is possible they may be
contained for a considerable period, depending on
environmental conditions and current natural exposure

levels (Stegeman and Teal, 1973). In any case, shellfish under conditions of chronic exposure, even at
low levels, are certain to contain some exogenous
hydrocarbon all the time.

As is expected from the uptake-release curves,
the efficiency with which oysters in the flow-through
system remove hydrocarbons from the water decreases
exponentially with increasing length of exposure
(Table 1). Efficiency here is taken as the observed
net hydrocarbon accumulated per unit time relative to
the total hydrocarbons in water theoretically being
passed over the gills, 1 liter/hr/gm wet weight
(Jørgensen, 1966). Even during initial stages of
exposure and under favorable conditions of flowing
water and phytoplankton availability, these oysters
removed hydrocarbons with an observed efficiency of
only 1%. Efficiencies were also calculated for
points on the uptake curve with net hydrocarbon up-
takes computed using the initial uptake rate and the

TABLE 1
*Efficiency of Petroleum Hydrocarbon Uptake by Oysters
During Prolonged Exposure*

Exposure Period at 106 µg/l	Increase in Hydro-carbon Concentration	Efficiency of Uptake
days 0-2	25 µg/g/day (0-49 µg/g)	1.0%
days 2-7	17 µg/g/day (49-133 µg/g)	0.7%
days 7-14	12 µg/g/day (133-218 µg/g)	0.5%
days 14-42	4 µg/g/day (218-328 µg/g)	0.16%
days 42-49	1 µg/g/day (328-334 µg/g)	0.04%

initial hydrocarbon release rate. The latter rate was estimated as the release of 6% of the current body burden per day. The calculated efficiencies fall on the same regression line (not shown) as the measured efficiencies when the former are calculated on a lipid weight but not a wet weight basis.

Perhaps more important than efficiency and route of entry or release of total hydrocarbons from the organism is the disposal efficiency of specific types of compounds. The duration of residence of particular hydrocarbons in the organisms will determine the composition of the total at a given time and, consequently, the potential for biological effect. Although a clear picture of mechanisms of hydrocarbon release by shellfish is not yet available, we can reach some tentative conclusions regarding relative disposal of various types of compounds.

Gas liquid chromotography (GLC) (Fig. 2), thin-layer chromatography and UV analyses were used to characterize hydrocarbon fractions A, B and C sequentially eluted from a silica/alumina column. Fraction A contained aliphatic, naphthenic, mono- and diaromatic hydrocarbons. Fraction B contained diaromatics, few monoaromatics and probably some higher molecular weight aromatics. Fraction C contained only the higher molecular weight aromatics. By comparing proportions of hydrocarbons in fractions A, B and C, it is readily apparent that the composition of the oil after being taken up by oysters is not the same as that of the contaminating oil (Table 2). The immediate observation that hydrocarbon fractions B and C represent a significantly greater portion of the total in the contaminated organisms than in the oil introduced indicates that aromatic hydrocarbons in these fractions are more readily accumulated by oysters than are aliphatic, napthenic and aromatic hydrocarbons in fraction A.

During the exposure period any variation in relative percentages of fractions A, B and C in oysters could not be correlated with either length of exposure or the tissue concentration of hydrocarbons.

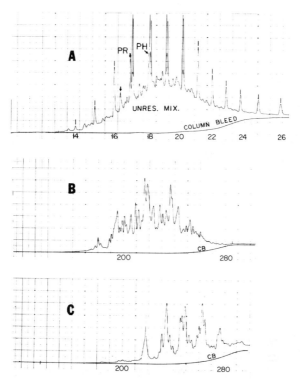

Fig. 2. Gas chromatograms of fuel oil hydrocarbons extracted from the water at point of input. A, B and C refer to three hydrocarbon fractions eluted from an alumina/silica column with three volumes of pentane (A), one volume of 10% benzene (B), and one volume of 20% benzene (C). In all three chromatograms temperature increases from left to right. In A the lines indicate n-alkanes and arrows branced alkanes (isoprenoids). PR refers to pristane and PH, phytane. The numbers refer to the chain lengths of the n-alkanes. Unresolved mixture refers to the complex mixture of napthenes, alkylated monoaromatics and diaromatics which elute too closely to be resolved. 200 and 280, temperature in GC column oven. Column Bleed, signal with no sample added.

337

TABLE 2
Percent of Total Hydrocarbon Contained in Column Chromatographic Fractions

Sample	Number	Fraction A	Fraction B	Fraction C
Input (100–900 µg/l)	7	85 (±5.4)	7 (±3.0)	7 (±3.4)
Oysters:				
Exposed 2–50 days* (50–334 µg/g tissue)	23	59 (±8.5)	20 (±5.8)	20 (±5.9)
Depurated 7 days (194 µg/g tissue)	1	63	18	19
Depurated 14 days (39 µg/g tissue)	1	69	18	11
Depurated 28 days (34 µg/g tissue)	2	83	9	7

*Exposed to 106 µg hydrocarbon/liter.

In depurating oysters, however, it is clear that the
fraction C aromatics, and to a lesser extent those in
fraction B (Table 2), are disposed of far more readily
than most hydrocarbons in fraction A. The hydrocarbons most rapidly lost are the straight and
branched alkanes in fraction A. The monoaromatic,
diaromatic and probably napthenic hydrocarbons in
this fraction have the longest half-life. Fraction A
constituted 85% of the 11 µg hydrocarbon/liter water
supplied to depurating oysters. Although this suggests equilibration is involved in establishing new
relative percentages, it does not require it.

In addition to the changes in aromatic content
relative to a contaminating oil, we find that proportions of the various compounds in a given column
chromatographic fraction also show changes. It is
apparent that the GLC patterns of fractions A, B and
C extracted from the organisms (Fig. 3) differ markedly from GLC patterns of the three fractions of the
input hydrocarbons (Fig. 2). Careful scrutiny reveals the patterns of peaks are qualitatively identical in each pair of fractions. In fraction A (Fig. 3),
however, the n-alkanes are dramatically reduced. In
fractions B and C the aromatic compounds eluting from
the GC column at lower temperatures are much reduced
and the compounds eluting at the higher temperature
are subsequently enriched relative to the input
hydrocarbon fractions. In all three fractions these
changes occur gradually during the exposure period
and are accelerated when oysters are transferred to
clean water.

Numerous reports concerning the composition of
hydrocarbons in various environmentally contaminated
shellfish (Blumer et al., 1970; Burns and Teal, 1971;
Ehrhardt, 1972; Farrington and Quinn, 1973) support
the theory that changes in hydrocarbon composition
similar to these observed in shellfish in the laboratory occur in nature as well. This is further exemplified by the fraction A hydrocarbons (Fig. 4) of
oysters sampled in Woods Hole seawater facilities
where the water source contained only those

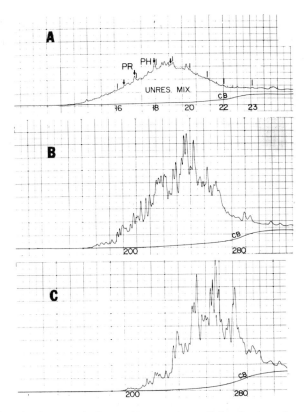

*Fig. 3. Gas chromatograms of hydrocarbons
extracted from oysters. Oysters sampled
after 50 days experimental exposure to
106 μg/liter fuel oil and 7 days depuration.
A, B and C refer to three fractions eluted
from an alumina/silica column with three
volumes of pentane (A), one volume of 10%
benzene (B), and one volume of 20% benzene
(C). Description of chromatograms of
fractions A, B and C as in Figure 2.*

hydrocarbons normally present. This included a probable input of petroleum-derived hydrocarbons from
ship operations close by but no experimental addition
of hydrocarbons. After several months these oysters
contained 10 ppm total hydrocarbons, which were

clearly petroleum derived, but very low relative concentrations of alkanes (Fig. 4) and also of compounds
in fractions B and C.

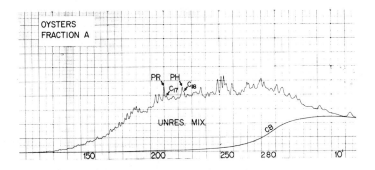

*Fig. 4. Gas chromatograms of hydrocarbons
extracted from oysters. Fraction A
hydrocarbons from oysters sampled after
extended residence in a seawater facility.
The water source for this facility was
not the same as that for the depurating
oysters in Table 2 and Figure 3. The
numbers refer to the temperature in the
gas chromatograph column oven. 10' refers
to the time held at 280°C. C_{17} and C_{18}
refer to n-heptadecane and n-octadecane,
respectively. Other descriptions are
as in Figure 2. Fraction A described
here equals ca. 90% of the total hydrocarbon in the sample.*

CONCLUSION

The uptake of complex hydrocarbon mixtures by
estuarine shellfish appears to be rather inefficient,
yet high tissue concentrations can be achieved even
when the concentration in the water is low. The
initial rate of uptake is linked to the concentration
in the water (Stegeman and Teal, 1973), while both
uptake and disposal rates must determine the net

341

accumulation during exposure. The rate of disposal appears to be a function of a given tissue concentration of hydrocarbons, although factors such as lipid character and hydrocarbon concentration in water certainly modify the relationship. Superimposed on the general pattern of hydrocarbon accumulation and disposal is the changing composition of oil in the tissues, in terms of total aromatic content as well as altered proportions of individual aromatic and aliphatic compounds present. Although change is occurring constantly, the rate of change in composition appears to be greater during a period of net loss of hydrocarbons. This probably results from a combination of factors including the lack of a continued input of all classes of compounds (Stegeman and Teal, 1973).

The results obtained here with chronic low level petroleum contamination agree with those of Anderson (1973), who reported that in acutely heavily contaminated systems lower molecular weight aromatics are retained best by shellfish, while higher molecular weight aromatics are more readily accumulated, but that both they and the paraffins are readily lost. The present results and those of Anderson differ in this regard from data obtained using very low levels of spin-labeled hydrocarbons (Roubal, 1974) in which it was reported that alkanes tend to occur in membrane interiors and aromatics on the surface. In Roubal's study paraffins were found to be much less subject to disposal than any aromatics. It seems most probable, however, that as cell membranes become saturated this pattern of retention would be masked by movement of hydrocarbon into more capacious intracellular lipid areas, and the patterns reported here and by Anderson (1973) would become dominant.

It is clear that classes of hydrocarbons and individual compounds in the complex mixture of oil hydrocarbons behave in a chemically distinct manner when in the oyster. The extent to which chemical or biological behavior of given compounds in an organism is modified by the presence or absence of other compounds is unknown. Yet, it is apparent that the total toxicological potential of an oil within the organism, as well as without, must be continually changing. In terms of potential biological effect the lower molecular weight aromatics should be considered as the most significant compounds. They are not only toxic but also apparently have a biological half-life longer than other hydrocarbons in shellfish.

A change in composition of oil within the tissues of shellfish and other estuarine organisms is probably not important in terms of such exposure effects as decreased biodeposition (Stegeman and Teal, 1973), and some behavioral effects (Jacobson and Boylan, 1973) which appear to be readily reversible. Effects of this type possibly result from interaction of hydrocarbons with surface membranes of ciliary or chemoreceptor cells and might be classed as narcotic in nature (Marsland, 1933; Seeman, 1972). Some other behavioral effects, such as those which involve membrane function in the central nervous system (Atema and Stein, 1972), could, however, be altered by changes in the retained hydrocarbons. A qualitatively changing complement of intracellular hydrocarbons should be important in modifying physiological effects as the alteration of normal lipid metabolism, observed in hydrocarbon contaminated in fin fish (Sidhu *et al.*, 1970). The potential for possible effects such as persistent histological changes, including neoplastic transformation and carcinogenesis, should also be modified as changes occur in the ratios of different types of hydrocarbons. Furthermore, it is likely that any cellular defense mechanisms and the susceptibility to pathogens should be affected. Clearly these latter possibilities would become more important as exposure periods became extended, and higher tissue concentrations of hydrocarbons were maintained for longer periods.

ACKNOWLEDGMENTS

This research was supported in part by NSF Grant GZ-1508 and the Woods Hole Oceanographic Institution, Contribution No. 3373. J. Ryther made available the aquaculture facilities. J. W. Anderson, J. Atema, J. W. Farrington, J. F. Grassle, J. M. Teal, and J. M. Vaughn commented on the manuscript. E. S. Stegeman assisted in preparing the manuscript.

LITERATURE CITED

Anderson, J. W. 1973. Uptake and depuration of specific hydrocarbons from fuel oil by the bivalves *Rangia cuneata* and *Crassostrea virginica*. In: *Background Papers for a Workshop on Inputs, Fates, and Effects of Petroleum in the Marine Environment*, Vol. 2, pp. 690-708. Washington, D. C.: U. S. National Academy of Sciences.

Atema, J. and L. Stein. 1972. Sub-lethal effects of crude oil on lobster *(Homarus americanus)* behavior. *Woods Hole Oceanogr. Inst. Tech. Report* 72-74.

Barbier, M., D. Joly, A. Saloit, and D. Tourres. 1973. Hydrocarbons from sea water. *Deep Sea Res. 20*:305-14.

Blumer, M., M. M. Mullen, and R. R. L. Guillard. 1970. A polyunsaturated hydrocarbon (3, 6, 9, 12, 15, 18-heneicosahexaene) in the marine food web. *Mar. Biol. 6*:226-35.

_____, G. Souza, and J. Sass. 1970. Hydrocarbon pollution of edible shellfish by an oil spill. *Mar. Biol. 5*:195-202.

Boehm, P. D. and J. G. Quinn. 1973. Solubilization of hydrocarbons by the dissolved organic matter in sea water. *Geochim. Cosmochim. Acta. 37*: 2459-77.

Burns, K. A. and J. M. Teal. 1971. Hydrocarbon incorporation into the salt marsh ecosystem from the West Falmouth oil spill. *Tech. Rep. Woods Hole Oceanogr. Inst.* 71-69. Unpublished.

Clark, R. C. Jr. and M. Blumer. 1967. Distribution of n-paraffins in marine organisms and sediments. *Limnol. Oceanogr.* 12:79-87.

_____ and J. S. Finley. 1974. Uptake and loss of petroleum hydrocarbons by mussels *(Mytilus edulis)* in laboratory experiments. Unpublished.

Ehrhardt, M. 1972. Petroleum hydrocarbons in oysters from Galveston Bay. *Environ. Pollut. 3:* 257-71.

Farrington, J. and J. G. Quinn. 1973. Petroleum hydrocarbons in Narragansett Bay. I. Survey of hydrocarbons in sediments and clams *(Mercenaria mercenaria)*. *Estuarine and Coastal Mar. Sci. 1:* 71-79.

Gilfillan, E. S. 1973. Effects of sea water extracts of crude oil on carbon budgets in two species of mussels. *Proc. Joint Conf. on Prevent. Control of Oil Spills*, pp. 691-95. Washington, D. C.: American Petroleum Institute.

Harvey, G. R., H. P. Miklas, W. G. Steinhauer, and V. T. Bowen. 1974. Observations on the distribution of chlorinated hydrocarbons in Atlantic Ocean organisms. *J. Mar. Res.* In press.

Jacobson, S. M. and D. B. Boylan. 1973. Seawater soluble fractions of kerosene: effect on chemotaxis in a marine snail, *Nassarius obsoletus*. *Nature 241:*213-15.

Johnson, R. W. and J. A. Calder. 1973. Early diagenesis of fatty acids and hydrocarbons in a salt marsh environment. *Geochim. Cosmochim. Acta 37:*1943-55.

Jørgensen, C. B. 1966. *Biology of Suspension Feeding*. New York: Pergamon Press.

Keizer, P. D. and D. C. Gordon, Jr. 1973. Detection of trace amounts of oil in sea water by fluorescence spectroscopy. *J. Fish. Res. Bd. Canada* *30*:1039-46.

Kotyk, A. 1973. Mechanisms of nonelectrolyte transport. *Biochem. Biophys. Acta 300*:183-210.

La Roche, G. 1973. Analytical approach in the evaluation of biological damage resulting from spilled oil. In: *Background Papers for a Workshop on Inputs, Fates, and Effects of Petroleum in the Marine Environment*, Vol. 1, pp. 347-74. Washington, D. C.: U. S. National Academy of Sciences.

Lee, R. F., R. Sauerheber, and A. A. Benson. 1972. Petroleum hydrocarbons: uptake and discharge by the marine mussel *Mytilus edulis*. *Science 177*: 344-46.

Love, R. M. 1970. *The Chemical Biology of Fishes*. London: Academic Press.

Mackay, D. and R. S. Matsugu. 1973. Evaporation rates of liquid hydrocarbon spills on land and water. *Canad. Jour. Chem. Engin. 51*:434-39.

Mackin, J. G. and S. H. Hopkins. 1961. Studies on oyster mortality in relation to natural environments and to oil fields in Louisiana. *Pub. Inst. Mar. Sci. 7*:3-131.

Marsland, D. 1933. The site of narcosis in a cell; the action of a series of paraffin oils on *Amoeba dubia*. *J. Cell. Comp. Physiol. 4*:9-33.

Meyers, P. A. and J. G. Quinn. 1973. Association of hydrocarbons and mineral particles in saline solution. *Nature 244*:23-24.

North, W. J., M. Neushul, Jr., and K. A. Clendenning. 1964. Successive biological changes observed in a marine cove exposed to a large spillage of mineral oil. *Symp. Pollut. Mar. par Microorgan. Prod. Petrol.*, Monaco. pp. 335-54.

Pelkonen, O. and N. T. Kärki. 1973. Effect of physicochemical and pharmacokinetic properties of barbituarates on the induction of drug metabolism. *Chem.-Biol. Interactions 7*:93-99.

Roubal, W. T. 1974. Spin-labeling of living tissue—
    a method for investigating pollutant-host
    interaction. This volume.
Sanders, H. L., J. F. Grassle, and G. R. Hampson.
    1972. The West Falmouth oil spill I. Biology.
    Woods Hole Oceanogr. Inst. Tech. Report 72-20.
    Unpublished.
Seeman, P. 1972. The membrane actions of anesthetics
    and tranquilizers. *Pharmacol. Rev. 24*:583-655.
Sidhu, G. S., G. L. Vale, J. Shipton, and K. E.
    Murray. 1970. Nature and effects of kerosene-
    like taint in mullet. *F.A.O. Tech. Conf. on
    Mar. Pollution.* Rome, December 1970.
Stegeman, J. J. and J. M. Teal. 1973. Accumulation
    release and retention of petroleum hydrocarbons
    by the oyster *Crassostrea virginica. Mar. Biol.
    22*:37-44.
Tenore, K. R. and W. M. Dunstan. 1973. Comparison
    of feeding and biodeposition of three bivalves
    at different food levels. *Mar. Biol. 21*:190-95.

THE EFFECT OF COLD SEAWATER EXTRACTS OF OIL FRACTIONS UPON THE BLUE MUSSEL, *MYTILUS EDULIS*

A. DUNNING and C. W. MAJOR

Department of Zoology
University of Maine
Orono, Maine   04473

Oil spills are a recurrent problem in marine systems and, while large spills are highly publicized, it is the sum of the innumerable small leaks and drops that probably constitute the major long range environmental problem. Despite the importance of the problem, there is surprisingly little data on the effects of oil on marine organisms. Most available information consists of field observations which have limited value. For example, they usually lack controls, results are often confounded with the effects of detergents used in the cleanup process, and the concentrations involved are, at best, conjectural. However, some laboratories studies on oil effects on organisms have been published.

Relatively low concentrations of crude oil kill the larvae of the copepod, *Anomolocera pattersoni*, and the hermit crab, *Diogenes pugilator*, within 3 days (Mileikousky, 1970). Adult copepods, such as *Acartia clausi*, *Paracalanus parus*, *Penelia avirostris*, *Centopages penticus*, and *Oithena nana* are killed relatively fast in concentrations of 0.0001 ml/l and

were killed in 1 day at the concentration of 0.1 ml/l
(Mironov, 1969). The amphipod, *Gammarus*, and the
freshwater planarian, *Dugesia*, cannot tolerate oil
in the sediments (McCauley, 1965).

Considerable confusion exists concerning the
effects of oil on other marine life. A number of
reports indicate that shellfish are highly resistant
to oil (Brisby, 1969; Mackin and Hopkins, 1962; and
Simpson, 1968). Corner, Southard, and Southard (1968)
state that barnacles are the most resistant of the
intertidal organisms, although they did note that
nauplii swimming activity was stopped in 3 hrs.
George (1961) earlier reported on the barnacle's
resistance, as well as that of the limpet, *Patella
vulgata*, in his studies of the Milford Haven spills.
Limpets were found in pools of oil immediately after
the Torrey Canyon spill showing no apparent adverse
effects from the oil (O'Sullivan and Richardson,
1967). *Patella* was previously reported to be able to
browse on oil-slicked rocks, ingest the oil, and pass
it through its digestive system with no adverse
effects noted (George, 1961).

Periodic immersion in oil has been reported to
have no noticeable effects on the oyster, *Crassostrea*
(Galtstoff *et al.*, 1935). Smith (1968) found *Mytilus
edulis*, the common blue mussel, to be quite resistant
to oil from the Torrey Canyon spill. Globs of oil
were even noticed in the mantle cavity thus sub-
stantiating the previous report of Mackin and Hopkins
(1962). Marx (1969) likewise substantiates the
mussel's hardiness in that he reported less than 10%
damage of mussel populations following the Union Oil
Blowout.

Many reports, on the other hand, have been
published showing that crude oil and its refined
products are toxic to marine organisms. A spill of
#2 oil off West Falmouth, Massachusetts, was reported
to "result in 95% dead or dying organisms in bottom
samples taken shortly after the spill" (Blumer *et al.*
1970). Immediate mortalities among barnacles, crabs,
mussels, and sea urchins were reported after the

Tampico Mara wreck in March 1957 off Baja, California
(Nelson-Smith, 1968a). Nelson-Smith (1968b) also
reported that *Monodonta*, *Littorina littorea*, *Littorina
obtusata*, and *Crassostrea virginica* were the most
sensitive to oil pollution. This work substantiates
the earlier work of Gowanlock (1935). He reported
that oysters grown in water which had been run through
crude oil showed 90% mortality in 3 months and 100%
mortality in 4 months, while controls showed only 17%
mortality. *Mytilus edulis* populations were reported
to be reduced to one-tenth their normal population by
oil from the Chrissi P. Goulandris in 1967 (Nelson-
Smith, 1968a). Rutzler and Sterrer (1970) reported
that populations of oysters, mussels, barnacles,
sponges, tunicates, and bryozoans were eliminated by
oil spilled from the tanker Bitwater.

Added to this increasing list of contradictory
results are scattered reports of sublethal effects of
oil. Concentrations of 0.1 ml/l and 1 ml/l of oil
impair the gripping ability of the tube feet of sea
urchins (North *et al.*, 1964). Mussels that were
juveniles during the West Falmouth spill were later
presumed to be sterile (Blumer *et al.*, 1970). Although oil was not found to be toxic to the limpet,
its shell was easily removed (O'Sullivan and Richardson, 1967). Oysters that lived under oil with no
apparent harmful effects showed depletion of glycogen
deposits (Galtstoff *et al.*, 1935).

Other problems with indirect effects are being
noted. DDT is one million times more soluble in oil
than water and thus extremely high concentrations of
this chlorinated hydrocarbon are found in the sediments when oil is present (Hartung and Klinger, 1970).
Oils also contain small amounts of carcinogenic substances (Bingham *et al.*, 1965; Carruthers *et al.*,
1967; Cook *et al.*, 1958).

Most oil "cleanups" are exclusively concerned
with the visible oil and are, to some extent, cosmetic,
in that what cannot be seen is presumed to be harmless.
In this study we have concentrated on those oil components most probably left in the water column.

MATERIALS AND METHODS

COLLECTION AND MAINTENANCE OF ANIMALS

All specimens of *Mytilus edulis* were collected at an uncontaminated site at Lamoine, Maine, in June, July, and August, 1971, from water ranging in temperature from $10°$ to $15°C$. The animals were returned immediately to the laboratory and kept in 10-gallon glass containers at a constant temperature of $12°C$. All seawater (32.5 gm/kg) used in this investigation was also collected from Lamoine. Distilled water was added to maintain a specific gravity of 1.025.

RESPIRATION STUDIES

Respiration was measured according to the method of Scott and Major (1972) using an Aminco refrigerated Warburg Respirometer. Since the respiration of *M. edulis* has been shown to vary with body size and temperature (Read, 1962), salinity (Schlieper, 1957), beach position (Baird and Drinnan, 1957), and seasonal change (Kruger, 1960), these variables were kept constant. All measurements were made at $10°C$. All animals chosen for the respiration studies were 1.3 ± 0.1 cm in length.

All animals were tested within 4 days of collection in standard Warburg flasks (20 ml). Each flask contained four animals, 0.1 ml of 15% KOH in the centerwell, and 4.15 ml of seawater or diluted extract. Seawater, seawater plus extract, and distilled water thermobarometers were employed.

Extract solution was added from the sidearms of the experimental flasks to give final concentrations in the flasks of 24% and 12% of the original extract. The respiration of five gill preparations (20 mg dry weight/flask) in 24% Esso Extra were compared with the respiration of five control gill preparations in pure seawater.

Finding that there was a gill respiratory effect and noting the excess visibility of mucus in exposed

352

animals, an experiment to measure clearance was performed. Four whole mussels were exposed to a 0.01%
neutral red solution in seawater and, in another container, four other mussels were similarly exposed to
the dye solution made up in 12% Esso Extra extract.
The mussels were selected to conform to those used in
the respiratory experiments.

EXTRACT PREPARATION

Esso Extra and #2 home heating fuel were obtained
from the coastal depot of the Webber Oil Company in
Bucksport, Maine. No. 6 industrial heating fuel
(Bunker C) was obtained from the Sprague Incorporated
Depot in Bucksport. These oils were transferred to
the laboratory in sealed glass containers. The extracts were prepared by the constant stirring, for
one-half hour, of equal volumes of oil and seawater
in sealed flasks at 12°C. The extracts were then
transferred to separatory funnels and allowed to
stand an additional half hour. The seawater layer
was then transferred to sealed flasks and used within
a half hour. This stock solution is referred to as
100% extract solution.

SURVIVAL IN EXTRACTS

Survival studies in 12% extracts of the three
oils were carried out in two series. In the first
series the experiment was carried out by exposing
mussels to 5 liters of 24% extract in aerated tanks
at a constant temperature of 12°C. Fifty animals
which were not fed were exposed to each of the three
extracts. In the second series 50 animals were
placed in 1 liter sealed flasks of the three 24%
extracts. These were kept at a constant temperature
of 12°C and were neither aerated nor fed.

SURVIVAL UNDER WHOLE OILS

The survival of *M. edulis* was investigated under
the influence of Esso Extra, #2 oil, and #6 oil. For

each oil, six 1-gallon tanks were used, one serving
as a control tank. Each tank contained 25 animals
and 2 liters of seawater. Oils were added to give
final concentrations of 2.5 ml/l, 1.5 ml/l, 1.0 ml/l,
0.5 ml/l, and 0.25 ml/l. All tanks were aerated and
kept at a constant temperature of 10°C. Glycogen
determinations were made on the animals under the
2.5 ml/l concentrations of the three oils according
to the method of Seifter *et al.* (1950).

STATISTICS

Wherever multiple comparisons were made, the
results were submitted to an analysis of variance.
When significant F values were obtained, the means
were compared by Duncan's Multiple Range Test to
determine those with significant deviations (P < .05).

RESULTS

RESPIRATION STUDIES

Figure 1 shows the mean results of respiration
studies under the influence of 24% and 12% extracts
of Esso Extra and #2 home heating oil. Analysis
between t = o and t = 60 min shows the Esso Extra
points are significantly different at the 0.05 level
of confidence while the #2 oil curve shows the same
significance for the 24% extract while the 12% extract
was more variable and significance was at the 0.1
level. The shells of animals in the 24% extracts
were closed at the termination of the experiment,
while in the 12% extracts they were open. Full
strength extracts of #6 industrial heating oil had
no measurable effects on respiration.

The results of a long-term respiration study of
28 animals in 12% Esso Extra extract are graphically
represented in Figure 2. The flasks were opened after
140 min and left in the refrigeration unit overnight.
In the morning the flasks were again closed and

354

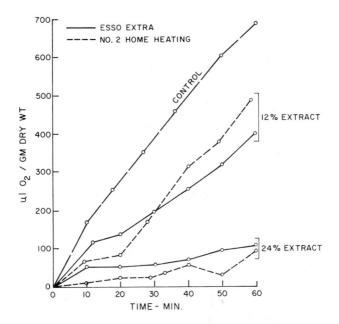

Fig. 1. The effects of water-soluble extracts of Esso Extra and #2 home heating oil on the respiration of Mytilus edulis *at 10°C, 56 animals in each case except the control which is the mean of 120 animals.*

respiration determined for 1 hr. At this time no significant difference was found between the respiration rates of the experimental and the control groups.

Figure 3 shows the results of respiratory studies of 20 animals with cut posterior adductor muscles maintained in seawater (control group) compared with 20 animals with cut posterior adductor muscles kept in 24% Esso Extra extract (experimental group). The control group with cut posterior adductor muscles did not differ significantly in oxygen consumption from intact animals. Whole animal respiration in 24% Esso Extra extract is included for comparison. Control animals placed down abyssal threads while the experimentals did not.

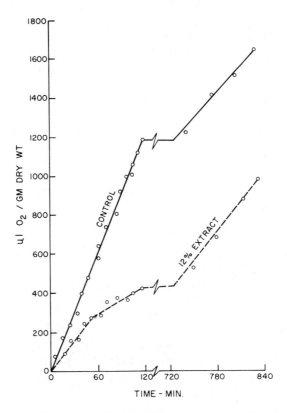

Fig. 2. The long-term effect of 12% extract of Esso Extra on the respiration of Mytilus edulis *at 10°C, 28 animals in each curve.*

Figure 4 shows the results of respiratory studies of isolated gill preparations in 24% Esso Extra extract. The respiratory depression of the experimental gill preparations is significant only over the first hour of measurement.

SURVIVAL IN EXTRACTS

No effects were noted on those animals placed in unsealed aerated tanks containing 12% extracts of the three oils.

356

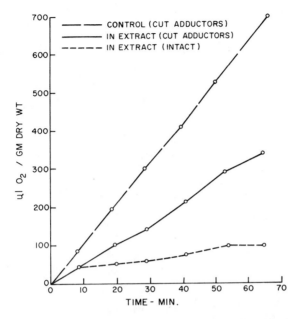

Fig. 3. The effect of 24% Esso Extra extract on the respiration of Mytilus edulis *with severed posterior adductor muscles. Each curve represents the mean of 28 animals at 10°C.*

Animals placed in 12% sealed extracts of #2 oil showed a decreased rate of shell closure by day 2 upon mechanical stimulation. Mantle retraction was not, however, noticeably impaired. The animals failed to place down any abyssal threads and failed to respond to any stimulation between day 5 and day 6, at which point they were declared dead.

All animals placed in 12% sealed extracts of Esso Extra also showed a decreased rate of shell closure by day 2 and a lack of thread formation. Further, after introduction into the extract, these animals remained almost fully closed for 30 hrs before they opened as fully as the controls. In contrast new animals placed in this old, 30-hr extract did not show any of these noted effects. Animals that had opened after being in the extract for 30 hrs closed

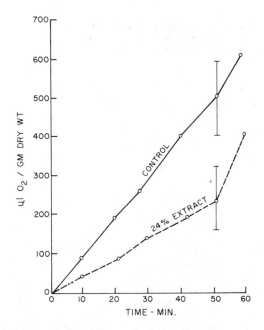

*Fig. 4. The effect of 24% Esso
Extra extract on the respiration
of isolated gill tissue of*
Mytilus edulis *at 10°C. Two
standard deviations are indicated
around each point at 50 min.*

for an additional 28 to 30 hrs if placed in fresh
extract. All animals died between day 4 and day 5.
All animals in controls, sealed seawater, were alive
and responsive up to day 12, at which point the
experiment was terminated.

All animals placed in 12% sealed extracts of
#6 oil showed a decreased rate of shell closure and
lack of thread formation. These animals likewise
died between day 5 and day 6.

SURVIVAL UNDER WHOLE OIL

All animals placed in concentrations of 2.5 ml/l
down to 1.0 ml/l of #2 oil showed decreased rate of

shell closure by day 2. Animals in the 0.5 ml/l concentration showed a decreased closure rate by day 4.
Animals in the lowest concentration, 0.25 ml/l,
showed no abnormal effects and did as well as the
controls throughout the experiments. Experimental
animals in the 2.5 to 0.5 ml/l concentrations were
not observed to put out their feet nor did they place
down threads. The glycogen content of six posterior
adductor muscles from animals in the 2.5 ml/l concentration of #2 oil was 14.79 ± 3 μg glycogen/mg wet wt
on day 4 compared with 26.62 ± 5.9 μg/mg wet wt for
controls on day 4. All animals in the 2.5 ml/l to
1.0 ml/l died between day 4 and day 5, while those in
the 0.5 ml/l concentration died between day 5 and
day 6. All animals in the 0.25 ml/l were alive on
day 12, at which point the experiment was terminated.

All animals under all five concentrations of
Esso Extra, 2.5 ml/l concentration showed an increased secretion of mucus not noted in the controls
or in animals kept in other oils. The glycogen content of the posterior adductor muscles for those
animals in the 2.5 ml/l concentration on day 4 was
27.8 ± 3.5 μg glycogen/mg wet wt. All animals were
alive on day 12, at which point the experiment was
terminated.

No noticeable effects of #6 oil were observed.
Glycogen content of the posterior adductor muscles on
day 4 in the 2.5 ml/l concentration was 30.61 ± 4.6
μg glycogen/mg wet wt.

DYE CLEARANCE

Animals in 12% Esso Extra extract took 200 min
to clear a standard neutral red extract which controls
cleared in 100 min.

DISCUSSION AND CONCLUSIONS

The immediate effect of water-soluble extracts
of Esso Extra and #2 oil is an almost complete

inhibition of oxygen uptake by 24% concentrations and partial inhibition at 12% concentrations. This response was found to be due to at least two factors. Firstly, animals with severed posterior adductor muscles, and therefore unable to close their valves, took up more oxygen in the 24% extracts (350 µl/gm/hr) than intact animals (90 µl/gm/hr). This amount of uptake was still well below control animals with cut adductors or intact controls in seawater (650 µl/gm/hr). Valve closure in response to the water-soluble components is therefore partly responsible for the lowered oxygen uptake.

Secondly, oxygen uptake also depends upon the rate at which water is moved over the gill surfaces by ciliary action and the rate that the oxygen is taken into the gill tissue. This latter factor is in turn regulated by the respiratory rate and rate of oxygen use by an organism. Animals were found to clear a 0.01% solution of neutral red in 12% Esso Extra extract at a much slower rate than did the control animals. This is probably an indication of reduced water movement due to reduced ciliary activity, but the possibility that this effect is due to an inability of the neutral red to adhere to the animals' tissues also exists.

Gray (1924) showed that when chloral hydrate was used to inhibit the ciliary activity of isolated *M. edulis* gill tissue there was a 50% reduction in oxygen uptake. A 50% reduction in oxygen uptake by isolated gill tissue was likewise found in these studies suggesting the possibility of an anesthetic action by the water-soluble components on ciliary activity. Additional weight is added to this hypothesis when one considers the nature of the compounds likely to be present in the extracts. First, the compounds probably are volatile as shown both by the rapid loss of toxicity on standing and the lack of effects if the animals are placed in open containers of the extracts. Secondly, the compounds in question must be at least slightly water soluble. Both of these facts point to the lower boiling point of

saturated hydrocarbons known to be abundant in the
oils. These compounds have also been shown to exert
a narcotic effect on a number of invertebrates
(Marshland, 1933; Goldacre, 1968). There has also
been at least one report (Galtstoff et al., 1935),
describing a narcotic effect of oil on the pumping
rate of the oyster, *Crassostrea*. These hydrocarbons,
which would be soluble in the lipid component of the
cellular membranes, have been found in high concentrations in the tissues of fish living in waste water
from petroleum complexes (Kiyoura, 1967).

The possibility that something is interfering
with oxygen uptake by the gills could be due to an
increase in mucus secretion noted in the whole oil
studies but not noticed in any of the extracts.
Aljarinskaya (1966) concluded that *M. galloprovincialis*
concentrated oil components in the mucous coverings
of the gills. An increased mucous secretion could
slow down ciliary activity, or vice versa. The water-
soluble components were removed, lost, or inactivated
with time. This was clearly shown by the disappearance of effects in old extracts, in which animals had
been sealed in for 30 hrs, on new animals.

The concentrations of oil used in the whole oil
studies were inadequate to cause respiratory depression via water-soluble components. Immediate aeration
of the tanks upon introduction of the oil also has-
tened evaporation of the volatile components. No. 2
oil prevented the secretion of abyssal threads which
is obviously detrimental to the survival of an intertidal organism exposed to tidal action. It has been
reported that a good guide to the condition of
M. edulis is its ability to reattach itself to the
substratum after disturbance (Smith, 1968; Swedmark
et al., 1971). Although adults move infrequently,
plantigrade larvae have specific light requirements
and move a number of times before they eventually
find a stable substrate (Verwey, 1957). Settlement
time from spawning to permanent beds may take 8 to
10 weeks (Seed, 1969). The fact that no foot pro-
trusion was observed explains the lack of thread

formation since the foot is necessary, not only for movement, but also for the placement of threads (Barnes, 1964).

Rate of shell closure under the influence of #2 whole oil and in the sealed extracts steadily decreased upon exposure time. Rate of shell closure has been used in other studies to indicate sublethal effects of toxic materials (Swedmark *et al.*, 1971). This depressed rate of shell closure could expose *M. edulis* to abnormally high predation. After exposure to oil, glycogen stores of the posterior adductor muscle were found to be depleted which Galtstoff *et al.* (1935) also found to be true for *Crassostrea*. This is particularly interesting as Daniel (1925) reported that even during periods of prolonged starvation *M. edulis* does not draw upon its glycogen stores, but retains them until the buildup of reproductive structures. The general weakened condition is not apparent in mantle retraction. Oil is then not uniformly effective on all muscles.

Concentrations of #2 whole oil down to 0.5 ml/1 cause a number of responses that were not evident at the 0.25 ml/1 concentration. Thus, a threshold of toxicity, for the time period studied, is between 0.5 and 0.25 ml/1, or 0.05% and 0.025%. The exact concentration is much lower than these figures, as most of the oils remain on the surface during the experiments. Additional experimentation for longer periods of time is needed as well as studies of the effects of these oils on the larval stages of *M. edulis*.

ACKNOWLEDGMENT

This research was supported by the Oil Research Fund.

LITERATURE CITED

Aljarinskaya, I. O. 1966. Behavior and filtering ability of the Black Sea mussel, *Mytilus galloprovincialis* in oil polluted water. *Zooch. Zh. (Russ.) 45*:998-1003.

Baird, J. A. and K. Drinnan. 1957. Ratio of shell to meat in *Mytilus* as a function of tidal exposure of air. *J. du Conseil. 22*:329-36.

Barnes, R. D. 1964. *Invertebrate Zoology.* Philadelphia: W. B. Saunders Co.

Bingham, E., A. W. Harton, and R. Tye. 1965. The carcenogenic potency of certain oils. *Arch. Environ. Health. 10*:449-51.

Blumer, M., G. Souza, and J. Sass. 1970. Hydrocarbon pollution of edible shellfish by an oil spill. *Mar. Biol. 5*:195-202.

Brisby, K. 1969. Oil slicks effects on Ricon Island. Amer. Pet. Inst. Spring Meeting, Proc. Coast District., Los Angeles, California, pp. 147-50.

Carruthers, W., H. N. Stewart, and D. A. Watkins. 1967. 1-2-Benzanthracene derivatives in Kuwait mineral oil. *Nature 213*:691-92.

Cook, J. W., W. Carruthers, and D. L. Woodhouse. 1958. Carcenogenicity of mineral oil. *Nature 213*:691-92.

Corner, E. D. S., A. J. Southward, and E. C. Southward. 1968. Toxicity of oil spill removers (detergents) to marine life: An assessment using the intertidal barnacle *Eliminius moclestus. J. Mar. Biol. Ass. U. K. 48*:29-47.

Daniel, J. 1925. Annual report for 1924 of Lancs. Sea-Fisheries Lab.

Galtstoff, P. S., H. F. Prytherch, R. O. Smith, and V. Koehring. 1935. Effects of crude oil pollution on oysters in Louisiana waters. *Bull. Bur. Fish. 45*:143-210.

George, M. 1961. Oil pollution of marine organisms. *Nature 192*:1209.

Goldacre, R. J. 1968. The effects of detergents and oil on the cell membrane. *Fld. Stud. 2*:131-37.

Gowanlock, J. N. 1935. Pollution by oil in relation to oysters. *Trans. Amer. Fish. Soc. 65*:293-96.

Gray, J. 1924. Mechanism of ciliary action. IV. Relation of ciliary activity to oxygen consumption. *Proc. Royal Soc. B 96*:95-114.

Hartung, R. and G. W. Klinger. 1970. Concentration of DDT by sedimented polluting oils. *Envir. Sci. Technol. 4*:407-10.

Kiyoura, R. 1967. Industrial waste water and hazards. Counter measures for waste water from petroleum complexes. *Kagaku Kagaku 31*:103-109.

Kruger, F. 1960. Zur Frage der Grässenabhängigkeit dis Savustaffverbrauches von *Mytilus edulis*. *Helgal. wiss. Mecresuntersuch.* 7:125-48.

Mackin, J. G. and S. H. Hopkins. 1962. Studies on oysters in relation to the oil industry. *Publ. Inst. Mar. Sci., University of Texas* 7:1-131.

Marshland, D. 1933. The site of narcosis in a cell: The action of a series of paraffin oils on *Amoeba dubia. J. Cell. Comp. Physiol.* 4:9-33.

Marx, W. 1969. Wildlife and the union oil blowout. *Underwater Nature* 6:32.

McCauley, R. N. 1965. The biological effects of oil pollution in a river. *Limnol. Oceanog. 11*: 475-86.

Mileikousky, S. A. 1970. The influence of pollution on pelagic larvae of bottom invertebrates in marine nearshore and estuarine waters. *Mar. Biol. (Berlin)* 6:350-56.

Mironov, O. G. 1969. Effect of oil pollution upon some representatives of Black Sea zooplankton. *Zool. Zh. 48*:980-84.

Nelson-Smith, A. 1968a. The effects of oil pollution and emulsifier cleaning on shore life in S. W. Britain. *J. Appl. Ecol.* 5:97-107.

_____. 1968b. Biological consequences of oil pollution. In: *The Biological Effects of Oil Pollution on Littoral Communities*, pp. 1-73, ed. by J. D. Cathy and D. R. Arthur. Field Studies Council.

North, W. J., M. Neushul, and K. A. Clendenning. 1964. Biological changes observed in a marine cove exposed to a large spillage of mineral oil. Symp. Poll. Mar. Micro-org. Prod. Petrol., Monaco, pp. 335-54.

O'Sullivan, A. J. and A. J. Richardson. 1967. The Torrey Canyon disaster and intertidal marine life. *Nature 214*:541-42.

Read, K. R. H. 1962. Respiration of the bivalved molluscs, *Mytilus edulis* and *Brachiodontes demissus*, as a function of size and temperature. *Comp. Biochem. Physiol.* 71:89-101.

Rutzler, K. and W. Sterrer. 1970. Oil pollution damage observed in tropical communities along the Atlantic seaboard of Panama. *Bioscience 20*: 222-24.

Schlieper, D. 1957. Comparative study of *Asterias rubens* and *Mytilus edulis* from the North Sea and the Western Baltic Sea. *Annee Biol.* 33:117-27.

Scott, D. and C. W. Major. 1972. The effect of copper on survival, respiration, heart rate, and electrocardiogram in the common blue mussel. *Biol. Bull. 143*:678-88.

Seed, R. 1969. The ecology of *Mytilus edulis* on exposed rocky shores. I. Breeding and Settlement. *Oecologia 3*:277-316.

Seifter, S., S. Dayton, B. Novice, and W. Muntwyler. 1950. The estimation of glycogen with the anthrone reagent. *Arch. Biochem.* 25:191.

Simpson, A. C. 1968. Oil, emulsifier and commercial shellfish. In: *The Biological Effects of Oil Pollution on Littoral Communities*, pp. 1-91, ed. by J. D. Carthy and D. R. Arthur. Field Studies Council.

Smith, J. E. 1968. *Torrey Canyon Pollution and Marine Life*. Boston, Mass.: Cambridge University Press.

Swedmark, M., B. Braaten, E. Emanuelsson, and A. Granmo. 1971. Biological effects of surface active agents on marine animals. *Mar. Biol. (Berlin) 9*:183-201.

Verwey, J. 1957. Discussion Coll. Intern. Biol.
     Mar., St. Roscoff. *Annee Biol.* *3*:238.

SPIN-LABELING OF LIVING TISSUE—
A METHOD FOR INVESTIGATING
POLLUTANT-HOST INTERACTION

WILLIAM T. ROUBAL

Northwest Fisheries Center
National Marine Fisheries Service
National Oceanic and Atmospheric Administration
Seattle, Washington 98122

Marine organisms the world over are exposed to many toxic and physiologically active substances, both natural and man-made. Those derived from man's activities are the land-based industrial and munici-pal wastes, pesticides, herbicides, forest fire re-tardants, and other substances which ultimately find their way into our streams, bays and oceans. Addi-tionally, there are the raw and processed petro-chemicals which enter waters from bilges of ships, oil cargo shipping accidents, and offshore drilling mishaps.

What happens to a living cell when a foreign substance enters it, and what becomes of the sub-stance itself when it interacts with cells and their constituents? The answers to these questions have been long in coming—admittedly they are difficult to answer. Only in recent years has the biochemical/biophysical know-how of science progressed to the point of making it possible to study pollutant-host interaction at the molecular level.

In this report I shall discuss some of our recent studies on uptake, transport, retention and interaction of hydrocarbons with lipoproteins, albumins, and membranes of fish. Studies of interaction, as monitored by pollutant-membrane investigations appear appropriate. Membranes play a dominant role in many life processes—neural function, osmoregulation, control of entry and exit of nutrients and metabolites, antibody function, and enzyme systems.

The application of radiotracer and spin-label methodology is discussed.

METHODS

IN VIVO *STUDIES—UPTAKE, TRANSPORT, AND RETENTION OF HYDROCARBONS*

General

Spin-labeled straight and branched-chain paraffins, spin-labeled aromatics, and spin-labeled long-chain fatty acids (Roubal, 1974) were incorporated into Oregon Moist Pellet hatchery food for feeding to fish. Fingerling Coho salmon were maintained in charcoal-treated, cold, running fresh water. The level of spin-label in the food ranged from 30 to 100 ppm. In conjunction with spin-labels, C^{14}-labeled hexadecane, heptadecane, nonane, benzene and naphthalene were also used. A total of 1 µC of radioactivity was administered via food to each fish over the feeding period.

Treated food was fed to fish over a 7- to 10-day period. The feeding regime was such that little of the treated food escaped the fish as they fed or browsed the bottom of the tanks. Blood constituents and various tissues (discussed later) were isolated according to the schemes of Freeman *et al.* (1963) and Roubal (1974) and were analyzed for radioactivity or hydrocarbon-membrane interaction as described by Roubal (1974).

368

Spin-labeling Theory—General

The theory and applications of spin-labeling are discussed in detail elsewhere (Roubal, 1972) and only a very abbreviated discussion is presented here. Spin-labeling is a spectroscopic technique, and, like all such techniques, is based on the absorption of energy of selected wavelengths by energy-absorbing substances.

Spin-labels belong to a class of compounds known as free radicals. Free radicals used here are the relatively stable nitroxide type (Hamilton and McConnel, 1968) and contain one unpaired electron per free radical molecule. The electron paramagnetic resonance (EPR) spectrometer is used for the measurement. The uniqueness of the spin-labeling method stems from the fact that when a spin-label is inserted into a biological system, the label gives a characteristic, environment-dependent spectrum—which will be discussed later.

INTERACTIONS OF HYDROCARBONS WITH NEURAL MEMBRANE OF COHO SALMON (IN VITRO STUDIES)

In vitro studies of membrane-pollutant interaction were performed on carefully excised tissue prelabeled with innocuous membrane-monitoring spin-labels (Roubal, 1974). Three conventional spin-labels (Fig. 1) were employed for prelabeling tissue

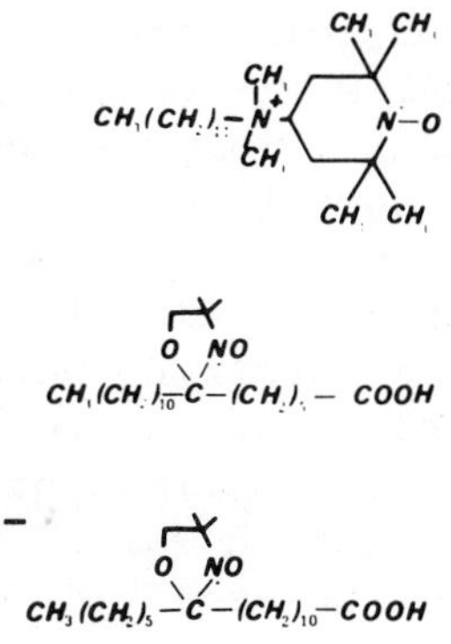

Fig. 1. Three common nitroxide spin-labels for studying membranes.
A. N,N-Dimethyl-N-dodecyl-N-tempoyl ammonium bromide.
B. "7-Nitroxide" derivative of stearic acid.
C. "12-Nitroxide" derivative of stearic acid.

prior to exposing tissue to low levels of hydrocarbons. Label A is described by Hubbell *et al.*
(1970), while labels B and C are synthesized by the
method of Keana *et al.* (1967) as described by Roubal
(1974).

Emphasis is on neural tissues since concentration of hydrocarbon is high in these tissues based on
weight. The discussion centers on spinal cord, although similar results are also obtained for lateral
line nerve, sheathed or desheathed.

RESULTS AND DISCUSSION

UPTAKE, RETENTION AND RELEASE OF HYDROCARBONS
(IN VIVO STUDIES WITH SPIN-LABELED AND C^{14}-LABELED
HYDROCARBONS)

Spin-labels fed to fish give rise to EPR activity in the blood. Invariably, when labeled hydrocarbons were fed, whole blood exhibited a simple,
three-line spectrum showing no environmental con-
straints imposed on the label by blood (Fig. 2A).
Figure 2B, on the other hand, indicates extensive
immobilization of label by the environment, and is
the type of spectrum recorded for protein-label complexes. This spectrum is obtained when long-chain
fatty acids and polar substances are fed to fish.

The data indicate that "insoluble" hydrocarbons
are transported by carriers in the blood in which the
hydrocarbon pollutant is able to intercalate into a
region of low viscosity (such as lipid). The data
also suggest (Fig. 2B) that long-chain polar substances associate with nonlipid carriers, such as
albumins.

In order to clarify the nature of the transport,
radioactive paraffins were injected into fingerling
fish and blood was collected 2 days later. The percent activity in lipoprotein fractions for heptadecane
C^{14}-injected fish was 37%, 46%, and 15% for chylomi-
crons/lipomicrons (S_f^o 400 - 10^5), S_f^o 0 - 400

lipoproteins, and HDL_2 + HDL_3 lipoproteins (78 – 100 Å), respectively. The ultracentrifugal residue, consisting mainly of albumin complexes, was low in hydrocarbon, and comprised the remaining 2% activity. In the test tube, labeled fatty acids plus blood albumins gave spectra of Figure 2B, and substantiates the albumin-pollutant transport hypothesis for polar materials.

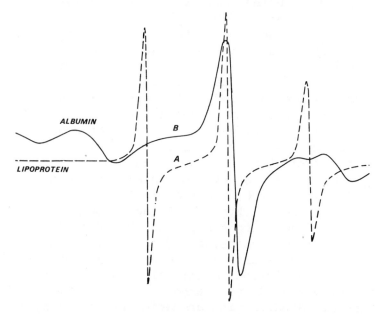

ALBUMIN

B

A

LIPOPROTEIN

Fig. 2. *Spin-label spectra of blood of fish fed hydrocarbons. A. Free tumbling condition; no label immobilization. B. Extensive immobilization.*

Replacement of label-treated food by label-free food (2-day feeding period) resulted in an abrupt drop in blood activity. At this time, however, tissues which were initially free of activity began to show a buildup of EPR signal (Roubal, 1974). In particular, the spinal cord, lateral line nerve, and brain exhibited EPR activity. Differences in EPR spectra were associated with this uptake. This is

not unexpected, however, as labels undoubtedly encounter a different environment when incorporated
into neural tissue.

Our original work showed that we could not
detect EPR activity in flesh muscle (Roubal, 1974).
It was suspected that label did not enter flesh, or
if it did, that metabolic activity here was sufficient to quickly reduce the free radical to a non-
radical entity. Using radiotracers it was later
shown that hydrocarbons do enter flesh.

Label retention appears to be somewhat dependent
on chain length, but more importantly it depends on
hydrocarbon class. Using radiotracers, activity in
tissue is raised and is maintained in an elevated
state just as long as food contains tracer. When
treated food is replaced by pollutant-free food, the
hydrocarbon content drops. In the case of nonane,
hexadecane, and heptadecane, initial levels in tissue
drop over a variant time span. Figure 3 shows a
typical time-course pattern for this. We see that
the total content of hydrocarbon in flesh is similar
to the rest of neural tissue. As an example, a
typical Coho fingerling's brain weighs about 150 mg,
while the flesh weighs about 2 grams (wet weight),
much of which is water. Absence of spin-label in
flesh was quite misleading initially, since radiotracer easily showed hydrocarbons to be present in
this tissue. Studies with aromatics are not complete,
but available data show that aromatics are rapidly
metabolized and/or excreted; with benzene, for
example, no radioactivity is detected in tissue at
the end of 2 to 3 days.

*HYDROCARBON-MEMBRANE INTERACTIONS (IN VITRO STUDIES
USING PRELABELED TISSUE FOR MEASURING INTERACTIONS IN
NEURAL TISSUE)*

Spin-labeling Measurements

Nitroxide-type spin-labels of the type used in
this study contain the unpaired electron density

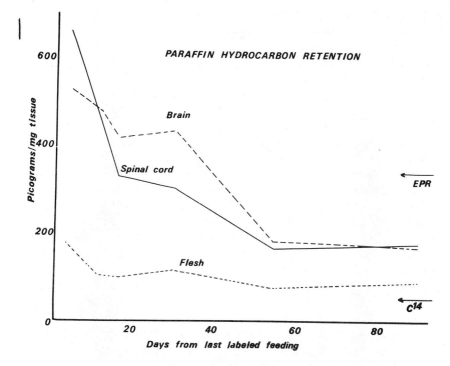

Fig. 3. Paraffin hydrocarbon content of neural tissue. The data indicate that concentrations do not all fall at uniform rates but that transport into and out of tissues occurs at different rates throughout the examination period. Arrows denote limits of detection by EPR and radiotracer.

localized in a p-orbital on nitrogen, the orientation of which is shown in Figure 4. Note that this orbital lies parallel to the amphiphilic axis in labels B and C. This orientation also aligns the long axis of the p-orbital perpendicular to the plane of the membrane surface. Since the interactions of the p-orbital with the surrounding environment are the governing factors for spectral line shapes and positions, and since the p-orbital is nonspherical, different spectra result when the p-orbital alignment changes with respect to a reference axis (here the plane of the membrane). Still another variable that contributes to the nature of the spectra is the rate (tumbling frequency) at which various orientations taken by the p-orbital change with respect to a reference

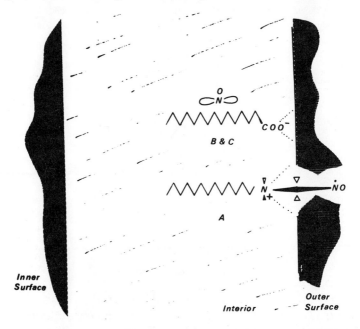

Fig. 4. Orientation of spin-label molecules and nitroxide p-orbitals in labeled tissue. Orientation of both the label amphipilic axis and p-orbital is as shown for labels B and C. Amphiphilic axis for A is as shown. The orientation of the p-orbital in A is not known with certainty.

point. Labels have characteristic EPR spectra from which the mobility of the p-orbital, and hence the fluidity of the membrane, may be monitored. Membrane perturbations which alter the alignment or tumbling rates (exposure of membrane to chemical or physical treatment) alter the EPR recordings.

The alignment of the p-orbital in A with respect to the rest of the molecule has not been made. The

label behaves similarly to labels with known align-
ments, however, and is useful for membrane studies.

Observe that label A intercalates into membrane
(Fig. 4) such that the p-orbital is positioned at or
near the membrane surface. Exposure of membrane to
surface treatments will change the environment of the
label in this portion of the membrane. Label B is
aligned as shown with the p-orbital somewhat below
the membrane surface. Changes in this region of the
membrane are detected by B. The p-orbital of label C
is even further removed from the surface, and reports
primarily on events deep within the hydrophobic
liquid-like lipid interior.

Effects of Treatments

Treatment of membrane by aromatics reduces the
immobilization at the surface, while treatment by
paraffins is less effective in this regard. Label B
experiments show a time-average orientation of the
p-orbital (so-called $2T_{11}$ anisotropy). Here aromatics are least effective in altering the control
situation, with the greatest effect resulting from
treatment by paraffins. Aromatics reduce some of the
anisotropy but paraffins effectively perturb this
region of the membrane and remove anisotropy altogether. The sharpening of the resonances indicate
that treatment produces a disordering on the membrane
in the region of B's p-orbital.

Label C control data show that anisotropic contributions in this lipid-rich portion of the membrane are
probably a reflection of the ordering of components by
cholesterol (cholesterol is a known ordering agent of
membranes and exists in high concentration in neural
tissue). The spectra indicate that both aromatics and
paraffins disorder the interior of the membrane. It
appears that the lipid-rich interior responds to all
treatments more readily than other regions. A critical examination of line widths indicate that paraffins are slightly more effective in this regard, however. Figure 5 shows spectra for labels A, B, and C.

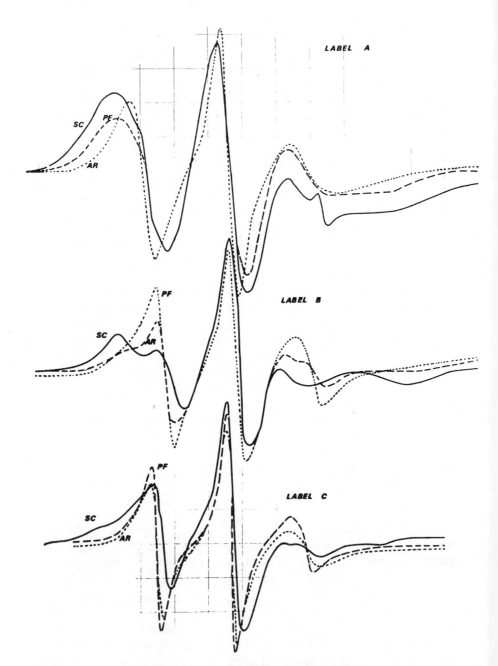

Fig. 5. Legend on next page.

Overall, the data indicate that aromatics, generally, alter surface properties while paraffins do not (Roubal, 1974). Evidently the pi-electron system characteristic of conjugated double bonds provides the necessary charge-transfer or other electron-mediated complexing capability, and directs the interaction to the polar membrane surface.

Differences in Membrane Perturbations and Possible Significance

The enhanced ability of aromatics to alter membrane surfaces may explain why these compounds are toxic to living organism. Membrane surfaces participate in ion exchange with fluids contacting the membranes. Organizational change here may alter the membrane's ion exchange capacity. Also, enzymes associated with membranes require certain membrane properties for their function. If the surface is altered, polar and other interaction between membrane and enzyme may alter enzyme activity. Membranes mediate osmoregulation and neural response, and any alterations of the surface structure may be considered detrimental.

It is true that paraffins penetrate membranes and alter internal organization (a region which is fluid from a molecular standpoint in native— untreated—tissue). Any additional perturbations by paraffins, however, may only be incidental to paraffin binding.

The short retention of aromatics in fish may also reflect surface-active properties of aromatics, whereby aromatics are exposed to enzyme systems which participate in detoxification. Paraffins, on the

---

Fig. 5. Spectra of prelabeled spinal cord before and after hydrocarbon treatment. Separate spinal cords were used for each test. SC, spinal cord; PF, spinal cord treated by paraffins; AR, spinal cord treated by aromatics.

other hand, intercalate with membrane interiors, out of contact with metabolically active processes. Hence they exhibit long residence times.

Spin-label studies, using new and better labels, and augmented with such techniques as electrophysiology and electronmicroscopy, will undoubtedly extend our knowledge of molecular processes which transpire when pollutants invade living tissue. In these beginning studies using EPR, we already see how spin-labeling fills a gap and provides information not obtainable by other methodologies.

ACKNOWLEDGMENT

The author wishes to thank Mr. Tracy K. Collier for preparing many of the samples for spin-labeling and for conducting the radiotracer study.

LITERATURE CITED

Freeman, N. K., F. T. Lindgren, and A. V. Nichols. 1963. The chemistry of serum lipoproteins. In: *Progress in the Chemistry of Fats and Other Lipids*, Vol. 6, pp. 216-50, ed. by R. T. Holman, W. O. Lundberg, and T. Malkin. New York: MacMillian Company.

Hamilton, C. L. and H. M. McConnell. 1968. Spin-labels. In: *Structural Chemistry and Molecular Biology*, pp. 115-49, ed. by A. Rich and N. Davidson. San Francisco: W. H. Freeman and Co.

Hubbell, W. L., J. C. Metcalfe, and S. M. Metcalfe. 1970. The interaction of small molecules with spin-labeled erythrocyte membranes. *Biochem. Biophys. Acta. 219*:415-27.

Keana, J. F. W., S. B. Keana, and D. Beetham. 1967. A new versitile ketone spin-label. *J. Am. Chem. Soc. 89*:3055-66.

Roubal, W. T.   1972.   Spin-labeling with nitroxide
    compounds.   A new approach to the *in vivo* and
    in vitro study of lipid-protein interaction.
    In: *Progress in the Chemistry of Fats and Other
    Lipids*, Vol. 13, pp. 61-88, ed. by R. T. Holman.
    London: Pergamon Press.
_____. 1974. *In vivo* and *in vitro* spin-
    labeling studies of pollutant-host interaction.
    In: *Mass Spectrometry and NMR in Pesticide
    Chemistry*, ed. by R. Haque and F. J. Biros.
    New York: Plenum Press. In press.

MULTIPLE ENVIRONMENTAL FACTOR EFFECTS ON PHYSIOLOGY AND BEHAVIOR OF THE FIDDLER CRAB, *UCA PUGILATOR*

WINONA B. VERNBERG,[1,2] PATRICIA J. DE COURSEY,[1,2,3] and JAMES O'HARA[1,4]

[1]*Belle W. Baruch Institute for Marine Biology and Coastal Research,*
[2]*Biology Department,* [3]*Psychology Department, University of South Carolina Columbia, South Carolina 29208* and [4]*Applied Biology, Decatur, Georgia 30030*

✗ In the last few years there has been increased awareness of the growing competition between industrial and recreational uses of estuaries. As a result, water quality control for multiple usage has become a vital problem. A key issue in assessing, and subsequently controlling, water quality has been the development of adequate monitoring measures. Bioassay methods offer possibilities but have not yet been widely exploited. Data presented in this report are the results of studies on a species characteristic of Atlantic estuaries, the fiddler crab, *Uca pugilator.* ✗To provide realistic water quality standards it is necessary to know synergistic effects between pollutant and normal environmental fluctuations, since one of the chief characteristics of estuaries is the large number and range of environmental variables throughout the year. Temperature,

381

salinity, humidity, and light intensity are only a
few of the factors which change from hour to hour.
While organisms living in an estuarine environment
can generally adapt to such physical changes, the
combination of a sublethal pollutant plus a stressful
environmental change may prove lethal. Accordingly,
these studies were undertaken to delineate the synergistic effects of sublethal mercury and cadmium pollutants and temperature-salinity fluctuations upon
both larval and adult *U. pugilator.* Mercury is used
in the manufacture of many industrial products,
including chlor-alkali and electroplating processes
and paper manufacturing. Cadmium is also used in
electroplating and other chemical processes and is a
byproduct of lead and zinc mining.

The fiddler crab is a major component of the
intertidal fauna. As is true for many marine species,
a sequence of developmental stages occurs and the
different stages occupy widely divergent habitat
niches. In *U. pugilator* the larvae are wholly planktonic while adults occupy the intertidal mudflat
regions. Since the presence of a species is dependent upon the successful completion of its life cycle,
it was considered crucial to assess both larval and
adult responses in these studies.

MATERIALS AND METHODS

Adult fiddler crabs, *Uca pugilator*, were collected at the Belle W. Baruch Coastal Research station near Georgetown, South Carolina. The animals
were brought into the laboratory where they were
maintained in plastic boxes containing approximately
100 cc filtered seawater. Boxes were tilted so that
crabs could freely select total or partial immersion.
Stock animals were routinely placed about 40 per box
at 25°C in 30 o/oo filtered seawater, on a 12L:12D
light schedule for at least 2 weeks before use in an
experimental program; they were fed on Clark's fish

pellets three times a week and the water changed after each feeding.

For survival studies the adult crabs were housed in plastic boxes containing 100 cc of the test solution; the solutions were changed three times weekly. Temperature and lighting schedules were maintained by an environmental control chamber. Crabs were checked daily to tally mortality until 50% had died.

For egg survival studies, crabs were brought into the laboratory in a nongravid state in summer, kept in control seawater and examined daily until the egg sponge appeared. Then they were placed individually in finger bowls with the desired solution for 7 days, and finally returned to control seawater. Daily checks were made to determine the condition of the egg sponge, as well as date of hatch and condition of zoeae upon hatching. Preliminary experiments with 32 crabs were followed by a definitive series using 20 experimental and 20 control crabs.

Metabolic rates for whole animals and also for hepatopancreas and gill tissues were measured in a Gilson apparatus. Acclimation and determination temperatures are presented in the observation section. Results are expressed as microliters of oxygen per hour per gram wet weight. Significant difference of means was calculated by the method of Simpson *et al.* (1960) for small samples.

For behavioral studies the activity of adults was measured quantitatively in various types of actographs. The chief type used was a plastic carousel suspended by delicate bearings on an axle; revolutions of the drum caused by activity of the crab activated a magnetic sensor. Activity counts were totaled on either an Esterline-Angus operations recorder or by numerical printout recorders. Printout data were used for calculating hourly and daily rates. For rhythmic analyses, the printout data were processed by a computer periodogram program (Suter and Rawson, 1968).

Tissue uptake of mercury was monitored in gill, hepatopancreas, muscle, and carapace tissues.

Tissues were removed from a group of crabs and frozen immediately. The concentration of mercury in each tissue was determined on a Perkin-Elmer MAS 50 or on a Coleman MAS 30. Techniques were based on the Environmental Protection Agency method, developed by the Analytical Quality Control Laboratory, using dilute nitric acid to digest the samples. In addition, the radioisotope ^{203}Hg was used for measuring uptake in gill and hepatopancreas. Cadmium uptake was determined in gill, hepatopancreas, green gland, and muscle, using the radioisotope ^{109}Cd. Tissues exposed to ^{203}Hg or ^{109}Cd were digested in soluene, and concentrations were determined by liquid scintillation on a Packard Tricarb Model 3320 counter.

For study of the effects of toxicants on the microanatomy of crab tissues, gill tissue from control and treated animals were examined under the electron microscope. The tissue was fixed in glutaraldehyde, then sectioned and examined with an electron transmission microscope.

To obtain larvae for the study, gravid females were collected from the intertidal flats of the Baruch Foundation property near Georgetown, South Carolina. The crabs were then brought into the laboratory, placed singly in finger bowls containing 30 o/oo filtered seawater, and maintained in a constant temperature box at 25°C with a 12:12 LD light schedule. In a series of preliminary experiments, considerable variation in viability of larvae and general vigor of different hatches was noted. Therefore, new hatches were qualitatively rated and only active, vigorous groups having less than 10% mortality at hatching were used for these experiments.

The larvae were reared in groups of ten per finger bowl under various temperature-salinity regimes with and without Hg or Cd. Solutions were changed daily, and the food source of newly hatched *Artemia* replenished at that time. Larvae were staged under 32 X magnification, using the morphological criteria of Hyman (1920).

In the survival studies newly hatched zoeae were set up under the specified temperature, salinity, and water conditions; mortality counts were made at the time of daily feeding and water change. Metabolic measurements were made on single larvae, using Cartesian Diver Respirometers with a total volume of 10 to 13 microliters. To determine rate and swimming pattern of the larvae, they were reared under specified conditions to the desired stage, then tested singly in swimming chambers. Larvae were transferred to a transparent 10 × 10 × 3 cm deep chamber having a grid of lines 0.5 cm apart etched on the baseplate. The number of lines crossed/minute served as an index of activity. Phototactic responses of larvae were measured in a light intensity gradient apparatus modified from Ryland (1960) and Bayne (1964). Intensity ranged from 1600 fc at the anterior end to 1300 fc at the posterior end.

After preliminary screening, a 10-min test interval was chosen. Kite diagrams (Bayne, 1964) were constructed from pooled data for all animals in a specific test condition. Phototactic responses were first determined for larval stages 1, 3, 5 and early crab stages at 25°C, 30 o/oo in either untreated seawater or 1.8 ppb Hg. Responses were also measured for first-stage zoea, which had been maintained 24 hrs under the various temperature-salinity regimes with and without the addition of mercury, and of third-stage larvae reared in 20°C regimes under control or Hg-treated conditions. It was not possible to measure third-stage responses in 30°C regimes due to high mortality rates at this temperature.

The toxicant solutions were made up in 30 o/oo filtered seawater. Mercury was provided in the experimental conditions as a known concentration of $HgCl_2$. A stock solution of $9 \times 10^{-3}M$ $HgCl_2$ in 30 o/oo filtered seawater was first prepared. This stock solution was checked at 4 to 6 week intervals with a Perkin-Elmer Atomic Absorption Spectrophotometer Model 303, and no changes in the mercury concentration were found. Shortly before use the stock was diluted to the three mercury test solutions: $9 \times 10^{-7}M$ $HgCl_2$ (0.18 ppm Hg), $9 \times 10^{-9}M$ (0.0018 ppm Hg), or $9 \times 10^{-11}M$ (0.000018 ppm Hg). The cadmium stock for all experiments was reagent grade $CdCl_2 \cdot 2.5\ H_2O$, made up to a solution of 1 mg Cd^{++}/ml H_2O. Dilution to the desired concentration was made using filtered seawater (30 o/oo) and/or distilled H_2O.

RESULTS

ADULT STUDIES

Survival

Preliminary studies established that under optimum conditions of temperature (25°C) and salinity (30 o/oo) adult crabs could survive for prolonged periods of time in seawater having an initial concentration of $9 \times 10^{-7}M$ $HgCl_2$ (0.18 ppm Hg). Under temperature and salinity stress, however, this concentration of mercury significantly shortened survival time (Vernberg and Vernberg, 1972). For example, under conditions of low temperature (5°C) and low salinity (5 o/oo), such as could occur following heavy winter rains, the crabs could not survive as long as under conditions of high temperature and low salinity. In winter animals without the added stress of a pollutant, 50% of the females died by day 20 and 50% of the males by day 7 (Fig. 1). With added Hg (0.18 ppm) 50% mortality of females was reached on day 8, and males by day 6. Under conditions of low salinity (5 o/oo) and high temperature (35°C), conditions very apt to occur following the heavy rains associated with a summer hurricane, both male and female *U. pugilator* can survive with very little mortality for at least 28 days (Fig. 1). With the addition of 0.18 ppm mercury, however, survival times of both males and females were reduced; 50% of the

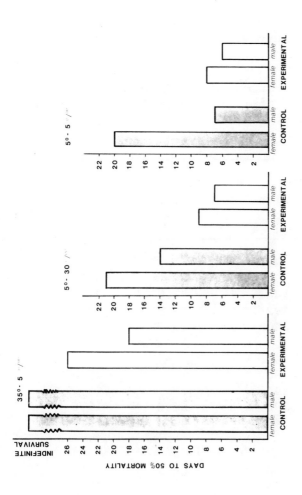

Fig. 1. *Mortality of adult Uca pugilator maintained under different temperature-salinity regimes with and without the addition of 0.18 ppm Hg.*

males had died by day 17 while 50% of the females survived to day 26.

The effect of Cd on the survival of adult fiddler crabs differed from the Hg picture (O'Hara, 1973). Table 1 shows the influence of temperature, salinity, and exposure time on the level of toxicant which kills 50% of the crabs. No differences were noted between survival rates of male and female crabs.

TABLE 1

Cadmium Concentrations (Cd^{++} in ppm) Lethal to 50% of Test Organisms (TLm) at Different Salinities, Times, and Temperatures

Salinity o/oo	Time hr	Temperature		
		10°C ppm	20°C ppm	30°C ppm
10	48	--	--	11.0
	96	--	32.2	6.8
	144	51.0	21.3	4.0
	192	28.5	18.0	3.0
	240	15.7	11.8	2.9
20	48	--	--	28.0
	96	--	46.6	10.4
	144	--	23.0	5.2
	192	52.0	16.5	3.7
	240	42.0	9.5	3.5
30	48	--	--	33.3
	96	--	37.0	23.3
	144	--	29.6	7.6
	192	--	21.0	6.5
	240	47.0	17.9	5.7

From O'Hara (1973).

The susceptibility of fiddler crabs to Cd was most pronounced in the thermosaline regime of 30°C and 10 o/oo. The concentration fatal to 50% of the organisms in 240 hrs (TLm-240 hrs) was calculated to be 2.9 ppm Cd^{++} under these conditions, whereas at 10°C, 30 o/oo the level was 47.0 ppm. Values generally were more influenced by temperature than by salinity within a given regime.

Egg survival studies compared hatch success of gravid females treated for 7 days in 0.18 ppm Hg with hatch success of females maintained in control seawater. In the preliminary experiments 36% of control females and 39% of the Hg-treated females dropped egg sponges prematurely. A hatching survival of 50% was realized for both groups in terms of total egg potential (including sponges dropped). In the final series 5% of control females died versus 20% of the Hg-treated ones; premature dropping of the egg sponge was 55% versus 80%. Therefore, total egg loss equaled 60% for controls and 100% for experimentals.

Thus considerable variability was seen between the two series in two successive years. Little difference between treated and untreated animals occurred in the first series, and the 50% overall hatching survival of both treated and control groups suggested that the egg stage, like the adult, might be less sensitive to Hg-poisoning than the larvae. In the second series, however, no treated eggs survived until hatching and the percentage of females which died in the Hg group was much higher than in the control group, suggesting considerable toxicity. The reason for the differences has not yet been resolved.

Uptake

Neither the male nor the female crabs survived well when exposed to mercury under conditions of environmental stress; it thus seemed possible that increased mercury uptake occurred in the tissues of the crabs under suboptimal temperature-salinity regimes. Therefore, the amount of mercury in selected tissues of crabs living under various conditions of temperature and salinity for 1, 3, 7, 14 and 28 days was determined. In five tissues assayed (gill, hepatopancreas, muscle, green gland, and carapace) the greatest accumulation occurred in gill tissue under optimum temperature and salinity (Vernberg and Vernberg, 1972). Uptake rates of mercury under sub-optimal conditions also were determined for gill and hepatopancreas tissue using ^{203}Hg (Vernberg and O'Hara, 1972). Gill tissue accumulated much greater amounts of mercury than did the hepatopancreas regardless of the experimental conditions. Over 82% of the mercury accumulation in the gill tissue occurred within the first 24 hrs in all thermal-salinity regimes, with only slight additions after 48- and 72-hr exposures in all experimental conditions except 5°C, 5 o/oo S regime. For this latter condition the mercury content of gill tissue declined slightly after 72 hrs, possibly due to necrosis and sloughing of the gill epithelium.

The amount of mercury accumulated in gill tissue within the first 24 hrs was significantly higher (at the 1% significance level or less) under the thermal-salinity regime of 5°C, 5 o/oo than in tissues of crabs maintained in any other set of experimental conditions. Under experimental conditions of 5°C, 30 o/oo, mercury accumulation in gill tissue was significantly greater than at 25°C, 30 o/oo, or at 33°C with either 5 o/oo or 30 o/oo. The least amount of mercury accumulated in gill tissue of crabs maintained at 33°C, 30 o/oo. Thus, low salinity increased mercury accumulation in gill tissue at all temperatures; mercury accumulation was further heightened at low temperature (Fig. 2).

In hepatopancreas tissue the concentration of mercury increased throughout the 72-hr exposure in all experimental conditions except 5°C, 5 o/oo, where mercury concentration was significantly lower than in crabs maintained under any of the other experimental regimes. After 72 hrs there was less mercury in the hepatopancreas of 5°C crabs regardless of the salinity

than in any other group. Crabs exposed to 33°C,
30 o/oo contained 14 times more mercury in the
hepatopancreas than crabs exposed to 5°C, 5 o/oo.

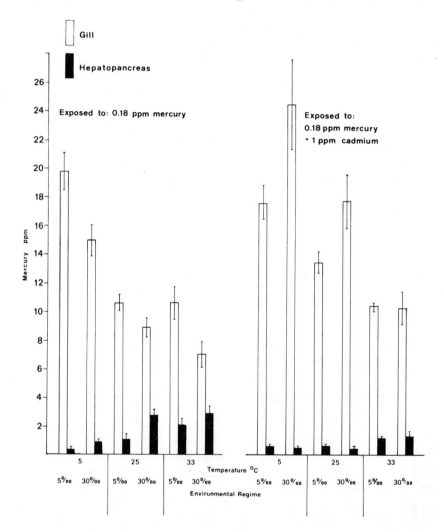

*Fig. 2. Uptake of mercury in gill and hepatopancreas
tissue of adult* Uca pugilator *after a 72-hr exposure
to 0.18 ppm Hg alone (left), or 0.18 ppm Hg in combination with 1 ppm Cd (right). Vertical lines indicate
± 1 standard error.*

391

However, the total body burden of mercury in the gills
and hepatopancreas, when calculated as mg metal present in both tissues, was relatively constant under all
experimental conditions (Fig. 3). Thus, the total
amount of mercury in the gill plus that in the hepatopancreas was essentially the same whether the crabs
were maintained at 5°C, 30 o/oo or 33°C, 5 o/oo, but

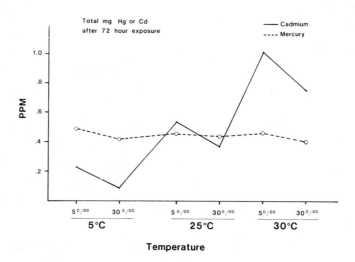

*Fig. 3. Total amount of Hg or Cd in gill
and hepatopancreas tissues of adult
Uca pugilator maintained under different
thermal-salinity regimes for 72 hrs in
0.18 ppm Hg or 1 ppm Cd.*

the relative amounts in each tissue were very different. These results suggest that the cause of death
at low temperatures may be related to the fact that
the crabs seem unable to transport mercury from gill
tissue to the hepatopancreas. There was no sex
difference in mercury uptake in tissue under either
optimal or suboptimal conditions.
    Preliminary screening with cadmium indicated that
a concentration of 1 ppm Cd was sublethal for adult
fiddler crabs maintained under optimum temperature-
salinity conditions (25°C, 30 o/oo). Both male and

female crabs survived under these conditions for prolonged periods of time. In a series of experiments
parallel to the Hg studies on uptake levels, cadmium
uptake rates were determined for the hepatopancreas
and gill tissue from crabs maintained under different
temperature-salinity ranges (O'Hara and Vernberg, in
manuscript).

In gill tissue, cadmium levels were essentially
the mirror image found for mercury levels. Whereas
the highest Hg levels in the gills were found at low
temperatures, with cadmium the highest levels occurred
at high temperature (Fig. 4). Cadmium was transferred
rapidly and in relatively high amounts to the hepatopancreas; under optimum conditions, the amount of
cadmium in the two tissues was approximately equal.
Relatively small amounts of cadmium were found in the
hepatopancreas of crabs under optimum salinity at
either high (33°C) or low (5°C) temperature. In
contrast to mercury, the total body burden level of
cadmium was dependent upon the environmental regime,
ranging from a low of 0.08 µg at 5°C, 30 o/oo to a
high of 1.2 µg at 33°C, 5 o/oo (Fig. 3).

Since these studies established that temperature
and salinity differentially affected the uptake of
cadmium and mercury in tissues of *U. pugilator*,
further work was initiated to consider the effect of
dual exposure to these two metals (O'Hara and
Vernberg, in manuscript). Mercury was added in the
form of $HgCl_2$ at an initial concentration of 0.18 ppm
Hg over a 72-hr period. Cadmium was added in the
form of $CdCl_2$ plus one µCi of ^{115}Cd to bring the total
initial concentration to 1 ppm Cd.

In both the gill and hepatopancreas, mercury
uptake was more influenced by the presence of cadmium
than was cadmium uptake by the addition of mercury
irrespective of thermo-salinity regime (Figs. 2 and
4). Generally, when a statistically significant
change in uptake did occur, the uptake of each metal
increased in the gills and decreased in the hepatopancreas. The presence of Cd also affected the distribution of Hg in the hepatopancreas and gill tissue.

When Hg alone was present in the water, the crabs were effectively able to transport the Hg from the gills to the hepatopancreas. For example, after

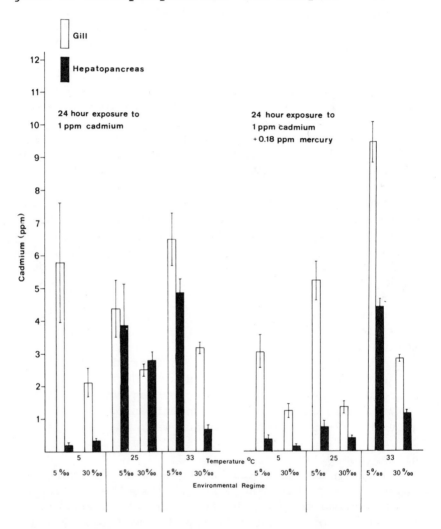

Fig. 4. Uptake of cadmium in gill and hepatopancreas tissue of adult Uca pugilator *after a 72-hr exposure to 1 ppm Cd alone (left), or 1 ppm in combination with 0.18 ppm Hg (right). Vertical lines indicate ± 1 standard error.*

after 72 hrs at 33°C, 30 o/oo, the percentages of Hg
in the gill and hepatopancreas were 35.1 and 64.9,
respectively. At 5°C, 5 o/oo, however, percentage
values were 96.3 and 3.7 (Vernberg and O'Hara, 1972).
The uptake of Cd in the gills was little affected by
the addition of Hg. Cd uptake was increased only at
high temperature and low salinity (30°C, 5 o/oo); at
low temperature and high salinity the rate of uptake
decreased, but the amount of decrease was statistically significant only after a 48-hr exposure.
Mercury significantly decreased Cd uptake in the
hepatopancreas at 25°C at both high and low salinity.
When both Cd and Hg were present, the crabs seemingly
lost their ability to transport the Hg from the gills
to the hepatopancreas, and the percentage of Hg in
the gills ranged from 90% to 98% regardless of the
temperature-salinity regime. Although TLm values
were not determined for the crabs, mortality rates in
the experimental boxes were considerably higher than
observed mortalities in crabs subjected to only one
metal. The inability of the crabs to survive for
long periods of time in the presence of both metals
is probably associated with the inability to transport
the Hg from the gills to the hepatopancreas.

Metabolism

In metabolic experiments with adult fiddler
crabs, rates were first established for males and
females at 25°C in 30 o/oo untreated seawater.
Uptake rates were essentially the same for both sexes.
After the base-line metabolism was determined, the
same animals were then maintained at 25°C in 30 o/oo
seawater with the addition of mercury, and metabolism
measured after 1, 3, 7, 14, 21, and 28 days' exposure.
Although a low level concentration of mercury was not
lethal to the crabs under optimum environmental conditions, metabolic rates were affected, especially
for males. The rate of oxygen uptake of males was
significantly lower than that of the females after
21 days in this sublethal concentration of mercury,

and the metabolic rate of the males had not returned to the base-line level by the end of the 28-day experimental period. Both males and females, however, continued to survive for another month under the mercury regime without any significant increase in mortality.

Under conditions of low temperature (5°C) and low salinity (5 o/oo) stress, females not only survived much longer than males but also maintained a steadier rate of oxygen uptake. The metabolic rates and patterns of the experimental female crabs were similar to those of the control female crabs. The metabolic rate of male experimental crabs was not significantly different from that of the female experimental or male and female control crabs after a 1-day exposure to mercury, but by day 3 the rate dropped markedly.

Oxygen uptake rates of female control crabs maintained in low salinity water (5°o/oo) and at high temperature (35°C) were relatively constant over a 28-day period and tended to be higher than that of control male crabs. The metabolic rates of mercury-treated female crabs remained fairly constant for the first 7 days and then declined rapidly. The uptake rates of experimental male crabs declined steadily from day 1 and tended to be lower than those of the females throughout the remainder of the time period (Vernberg and Vernberg, 1972).

To pinpoint why Hg-treated males died sooner than females and did not respond metabolically in the same manner, tissue metabolism studies were carried out. Crabs were collected in winter or early spring, then either warm-acclimated (25°C) or cold-acclimated (10°C) in the laboratory for a minimum of 2 weeks. Each group was then subdivided into temperature-salinity groups:

    5°C, 5 o/oo
    5°C, 30 o/oo
    33°C, 5 o/oo
    33°C, 30 o/oo
    25°C, 30 o/oo (warm-acclimated crabs only)

In crabs that were cold-acclimated, there were no differences between males and females in the metabolism of isolated tissues regardless of the temperature-salinity regimes; furthermore, there was no difference between metabolic rate of hepatopancreas tissue from warm-acclimated males and females. Males failed to survive at 5°C, 5 o/oo long enough for experimentation. There was, however, a marked difference in metabolic response of gill tissue from warm-acclimated male and female crabs subjected to low temperature in combination with high salinity (5°C, 30 o/oo) (Fig. 5). Under such conditions the metabolic rate of gill tissue from control and

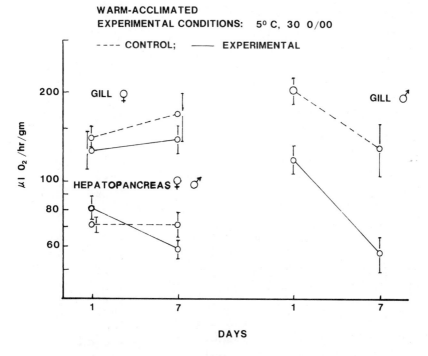

Fig. 5. *Metabolic rates of gill and hepato-pancreas tissue from warm-acclimated adult crabs maintained at 5°C, 30 o/oo, with and without 0.18 ppm Hg. Vertical lines indicate ± 1 standard error.*

experimental female crabs remained relatively constant through day 7, while gill rates from control
and experimental male crabs decreased sharply over
the same time period. Furthermore, in contrast to
females, the oxygen uptake rates of gill tissue from
mercury-exposed males decreased more rapidly and was
significantly lower than that of control male gills.

Behavior

Qualitative observations of fiddler crabs in
sublethal concentrations of mercury indicated that
after several weeks' exposure, sluggishness and lack
of responsiveness set in, followed by a state of
torpor before death (DeCoursey, unpublished data).
Experiments were undertaken using locomotor activity
as a measure of sublethal Hg effects on normal functions at optimum temperature-salinity conditions.
The total amount of daily activity was first considered. Several difficulties were encountered;
feeding disturbances often induced large activity
artifacts. Even after changes in feeding techniques
proved successful in reducing these artifacts, the
large interindividual differences in amount of activity were problematical. The comparison of control
and experimental groups would require a large sample
size to overcome the variability that might mask
treatment differences. On the other hand, consecutive daily totals for individuals seemed stable
enough to serve as a control for a subsequent Hg-
exposure period. Therefore, a series of experiments
was initiated consisting of a 2-week control period
in 30 o/oo untreated seawater, followed by a 2-week
period in 0.18 ppm Hg. Mean activity levels in one
example (Fig. 6) indicate a 38% reduction of activity
following mercury treatment. The average change in
mean activity for individual crabs during Hg treatment ranged from 5% increase to 38% reduction; mean
reduction for the entire group was 23%.
Secondly, the daily patterns of activity were
considered. Print-out records were used to construct

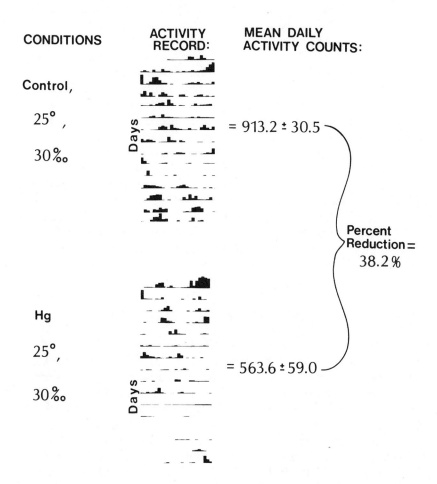

CONDITIONS   ACTIVITY RECORD: MEAN DAILY ACTIVITY COUNTS:

Control, 25°, 30‰   = 913.2 ± 30.5

Percent Reduction = 38.2%

Hg 25°, 30‰   = 563.6 ± 59.0

Fig. 6. Activity of one adult Uca pugilator *in a revolving drum actograph for 15 days in untreated seawater under constant dim illumination (above), followed by 16 days in 0.18 ppm Hg (below). Actograms are constructed with 24-hr scans mounted vertically in chronological sequence. Mean daily counts (right) indicate one-half revolutions of the 6-inch drum.*

actograms of activity with hourly values arranged in daily scans (Fig. 6). Under laboratory conditions of constant temperature (25°C) and continuous dim lighting, in untreated seawater (30 o/oo), activity scans were highly variable. Often no clear-cut marker point was discernable to use for frequency analysis. Instead, a high level of background activity often obscured rhythmic events. Some records suggested tidal components (Fig. 6), other crabs appeared circadian in frequency, while some were almost continuously active (Rawson and DeCoursey, 1974). Therefore, a computer program for periodogram analysis (Suter and Rawson, 1968) modified from Enright (1965) was used to detect dominant frequencies of the activity rhythm during the control and mercury periods. When a rhythm was detectable during the control period, the periodogram analysis indicated that no change in frequency was induced by Hg treatment.

Electronmicroscope Study of Tissue Anatomy

Electron microscope studies of the tissues of adults treated for 42 days in 9×10^{-7}M $HgCl_2$ revealed the sites of concentration and tissue damage. Mercury was found primarily in the gills, green glands, and hepatopancreas, with highest concentration in the gills. The mercury caused extensive alteration of the ultrastructure of the gill filaments (Fig. 7A). Normally, the filaments are characterized by tightly packed epithelial cells (Fig. 7B). The basal plasma membranes are thrown into folds that penetrate the cell almost to their apical surfaces. Numerous mitochrondria are found localized within the folds. In gill tissue from crabs maintained in the sublethal concentration of mercury, the filaments showed less cytoplasmic protein, disappearance of membrane folds, and decreased number of mitochrondria. Swelling and loss of the cristae of the mitochondria were also observed (N. Watabe and W. B. Vernberg, unpublished).

400

LARVAL STUDIES

Larval studies centered on the synergistic effects of Hg or Cd combined with temperature and salinity stress. Zoeal Stages I, III, V and megalopa stages were selected for study; measurement parameters included survival, metabolism, and behavior.

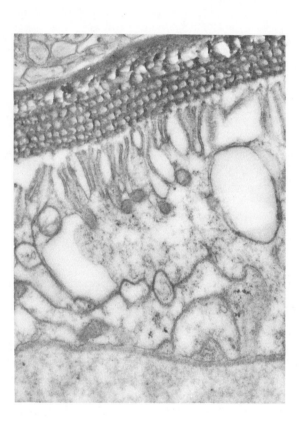

Fig. 7A. Ultrastructure of gill filaments of adult Uca pugilator, *maintained for 6 weeks under an optimal thermal-salinity regime in seawater containing 0.18 ppm Hg.*

B

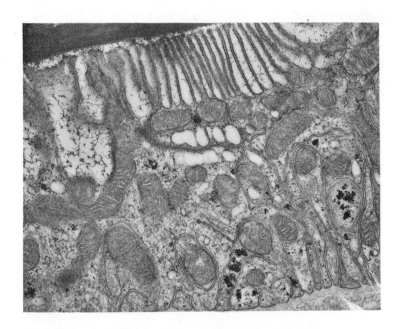

*Fig. 7B. Ultrastructure of gill filaments
of adult* Uca pugilator, *maintained for
6 weeks under an optimal thermal-salinity
regime in untreated seawater.*

Effect of Hg at Three Sublethal Concentrations under Optimum Conditions

Survival experiments were first carried out with three concentrations of mercury (DeCoursey and Vernberg, 1972) in an optimum salinity-temperature regime (25°C, 30 o/oo). The highest concentration selected, 9×10^{-7}M $HgCl_2$ (0.18 ppm Hg), was sublethal to adults over a period of 6 weeks; the two lower concentrations, 9×10^{-9}M $HgCl_2$ (1.8 ppb Hg) and 9×10^{-11}M $HgCl_2$ (0.0018 ppb Hg) are concentrations within the range reported for certain polluted estuaries (Klein and Goldberg, 1970). One hundred zoeae were set up in finger bowls for each of the three concentrations. A concentration of 9×10^{-7}M $HgCl_2$ quickly proved fatal to Stage I zoeae; 50% survival time was

less than 24 hrs. The two lower concentrations of
mercury also markedly affected survival, for survival
time was considerably reduced in comparison to control values: 8 days in $9 \times 10^{-9}M$ $HgCl_2$, or 11 days
in $9 \times 10^{-11}M$ $HgCl_2$, in contrast to 18 days in untreated seawater (Fig. 8).

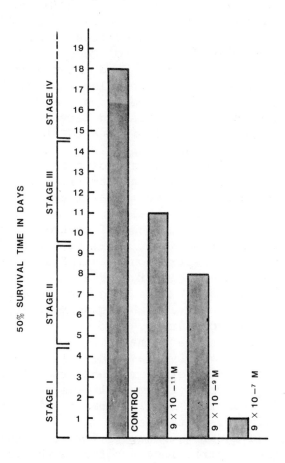

Fig. 8. *50% survival time values for zoeae*
reared in untreated seawater or in three
concentrations of HgCl_2. The approximate
number of days for each stage of larval
development are indicated by brackets.
(From DeCoursey and Vernberg, 1972.)

Additional viability experiments confirmed the findings of the above results (Fig. 8). Experiments with 500 newly hatched (Stage I) zoeae from different hatches suggested considerable variability in survival time in 9×10^{-7}M $HgCl_2$, ranging from 5 to 72 hrs. Furthermore, when control-reared Stage III or Stage V zoeae were placed in 9×10^{-7}M $HgCl_2$, they appeared even more sensitive. Whereas a few first stage and occasional third stage zoeae were able to survive for the 24-hr period, no Stage V larvae were alive after 24 hrs; most were dead after 6 hrs.

Rearing success statistics also support the above results. Survival to megalopa stage was definitely reduced for mercury-treated larvae: 6 of 100 in 9×10^{-9}M $HgCl_2$, 3 of 100 in 9×10^{-11}M $HgCl_2$, but 20 of 100 in the control group.

There was no immediate metabolic response in Stage I zoea to any of the concentrations of mercury. Oxygen uptake rates of these larvae 1 hr after exposure to the three experimental concentrations of $HgCl_2$ were unchanged from those of control larvae. A slightly longer exposure of 6 hrs to 9×10^{-7}M $HgCl_2$ markedly depressed respiration rates of all stages tested, although the greatest decrease in metabolic rate occurred in Stage V zoea (Fig. 9A), where the rate of mercury-exposed larvae was approximately one-third that of control larvae.

Observations on the normal swimming behavior of Stages I, III, and V served as a base line for detecting effects of mercury exposure (DeCoursey and Vernberg, 1972). Using the maxillipeds, the control zoeae usually swam in a fairly straight line. This type of swimming was interspersed with a variable amount of "tail-lashing maneuvers," which resulted in a rapid change of direction, or often a whirling type of locomotion. Stage I zoeae usually swam in a start and stop fashion, while Stage III zoeae, with a marked increase in size and complexity of the maxillipeds, were strong, steady swimmers. Stage V zoeae, which had increased greatly in weight with little further development of the maxillipeds, were relatively

404

slow, sluggish swimmers, often hovering close to the substrate. Such stage-dependent differences are reflected quantitatively in the rate of swimming (Fig. 9B).

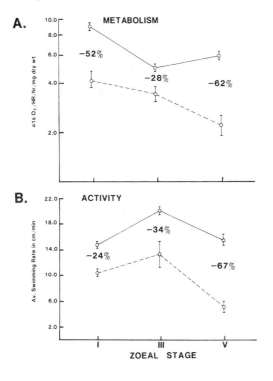

Fig. 9. Percent change in metabolic rates (A), and swimming activity (B) of zoeal Stages I, III, and and V after 6-hr exposure to a mercury concentration of 9 × 10^{-7}M $HgCl_2$. ● = control zoeae, O = experimental zoeae. Vertical lines indicate ± 1 standard error. (From DeCoursey and Vernberg, 1972.)

When zoeae were first introduced into the mercury, they often remained motionless for several seconds on the bottom of the dish, then darted about erratically with considerable tail lashing for several minutes before adopting a characteristic swimming pattern. As the effects became more pronounced with time, the zoeae manifested marked swimming abnormalities, such as spiral swimming, swimming on their sides, or settling slowly to the bottom, followed by disoriented twitching movements. This was reflected by total swimming distance/unit time (Fig. 10).

The effect of mercury solutions on normal activity was assayed at regular intervals after the start of exposure by tracking the actual swimming path in order to determine rate of activity. Exposures of 6 hrs to 9×10^{-7}M $HgCl_2$ reduced the swimming rate of all larvae, with a greater effect on Stage V larvae.

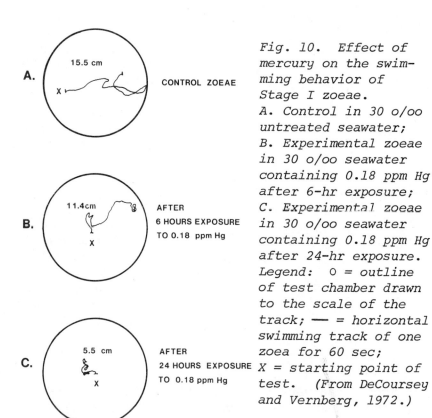

A. 15.5 cm — CONTROL ZOEAE
X

B. 11.4cm — AFTER 6 HOURS EXPOSURE TO 0.18 ppm Hg
X

C. 5.5 cm — AFTER 24 HOURS EXPOSURE TO 0.18 ppm Hg
X

Fig. 10. Effect of mercury on the swimming behavior of Stage I zoeae.
A. Control in 30 o/oo untreated seawater;
B. Experimental zoeae in 30 o/oo seawater containing 0.18 ppm Hg after 6-hr exposure;
C. Experimental zoeae in 30 o/oo seawater containing 0.18 ppm Hg after 24-hr exposure.
Legend: O = outline of test chamber drawn to the scale of the track; ▬ = horizontal swimming track of one zoea for 60 sec; X = starting point of test. (From DeCoursey and Vernberg, 1972.)

As with metabolism, activity measurements suggest that the older stages are more sensitive to mercury than the newly hatched larvae (Fig. 9B). For the two more dilute mercury solutions, the rate of swimming was changed relatively little; it was depressed to some extent for Stage I and elevated slightly for the older stages.

Interactions of Hg, Temperature and Salinity

Since the larvae inhabit estuaries where temperature-salinity conditions are often suboptimal, the next series of experiments considered effects of temperature-salinity stress within the range encountered in nature, with and without a low level concentration of mercury on three parameters: survival, phototactic response, and O_2 consumption of the larvae (Vernberg et al., 1973).

After hatching, groups of control and Hg-treated larvae ($9 \times 10^{-9}M$ $HgCl_2$ = 1.8 ppb Hg) were reared in each of the following temperature-salinity conditions:

1. $30^\circ C$, 30 o/oo
2. $30^\circ C$, 20 o/oo
3. $20^\circ C$, 30 o/oo
4. $20^\circ C$, 20 o/oo

Viability of larvae under the various regimes was evaluated by means of 96-hr mortality rates. A total of 2400 larvae were used with 300 in each of the eight regimes, with mortality checked daily. The 96-hr mortality studies are summarized in Figure 11. In all of the environmental regimes except $30^\circ C$, 30 o/oo, suboptimal regimes caused an increase in mortality over that of larvae maintained under optimal temperature-salinity conditions; high temperatures were especially stressful. Under conditions of high temperature ($30^\circ C$) and optimum salinity, there was no difference between the controls and mercury-exposed zoeae in the 96-hr survival. However, larvae maintained at high temperature and in low salinity showed a 27% increase in mortality over larvae not exposed to mercury. At low temperature there was a marked increase in mortality at both optimal and low salinity with the addition of mercury. Percentage mortality data were analyzed by means of a factorial design with three factors: temperature (T), salinity (S), and mercury (Hg). Specifically, the design was a 2^3 factorial with three replications of 100 larvae

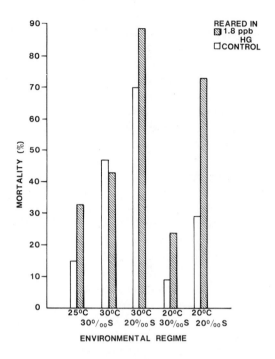

Fig. 11. Effects of optimal and suboptimal temperature-salinity regimes with and without added Hg on the 96-hr survival time of larval Uca pugilator. *(From Vernberg et al., 1973.)*

each, making a total of 24 observations. The levels of the factors were: temperature, 20°C and 30°C; salinity, 20 o/oo and 30 o/oo; and mercury, 0 ppb and 1.8 ppb. Thus, the 24 observations may reasonably be considered as continuous responses of a function of the three factors and interactions. Since the observations are treated as percentage measurements generated by data from binomial populations, the transformation $y = \arcsin \sqrt{x}$, where x is observed percent mortality, is appropriate to stabilize variances (Mendenhall, 1968).

The analysis of variance for this data indicated the following factors had significant effects at the 5% level:

$$T, \; S, \; Hg, \; T \times Hg, \; S \times Hg$$

The interaction of $T \times S$ was not significant.

The metabolic rates of first-stage larvae reared under less than optimal regimes without added mercury were significantly decreased over that of larvae reared under optimal conditions (Fig. 12). The addition of mercury markedly affected the respiration rate of larvae reared under all temperature-salinity regimes, including optimal ones, except at 30°C, 30 o/oo. The effect seemed to be temperature dependent: at the higher temperatures, 25°C and 30°C,

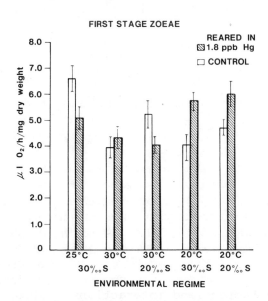

FIRST STAGE ZOEAE

Fig. 12. Metabolic rates of first-stage zoeae of Uca pugilator *reared under optimal and suboptimal temperature regimes with and without addition of Hg. Vertical lines = ± standard error. (From Vernberg et al., 1973.)*

409

mercury depressed metabolic rates; at 20°C, mercury enhanced metabolism in either optimal or suboptimal salinities.

Larvae reared to the third stage under an environmental regime of 20°C, 30 o/oo did not significantly differ metabolically from larvae reared under an optimum temperature-salinity regime (Fig. 13).

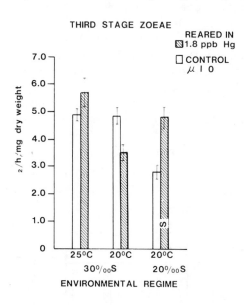

Fig. 13. Metabolic rates of third-stage zoeae of Uca pugilator *reared under optimal and suboptimal temperature regimes with and without addition of Hg. Vertical lines = ± standard error. (From Vernberg et al., 1973.)*

When the low level concentration of mercury was added, however, the metabolic rate of larvae reared under optimal temperature and salinity increased significantly, while larvae reared under a suboptimal regime showed a significant decrease in metabolism. The metabolic rate of larvae reared under the suboptimal temperature-salinity regime of 20°C, 20 o/oo decreased 56% over that of larvae reared under optimum

410

conditions; with the addition of mercury to the suboptimal regime, oxygen consumption rates of the larvae
were heightened markedly. Too few zoea survived at
30°C to permit experimentation.

The phototactic responses of Stages I, III, V
zoeae, and also megalops plus early crab stages were
measured in the light gradient apparatus. Under optimum conditions (25°C, 30 o/oo salinity, untreated
seawater), all zoeal stages tested were markedly
photopositive; most megalops were also photopositive,
but early crab stages appeared to be indifferent to
light (Vernberg et al., 1973). Paralleling the 96-hr
mortality and metabolism studies on synergistic
effects described immediately above, groups of larvae
were reared in the specified temperature-salinity Hg
conditions, and the phototactic response then measured. Considerable variability was found for the
Stage I groups. Therefore, the data were analyzed
statistically using 2^3 factorial design with the same
factors (temperature, salinity, mercury) and levels
as described above for the 96-hr mortality study.
The same assumptions are reasonable and the response
measured was the percent photopositive. All animals
in the anterior five segments of the gradient tube
were considered photopositive, and those in the
posterior five segments, photonegative. The observed
percentages were again transformed by $y = \arcsin x$,
where x was the observed percentage. The analysis of
variance for this data indicates that the following
effects were significant:

At the 5% level: $T \times Hg$, $S \times Hg$
At the 10% level: Hg (and $T \times Hg$, $S \times Hg$)

The interaction $S \times T$ was not significant.

Larvae reared to the third stage under suboptimal
conditions showed marked changes in phototactic response. At 20°C, in a salinity of either 20 o/oo or
30 o/oo, the photopositive response was sharply
decreased (Fig. 14). Mercury-exposed larvae reared
under these regimes were much more photopositive than
control ones. Too few Stage III experimental units

were available for a complete analysis of variance.
Chi-square analysis for the two levels of temperature
and the two levels of Hg (at a salinity of 30 o/oo)
indicates an interaction between T and Hg at the
5% significance level. At a temperature of $20°C$, a
repetition of the analysis at the two levels of Hg
(0 ppb and 1.8 ppb) and the two levels of salinity
(20 o/oo and 30 o/oo) indicated that the interaction
of S with Hg was not significant.

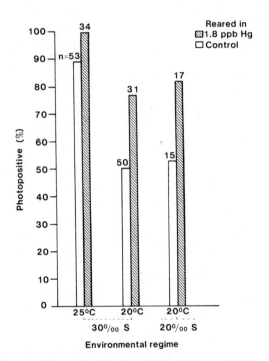

Fig. 14. *Phototactic response of third-stage zoea of* Uca pugilator *with and without addition of Hg.* n = number of zoea. *(From Vernberg* et al.*, 1973.)*

Interactions of Cd, Temperature and Salinity

The final part of the larval studies concerned
the effect of rearing in sublethal cadmium (1 ppb) on

survival, metabolism and swimming rate of zoeae. The
synergistic effects of temperature, salinity, and
cadmium on 96-hr mortality of zoeae were evaluated.
Estimation of percentage mortality of *U. pugilator*
was based on a response surface fitted to observed
mortality under 13 combinations of salinity and temperatures with and without the addition of cadmium.
Tolerance levels of the cadmium-exposed zoea were
narrowed (Fig. 15A and B); salinity tolerances were
not as great and the temperature range was reduced
and shifted downward. After 96 hrs, 80% of the control larvae survived temperatures of 21-28°C and
salinities of 23-34 o/oo. In experimental larvae,
these ranges were 18-23°C and 24-29 o/oo (Fig. 15A
and B). In the predictive equation where temperature
and salinity were used as predictors of mortality,

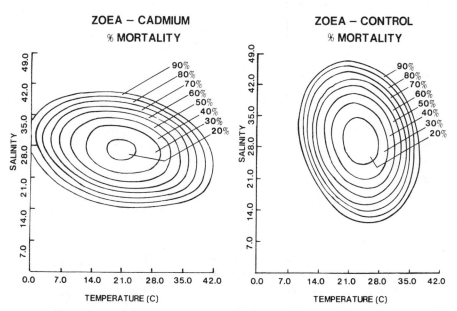

*Fig. 15. Estimation of percentage mortality of
first-stage* Uca pugilator *zoea based on response
surface fitted to observed mortality under
13 combinations of salinity and temperature
(a) with, and (b) without the addition of 1 ppb Cd.*

temperature (T_1 and T_2) was the most important cause
of death among control zoeae, but salinity (S_1 and S_2)
was the most important factor among the experimental
ones.

The effect of 1 ppb cadmium upon metabolism of
zoeae under optimum temperature-salinity conditions
is summarized in Figure 16A. Cadmium elevated respiration of Stages I and III zoeae but greatly depressed
the rate in Stage V zoeae.

Swimming rates of control and Cd (1 ppb) reared
zoeae at optimum salinity and temperature conditions
were determined for zoeal Stages I, III, and V.
Cadmium resulted in a decrease of activity at all
stages, but the difference was significant only for
Stage I larvae (Fig. 16B). Control rates paralleled
the earlier Hg controls with an increase in average
rate in Stage II, and gradual decrease to Stage V as
the larvae grew larger.

DISCUSSION

The sea contains trace amounts of many heavy
metals, some of which are essential for normal growth
in marine organisms. However, in higher concentrations certain metals, including mercury and cadmium,
are highly toxic. The serious consequences of environmental contamination with mercury have been recognized only within the past few years. Events such as
the Minimata tragedy in Japan in the 1950s (Kurland
et al., 1960) and the linkage of mercury poisoning
with declining bird populations in Sweden (Otterlind
and Linnerstedt, 1964) focused attention on the very
real problem of mercury pollution in the ecosystem.
More recently a comprehensive review of the occurrence and general effects of mercury in the ecosystem
has been published by Peakall and Lovett (1972).
These authors have speculated that, since the ocean
contains approximately four orders of magnitude more
mercury than man adds annually, the concentration in
the ocean is unlikely to be affected. Freshwater

414

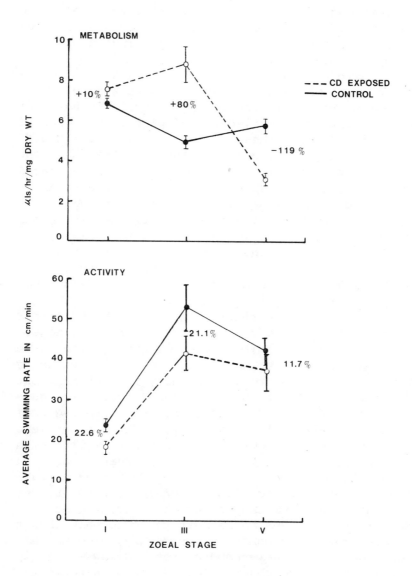

Fig. 16. *Percent change in metabolic rates (A), and swimming activity (B) of zoeal Stages I, III, and V of* Uca pugilator *after rearing in 1 ppb cadmium.*

systems and estuaries, however, receive much of the
mercury released from industrial processes, including
chlor-alkali production, electroplating and paper
manufacturers.

Toxic cadmium salts which contribute to the widespread environmental pollution (McKee and Wolf, 1963)
are also frequently found in the waste discharge of
electroplating plants, chemical plants, or lead and
zinc mines. From these sources cadmium is flushed
down to the estuarine and marine environments. The
importance of cadmium pollutants has been indicated
by its association with the crippling "itai-itai"
disease of Japan (Kobayashi, 1971). Although the
effects of cadmium have been investigated for numerous
freshwater organisms (Doudoroff and Katz, 1953; Ball,
1967; Mount and Stephan, 1967), only Gardner and
Yevich (1970), Jackim *et al.* (1970), and recently
Eisler (1971) have examined the effects of cadmium on
estuarine forms. Eisler alone has reported the
effects of normal variations in salinity and temperature on the toxic effect of cadmium in mummichogs.

Adult crabs can tolerate relatively high concentrations of mercury and cadmium for long periods of
time when temperature and salinity are optimal, but
under a stressful temperature-salinity regime, survival time is considerably shortened (Vernberg and
Vernberg, 1972). In this study, fiddler crabs were
found to withstand mercury in combination with high
temperature and low salinity better than in combination with low temperature and low salinity; mortalities at both low and high temperatures were greater
than under optimum temperature-salinity regimes.
Mercury also proved to be more toxic to male than to
female crabs. In contrast to mercury, cadmium was
most toxic at higher temperatures and low salinities,
and there were no observed differences in toxicity
between males and females. In other species of animals the influence of temperature on toxicity of
pollutants is not clear. Sprague (1970) stated that
"no assumptions should be made about temperature
effects on toxicity." Only slight temperature effects

have been reported for toxicity of naphthenic acid
and zinc to fish (Cairns and Scheier, 1957, 1962).
On the other hand, there appears to be an inverse
correlation between temperature and toxicity with
animals exposed to chlorinated hydrocarbons (Elson,
1967; Anderson, 1968).

With both metals, death of the fiddler crabs
probably is related to the accumulation of metal in
the gills with an associated breakdown in osmoregulatory or respiratory functions. There are, however,
major differences between the uptake and accumulation
of these two metals. Total mercury uptake is independent of the thermal-salinity regime, but at low
temperature mercury is not translocated away from the
gills to the hepatopancreas, thus leaving high mercury
residues in the gill. Similarly, mercury was not
translocated from the gills to the hepatopancreas
under any temperature-salinity conditions in crabs
that were simultaneously exposed to mercury and cadmium. Cadmium uptake levels, on the other hand, are
strongly dependent on temperature-salinity conditions.
At each temperature, crabs in low salinity water accumulated more cadmium than those exposed to high
salinity (Table 1).

One of the most intriguing puzzles in these studies has been the greater sensitivity of the male to
mercury poisoning in comparison to the female, both
at high and low temperature. It cannot be explained
on the basis of greater mercury uptake by tissue of
male crabs under stressful environmental conditions,
for these rates were essentially the same in tissues
of male and female crabs (Vernberg and O'Hara, 1972).
However, one clue lies in the difference in metabolic
responses of male and female crabs, both at the whole
animal and the tissue level. Under conditions of
thermal and salinity stress without the addition of
mercury the metabolic rate of female crabs tended to
be more stable and less depressed than the rate of
male crabs. The addition of mercury to the already
stressful conditions accentuated these differences.
The differential response of males and females to the

417

temperature-salinity-mercury stress may reflect neuro-endocrine differences.

At the tissue level, metabolic measurements of the tissues from crabs exposed to different combinations of temperature and salinity again indicate that high temperature and low salinity regimes are less stressful than low temperature and low salinity. Although mercury does alter to some extent the metabolic response of tissues from crabs exposed to mercury in combination with high temperature and low salinity, these alterations do not appear until after a relatively long exposure. In contrast, at low temperature in combination with either low or optimum salinities, alterations in tissue metabolic response are frequently apparent after a 1-day exposure to mercury.

The high variability of activity in adult fiddler crabs, and the time and expense involved in semi-automatic recording, raise serious problems in development of an activity bioassay. Probably the most feasible method of activity assaying would be to use large numbers of crabs maintained for a fixed period of time in either control seawater or treated seawater. Activity could then be measured for a relatively short period in the actographs and mean values obtained for the control and experimental groups. This approach has not yet been tested.

Larval stages of *U. pugilator* were several orders of magnitude more sensitive to mercury than was the adult stage. At a concentration of Hg sublethal to the adults, larvae could survive a scant 24 hrs. Sensitivity of the larvae to acute mercury exposure increased as the larvae developed. Although a few of the larvae were able to live and grow in dilute mercury solutions with only slight changes in activity or metabolism, the viability tests showed that the normal attrition rate of developing control zoeae was greatly accelerated in mercury-stressed larvae. The small proportion of larvae to reach Stage V in 9×10^{-9} (1.8 ppb Hg) or 9×10^{-11}M $HgCl_2$ (0.0018 ppb Hg) doubtless represent highly resistant individuals. The

locomotor behavior of larval *Uca pugilator* was also
modified by mercury. With three concentrations of
mercury, swimming activity of all stages was modified
in direct relationship to the concentration and duration of exposure. Such results suggest that marine
invertebrate larvae may be considerably more sensitive
to mercury pollution than previously suspected.
Similarly, differences in tolerance to mercuric acetate have been found in stages of the life cycle of
the fish *Oryzias latipes* where embryos were more sensitive than larvae, and these in turn were more sensitive than adults; the least sensitive was the egg
stage (Akiyama, 1970).

Despite the fact that there are wide fluctuations
in temperature and salinity in the estuary, numerous
mortality studies have shown that most temperate zone
species of crustacean larvae develop over a rather
limited temperature-salinity range (Costlow and
Bookhout, 1962, 1971; Costlow *et al.*, 1966; Vernberg
and Vernberg, 1974). Our survival data on larvae of
U. pugilator reared under suboptimal regimes demonstrated that warm water and low salinity were particularly harmful to developing larvae. Mortality at
high temperature (30°C) was greatly increased over
25°C (optimum) values, and only a few *U. pugilator*
larvae underwent development to the crab stage under
low salinity regimes regardless of the temperatures.
At low temperatures, mercury sharply increased mortality rates. However, at high temperatures control
mortalities were so great without mercury that the
added stress of mercury did not appreciably alter
mortality levels.

Physiological responses of larvae also reflect
the stresses of suboptimal conditions. The immediate
effect of reduced temperature (20°C) with optimum
salinity was to depress the metabolic rate of the
zoeae, although the animals did acclimate with time
to the lower temperature. This acclimation is evi-
denced by the fact that the metabolic rate of third-
stage larvae reared at 20°C was the same as the rate
in these larvae reared at 25°C. In a combination of

low temperature and low salinity, however, the zoeae did not show metabolic acclimation. These studies can be correlated with the survival studies which indicated that low temperature and low salinity regimes were especially stressful. The marked alteration of respiration rates in larvae exposed to mercury could appreciably affect the ability of the larvae to compete in the estuarine environment. For example, the depressed respiratory rate of mercury-exposed Stage I zoea could be reflected in decreased ability to capture food or escape predators.

The phototactic response of Stage I *Uca* larvae was strongly photopositive regardless of the salinity-temperature-Hg regime, while the phototactic response of Stage III larvae depended upon the environmental regime. In third-stage larvae reared at 20°C, the photopositive response of the control group was considerably reduced compared to the mercury-treated group. Salinity was not a critical factor since the phototactic response was essentially the same in larvae reared at either 30 o/oo or 20 o/oo. The phototactic responses of the larvae of many intertidal zone animals are highly adaptive, bringing them into phytoplankton rich areas for growth and development (Thorson, 1964). In view of the importance of the phototactic response in larval development, any modifications might seriously impair chances for survival of the larvae.

Although cadmium did increase mortality of zoeae over those of control larvae, it was not as toxic as mercury. Under optimum conditions cadmium-exposed zoeae survived to the megalops stage in the same number as control ones, and it was only under suboptimal conditions that mortalities increased.

Curiously, under optimum temperature-salinity conditions the metabolic rates of zoea reared in cadmium were more affected than were those of zoea reared in mercury. The metabolic rates of mercury-reared zoeae were essentially the same as those of control zoeae. In cadmium-reared larvae, however, both Stage III and Stage V zoea were markedly different from

420

that of control larvae. Such results suggest again
that pollutants can modify normal physiological func-
tioning, thereby reducing changes for survival in
nature, but the mode of action of a heavy metal contaminant may well vary with the metal involved.

SUMMARY

The effect of both cadmium and mercury on *Uca
pugilator* depends on a number of factors including
stage of the life cycle, sex, thermal history, and
environmental regime. The two metals do not neces-
sarily affect *Uca* in the same manner. Larvae are
several orders of magnitude more sensitive to mercury
than are adult stages, and adult males are more sensitive to mercury stress than are females. Mercury
is most toxic at low temperatures and low salinity.
Warm-acclimated animals (summer animals) are less
tolerant of mercury at low temperatures than cold-
acclimated (winter) ones, and concentrations of mercury that are sublethal under optimum conditions of
temperature and salinity become lethal when
temperature-salinity regimes become stressful. Distribution of mercury in the tissues of the crab is
dependent on the environmental regime, but the total
body burden is not.
    Adult crabs are also less sensitive to cadmium
poisoning than are larvae, but there were no observed differences in mortalities between males and
females. Cadmium is most toxic at high temperature
and low salinity, and both the distribution and total
body burden of cadmium is dependent upon environmental
conditions.

ACKNOWLEDGMENTS

These studies were supported in part by Grants
18050FYJ and 18080FYI from the Environmental Protection Agency and also by the Belle W. Baruch Institute

421

for Marine Biology and Coastal Research. The authors
express appreciation to the staff of the Institute
for dedicated care and assistance in the conduct of
the experiments and the preparation of the manuscript.

LITERATURE CITED

Akiyama, A. 1970. Acute toxicity of two organic
   mercury compounds to the teleost, *Oryzias
   latipes*, in different stages of development.
   Bull. Jap. Soc. Sci. Fish. 36:563-70.
Anderson, J. M. 1968. Effect of sublethal DDT on
   the lateral line of brook trout *Salnelinus
   fontinalis*. *J. Fish. Res. Bd. Canada 25*:2677-82.
Ball, I. R. 1967. The toxicity of cadmium to
   rainbow trout (*Salmo gairdnerii* Richardson).
   Water Res. 1:805-806.
Bayne, B. L. 1964. The responses of the larvae of
   Mytilus edulis (L.) to light and to gravity.
   Oikos 15:162-74.
Cairnes, J. and A. Scheier. 1957. The effects of
   temperature and hardness of water upon the
   toxicity of zinc to the common bluegill (*Lepomis
   macrochirus* Raf.). *Notulae Natur. 299*:1-12.
_____ and _____. 1962. The effects of
   temperature and water hardness upon the toxicity
   of naphthenic acids to the common bluegill
   sunfish *Lepomis macrochirus* Raf., and the pond
   snail *Physa heterostropha* Say. *Notulae Natur.
   353*:1-12.
Costlow, J. D. Jr. and C. G. Bookhout. 1962. The
   larval development of *Sesasma reticulatum* Say
   reared in the laboratory. *Crustaceana 4*:281-94.
_____ and _____. 1971. The effect of
   cyclic temperatures on larval development in the
   mud-crab *Rhithropanopeus harrisii*. In: *Fourth
   European Marine Biology Symposium*, pp. 211-20,
   ed. by D. J. Crisp. London: Cambridge University Press.

_____, _____, and R. J. Monroe. 1966.
Studies on the larval development of the crab,
Rhithropanopeus harrisii (Gould). I. The effect
of salinity and temperature on larval development. *Physiol. Zool. 39*:81-100.

DeCoursey, P. J. and W. B. Vernberg. 1972. Effect
of mercury on survival, metabolism, and behavior
of larval *Uca pugilator. Oikos 23*:241-47.

Doudoroff, D. and M. Katz. 1953. Critical review of
literature on the toxicity of industrial wastes
and their components to fish. II. The metals,
as salts. *Sewage Ind. Wastes 25*:802-39.

Eisler, R. 1971. Cadmium poisoning in *Fundulus
heteroclitus* (Pisces: Cyprinodontidae).
J. Fish. Res. Bd. Canada 28:1225-34.

Elson, P. F. 1967. Effects on wild young salmon of
spraying DDT over New Brunswich forests.
J. Fish. Res. Bd. Canada 24:731-67.

Enright, J. T. 1965. The search for rhythmicity in
biological time series. *J. Theor. Biol. 8*:
426-68.

Gardner, G. R. and P. O. Yevich. 1970. Histological
and hematological response of an estuarine
teleost to cadmium. *J. Fish. Res. Bd. Canada 27*:
2185-96.

Hyman, O. W. 1920. The development of *Gelasimus*
after hatching. *J. Morp. 33*:484-525.

Jackim, E., J. M. Hamlin, and S. Sonis. 1970.
Effects of metal poisoning on five liver enzymes
in the killifish *(Fundulus heteroclitus).
J. Fish. Res. Bd. Canada 27*:383-90.

Klein, D. H. and E. D. Goldberg. 1970. Mercury in
the marine environment. *Environmental Science
and Technology 4*:765-68.

Kobayashi, J. 1971. Relation between "itai-itai"
disease and the pollution of river water by
cadmium from a mine. In: *Advances in Water
Pollution Research, 1970*, Vol. 1, pp. 1-25,
ed. by S. H. Jenkins. Oxford: Pergamon Press.

Kurland, L. T., S. N. Faro, and H. S. Siedler. 1960.
Minimato disease. *World Neurol. 1*:320-25.

McKee, J. E. and H. W. Wolf. 1963. *Water Quality Criteria*, 2nd ed. California State Water Quality Control Board, Publ. 3-A.

Mendenhall, W. 1968. *Introduction to Linear Models and the Design and Analysis of Experiments.* Belmont, Calif.: Wadsworth Publishing Company.

Mount, D. I. and C. E. Stephan. 1967. A method for detecting cadmium poisoning in fish. *J. Wildl. Manage. 31*:168-72.

O'Hara, J. 1973. The influence of temperature and salinity on the toxicity of cadmium to the fiddler crab, *Uca pugilator. Fish. Bull. 71*: 149-53.

_____ and W. B. Vernberg. 1973. Effect of simultaneous exposure to cadmium and mercury on uptake rates in *Uca pugilator*. In manuscript.

Otterlind, G. and I. Linnerstedt. 1964. Avifauna and pesticides in Sweden. *Var. Fagelvarld. 24*: 363-415.

Peakall, D. B. and R. J. Lovett. 1972. Mercury: Its occurrence and effects in the ecosystem. *Bio. Sci. 22*:20-25.

Rawson, K. S. and P. J. DeCoursey. 1974. A comparison of the rhythms of mice and crabs from intertidal and terrestrial habitats. In: *Biological Rhythms in the Marine Environment*, ed. by P. J. DeCoursey. Columbia, S. C.: University of South Carolina Press. In press.

Ryland, J. S. 1960. Experiments on the influence of light on the behavior of polyzoan larvae. *Exp. Biol. 37*:783-800.

Simpson, G. G., A. Roe, and R. C. Lewontin. 1960. *Quantitative Zoology.* New York: Harcourt Brace Company.

Sprague, J. B. 1970. Measurements of pollutant toxicity to fish. II. Utilizing and applying bioassay results. *Water Res. 4*:3-32.

Suter, R. B. and K. S. Rawson. 1968. Circadian activity rhythm of the deer mouse, *Peromyscus*: effect of deuterium oxide. *Science 160*:1011-14.

424

Thorson, G. 1964. Light as an ecological factor in the dispersal and settlement of larvae of marine bottom invertebrates. *Ophelia 1*:167-208.

Vernberg, F. J. and W. B. Vernberg. 1974. Adaptations to extreme environments. In: *Physiological Ecology of Estuarine Organisms*, ed. by F. J. Vernberg. Columbia, S. C.: University of South Carolina Press. In press.

Vernberg, W. B., P. J. DeCoursey, and W. J. Padgett. 1973. Synergistic effects of environmental variables on larvae of *Uca pugilator* (Bosc). *Mar. Biol. 22*:307-12.

_____ and J. O'Hara. 1972. Temperature-salinity stress and mercury uptake in the fiddler crab, *Uca pugilator. J. Fish. Res. Bd. Canada 29*: 1491-94.

_____ and F. J. Vernberg. 1972. The synergistic effects of temperature, salinity and mercury on survival and metabolism of the adult fiddler crab, *Uca pugilator. Fish. Bull. 70*:415-20.

SOME PHYSIOLOGICAL CONSEQUENCES OF POLYCHLORINATED BIPHENYL- AND SALINITY-STRESS IN PENAEID SHRIMP

D. R. NIMMO and L. H. BAHNER

Environmental Protection Agency
Gulf Breeze Environmental Research Laboratory
Gulf Breeze, Florida 32561

Estuaries are dynamic environments where there are many factors that fluctuate, such as temperature, salinity, currents, hydrostatic pressure, and oxygen or carbon dioxide concentrations. Unfortunately, domestic sewage (nutrients), oils, industrial chemicals, pesticides, metals, or altered temperatures are an influence in estuaries. The combined effects of the natural and man-introduced factors are largely unknown. In contrast, these interactions could adversely affect the biota of an estuary before such a trend was recognized. Therefore, one of the major problems facing us today is understanding and predicting the interactions of pollutants and natural stresses.

It is common knowledge that the commercial shrimps along the Gulf Coast undertake distinct euryhaline migrations. After adult shrimp spawn in the open Gulf from spring to fall, the post-mysids and juveniles migrate into the fresher waters of bays where they grow rapidly to adulthood before returning to the Gulf. Obviously, these stages of

shrimp must be able to adjust to the changing salinities encountered in the estuary, and any factor
diminishing the ability of the shrimp to adjust physiologically to these changes would have a detrimental
effect on them.

One group of chemicals introduced by man that
has recently been of concern to many ecologists is the
PCBs, or polychlorinated biphenyls. In 1969, a PCB,
identified as Aroclor ® 1254[a] was discovered as a contaminant in water, sediment, and fauna of Escambia Bay,
Florida (Duke *et al.*, 1970). An early survey indicated
that whole body residues of the chemical in feral
shrimp were as high as 14 mg/kg whole body (Nimmo
et al., 1971a). Subsequent toxicity tests on juvenile
pink shrimp *(Penaeus duorarum)* revealed that about
1.0 µg/ℓ in the water would kill 50% of the experimental animals within 15 days (Nimmo *et al.*, 1971b).

While conducting bioassays at our laboratory we
noted on several occasions that salinity appeared to
affect toxicity. In one instance, adult pink shrimp
were exposed chronically to a sublethal concentration
of the chemical (about 1.0 µg/ℓ). The purpose of the
test was to determine whether structural damage might
occur in gill tissue. On day 27 of exposure at which
time we had recorded no previous deaths from the PCB,
the salinity of the incoming water decreased from
20 o/oo to 11 o/oo within 4 hrs due to rain, tides and
wind. As a result, ten experimental shrimp died before
the salinity had returned to 20 o/oo. During the next
2 days, the salinity was lowered again by aberrant
tides and climatic conditions and more experimental,
but not control, shrimp died. We, therefore, became
interested in the possible interaction of Aroclor ®
1254 and environmental stress, particularly the effect
of PCB on the ability of shrimp to regulate osmotically
and ionically at reduced salinities.

---

[a]Mention of commercial products does not constitute endorsement by the U. S. Environmental Protection
Agency.

MATERIALS AND METHODS

Adult brown shrimp *(Penaeus aztecus)*, 11.5 to 13.7 cm, rostrum to telson, were captured near Gulf Shores, Alabama, and used in our studies. Approximately equal numbers of both sexes were used and the methods of exposure to the Aroclor ® 1254 were similar to those reported previously (Nimmo *et al.*, 1971a) except that the shrimp were maintained at 30 ± 1 o/oo S, 25 ± 2°C, and the exposures to the chemical were "sublethal" and lasted but 7 days. Three µg/ℓ were chosen as the test concentration because previous tests with adult brown shrimp, as well as adult pink shrimp *(P. duorarum)*, demonstrated that this concentration would cause 50 o/oo mortality within 30 days.

Following exposure to PCB, equal numbers of PCB-exposed and control shrimp were transferred to separate aquaria. The experimental procedure is shown in Figure 1. Since the possibility existed of physiological stress from handling or inherent in the experimental design, both PCB-exposed and control shrimp were analyzed for osmotic and ionic concentrations after being subjected to the procedure without external salinity change (30 o/oo). While temperature was kept constant, the salinity in each aquarium was gradually lowered during 8 hrs to a predetermined level. For the first group, the salinity was maintained at 30 o/oo for 8 hrs; the second, salinity was lowered from 30 o/oo to 22 o/oo; the third, from 30 o/oo to 10 o/oo; and the fourth, from 30 o/oo to 7 o/oo. Although there was a time differential between groups of shrimp, and, therefore, a possible difference in test animals due to a slight loss of PCB, analyses for the chemical revealed no significant difference in whole body concentrations among groups (Table 1). As in earlier studies, there was a wide range in individual concentrations of PCB (Nimmo *et al.*, 1971b).

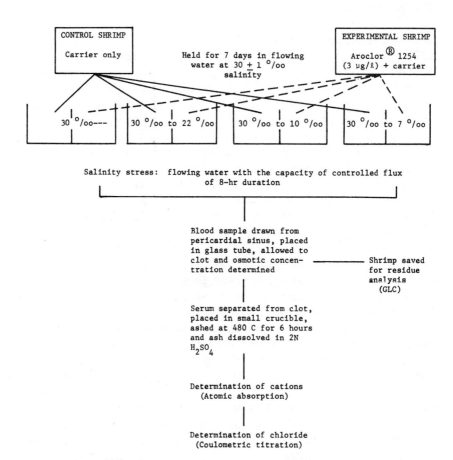

Fig. 1. Flow sheet for experimental procedure.

Samples of whole blood (hemolymph) for osmotic concentration were removed from each shrimp by pericardial puncture. A ground-glass syringe (1 ml), fitted with a #22 gauge stainless steel needle, was inserted into the animal at an oblique angle to obtain at least 0.3 ml of blood from each animal. The blood was immediately transferred to an osmometer tube and the osmotic pressure determined with a Fisk ® Model G-66 osmometer. Since the blood clotted quickly, it was difficult to determine whether actual

TABLE 1
*Whole Body Concentrations of Aroclor ® 1254 in PCB-Exposed Shrimp**

Salinity o/oo	mg/kg Average	Range
30	9.6	3.0-13.7
22	7.8	1.9-18.8
10	7.1	3.9-14.0
7	8.5	2.8-15.0

*Concentration of Aroclor ® 1254 in the test water was 3 ppb; length of exposure, 7 days. Control shrimp had less than 0.1 ppm of Aroclor ® 1254.

osmotic pressure was measured on whole blood or on serum, but an individual analysis showed no significant difference in osmotic concentrations. Replicate determinations of 10 separate aliquots of pooled sera from several shrimp yielded a standard error of 1.2 (mean concentration ∿ 629 mOs).

Analyses of ions were performed on ashed sera. To prepare the sample, the clot contained in the osmometer tube was squeezed with a small glass rod, the clot was removed and 0.2 ml of the serum was transferred to a small crucible. The crucible was placed in an oven and the contents ashed at 480°C for 6 hrs, cooled and the ash was dissolved in 2N H_2SO_4. Analysis of chloride was performed with a Buchler-Cotlove Chloridometer ® and cations were determined on a Model 403 Perkin-Elmer ® atomic absorption spectrophotometer equipped with a deuterium arc-background corrector and an HGA-70 heated graphite atomizer. Cations in standard solutions were in the same proportions as those in the sera. As a check on our methods, an analysis of a single aliquot of serum by emission and atomic absorption yielded identical results for potassium. Five replicates of pooled sera from several shrimp yielded a standard error of 0.24 mEq/l for Cl (mean = 257 mEq/l). Replicate analyses on serum aliquots yielded standard errors in mEq/l of 8.26 for Na (mean = 324.9), of 0.04 for Mg (mean = 16.0), of 0.07 for K (mean = 8.1), of 0.19 for Ca (mean = 4.74), and of 0.10 for Cu (mean = 2.84).

The 95% confidence interval was used to evaluate significance of differences in the data. The 95% intervals are indicated in the graphs by vertical bars on each datum and are listed in each table. Since it is sometimes difficult to relate "osmolality" or "osmotic concentration" to the environment, we expressed the concentration of the environment as salinity. The relationship between mOs and salinity is indicated along the X axis in Figure 2.

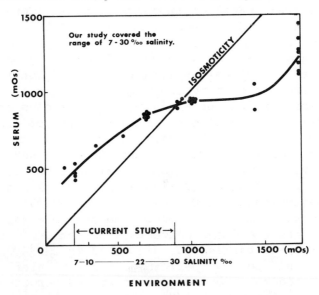

Fig. 2. *Osmotic concentration: Serum-environment in* Penaeus aztecus *(after McFarland and Lee, 1963). Our study covered the range of 7 o/oo to 30 o/oo salinity.*

Concentrations of Aroclor ® 1254 were determined on individual shrimp by gas chromatography, using procedures summarized earlier (Nimmo *et al.*, 1971b).

RESULTS

The most significant result of this study was the discovery that a sublethal concentration of Aroclor 1254 at constant salinity for 7 days became lethal when the species was subjected to a gradual decrease in salinity over an 8-hr period. Usually the shrimp exhibited increased swimming activity while the salinity changed from 30 o/oo to 20 o/oo; this was the only observed behavioral aberration. Experimental shrimp began to die at a salinity of about 12 to 13 o/oo. After 8 hrs of exposure to 10 o/oo and 7 o/oo salinity, mortality of experimental shrimp was nearly 50%. When 50% of the experimental shrimp had become moribund or had died, living PCB-exposed shrimp were taken for analyses of osmotic concentrations and ion determinations on sera.

The results of these analyses indicated that concentrations of most major ions in the sera of PCB-exposed shrimp became significantly less as the ambient salinity decreased, the sum of major ions (*e.g.*, Na, Ca, Mg, K, Cu, and Cl) was 18% less after ambient salinity reached 10 o/oo or 7 o/oo (Fig. 3). Of this total, sodium was 16% less (Fig. 4), chloride, 19% (Fig. 5) and calcium, 25% (Fig. 6). There was some indication that magnesium decreased, although the loss was not statistically significant (Table 2). No apparent differences in potassium (Table 3), or copper (Table 4) were noted.

Data for iron (Table 5) are not included in the totals in Figure 3 because we could not distinguish the divalent from the trivalent form.

Despite significant alterations in the major ion complement or in some major ions, osmotic concentration was not significantly affected by PCB and salinity stress (Table 6). Seemingly, osmotic pressure was less in PCB-exposed shrimp at 10 o/oo or 7 o/oo salinity, but individual variation was too great to show a significant difference from controls.

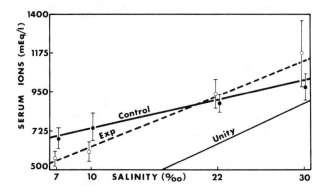

Fig. 3.   Total ions in brown shrimp serum in relation to salinity.

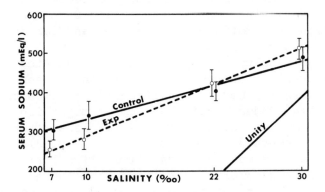

Fig. 4.   Serum sodium in brown shrimp in relation to salinity.

DISCUSSION

Knowledge of interactions between toxic compounds and environmental factors is essential for predicting their effects on ecosystems or species. Examples of this need have been demonstrated in both fresh and marine investigations. In fresh water,

434

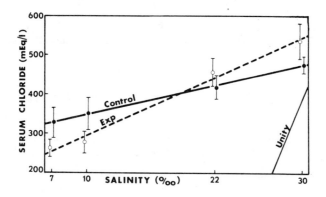

Fig. 5. Serum chloride in brown shrimp in relation to salinity.

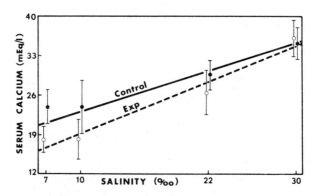

Fig. 6. Serum calcium in brown shrimp in relation to salinity.

low-level chronic exposure of the darter *(Etheostoma nigrum)* to dieldrin greatly affected its ability to survive thermal stress (Silbergeld, 1973). The sub-lethal effects of mercury on fiddler crabs *(Uca pugilator)* reduced survival times when crabs were placed under temperature and salinity stress (Vernberg and Vernberg, 1972). Mortality of fiddler crabs that were previously exposed to cadmium was greatest at high temperatures and low salinities (O'Hara, 1973).

435

TABLE 2

Average Serum Concentrations of Magnesium in Brown Shrimp in Relation to Salinity

Salinity	mEq/l			
	Control	95% Conf. Interval	Experi-mental	95% Conf. Interval
o/oo				
30	20.1	16.7-23.5	17.3	12.7-21.8
22	16.3	14.2-18.4	14.6	12.3-16.9
10	14.8	12.4-17.2	13.0	11.0-15.0
7	15.0	11.7-18.3	11.5	8.9-14.2

TABLE 3

Average Serum Concentrations of Potassium in Brown Shrimp in Relation to Salinity

Salinity	mEq/l			
	Control	95% Conf. Interval	Experi-mental	95% Conf. Interval
o/oo				
30	14.2	13.5-14.9	14.7	13.6-15.7
22	11.7	11.1-12.3	12.1	11.2-12.9
10	10.4	9.8-11.2	10.1	8.7-11.4
7	9.6	8.4-10.7	10.0	8.4-11.6

In our studies Aroclor ® 1254 possibly interfered with the adenosine triphosphatase (ATPase) activity in gills of shrimp. ATPase activity is associated with active ion transport (Tanaka, Sakamoto, and Sakamoto, 1971). Polychlorinated insecticides and the related polychlorinated biphenyls have been shown by several "in vitro" assays to inhibit ATPase in the tissues of fishes (Davis and Wedemeyer, 1971a, 1971b; Davis, Friedhoff and Wedemeyer, 1972; Cutkomp *et al.*, 1972; Yap *et al.*, 1971) and in the nerves of lobsters (Matsumura and Narahashi, 1971). The need for greater efficiency

TABLE 4
Average Serum Concentrations of Copper in Brown Shrimp in Relation to Salinity

Salinity	mEq/l			
	Control	95% Conf.	Experi-	95% Conf.
o/oo		Interval	mental	Interval
30	4.7	3.6-5.9	5.3	4.4-6.3
22	4.0	3.4-4.6	5.9	4.8-7.0
10	5.2	4.0-6.4	4.6	3.8-5.3
7	4.8	3.9-5.7	5.7	5.0-6.4

TABLE 5
Average Serum Concentrations of Iron in Brown Shrimp in Relation to Salinity

Salinity	mM/l			
	Control	95% Conf.	Experi-	95% Conf.
o/oo		Interval	mental	Interval
30	.23	.17-.29	.31	.15-.48
22	.28	.15-.35	.18	.14-.22
10	.20	.15-.25	.26	.17-.35
7	.24	.14-.33	.33	.24-.41

TABLE 6

Average Serum Osmotic Concentrations in Brown Shrimp in Relation to Salinity

Salinity o/oo	Control	Milliosmoles 95% Conf. Interval	Experimental	95% Conf. Interval
30	749	728-771	752	726-779
22	687	665-708	688	658-718
10	551	508-593	523	500-546
7	547	521-572	516	495-537

or capacity of ATPase in marine organisms can be inferred from the results of Pfeiler and Kirschner (1972), who showed that gill ATPase activity of rainbow trout adapted to salt water, was greater than in fish adapted to fresh water.

Polychlorinated hydrocarbons have interfered with either osmo- or ionic-regulation in aquatic animals (Eisler and Edmunds, 1966; Grant and Mehrle, 1970; Kinter *et al.*, 1972; Nimmo and Blackman, 1972). The physiological relationship of ionic effects to that of ATPase activity was first reported by Kinter *et al.* (1972), who postulated that lipophilic agents such as DDT and PCBs, might interact with the phospholipid-activating components of the lipoprotein enzyme. The effect of dieldrin on ion movement in the nervous system of cockroaches showed that dieldrin inhibited binding of calcium to the phospholipid moiety of the enzyme, thus inhibiting the movement of calcium across the nerve membrane (Hayashi and Matsumura, 1967). Calcium salts in fresh water greatly increased the ability of marine and euryhaline animals to survive in that medium (Black, 1957). Toxic symptoms of DDT poisoning in freshwater fish could be alleviated by the addition of calcium salts (Keffler, 1972). In

our studies calcium was lowered 25% in the sera of
PCB-exposed shrimp at 7 o/oo salinity, and it may be
that this reduction was responsible in part for the
observed decrease in sodium, chloride, and other ions.

Field observations of juvenile and subadult brown
shrimp by several investigators indicate that those
shrimp tolerate a wide range of salinities:

Salinity	Location	Observer
0.22-0.36 o/oo	St. Lucie estuary, Florida	Gunter and Hall (1963)
0-1.0 o/oo	Mobile Bay, Ala.	Loesch (In Gunter et al., 1964)
0-2.0 o/oo	Choctawhatchee Bay, Florida	Nimmo (unpubl. data)
69 o/oo	Laguna Madre, Texas	Simmons (1957)

Gunter, Christmas, and Killebrew (1964) found
that in Texas bays young brown shrimp were most
abundant at 10 to 30 o/oo, with greater abundance
above 20 o/oo. Zein-Eldin and Aldrich (1965) found
that postlarval brown shrimp withstood a wide range
of salinity-temperature combinations.

It is evident that adult shrimp are osmoregula-
tors at all but extremes of salinity (Fig. 2).
Williams (1960) gives the isosmotic point of shrimp
hemolymph at 26.5 o/oo (788 mOs), as compared to our
calculation of 23.4 o/oo (694 mOs) in control shrimp.
Nevertheless, there was no significant difference
between PCB-exposed and control shrimp (Table 6).
Obviously, a slight change in environmental osmotic
pressure would not be critical to the osmoregulatory
ability of the animals at isosmoticity, but was
critical in the dilute environment. Therefore, the
salinities where the PCB exerted its greatest effect
in the laboratory (as judged from mortality of shrimp)
were well within the range in which brown shrimp
occur in nature.

Although there was no appreciable difference in
osmotic concentration between control and PCB-exposed
shrimp, there was significant difference in major
ions in the sera of PCB-exposed shrimp (Fig. 3).

McFarland and Lee (1963) found that the point of convergence of the total ions in the hemolymph of feral
shrimp to that of the environment was 27 o/oo
(800 mOs). In our study, by extrapolation, convergence occurred at 34 o/oo (1000 mOs) in controls,
whereas in PCB-exposed shrimp showed no convergence
and total ions paralleled unity.

In surveys conducted soon after the PCB was
first discovered in the Pensacola estuary, the distribution of shrimp in relation to salinity seemed to be
related to the amount of the chemical in the animals
(Nimmo *et al.*, 1971b). Of three species captured,
brown shrimp had the highest whole-body residues
(14 ppm) although most samples were lower. Seemingly,
a concentration of 14 ppm PCB in feral shrimp would
have been lethal if the animals were subjected to
salinity stress such as imposed by our experimental
procedure. We have unpublished data that suggest
existence of a threshold in average whole-body concentration of PCB (5.6 to 7.8 mg/kg) in pink shrimp
(P. duorarum) that would be lethal when superimposed
on salinity stress caused by our procedure. However,
a recent survey of feral shrimp from the Penacola
estuary showed that young adult shrimp now have only
a fraction of PCB concentrations found in 1969/70
periods. For example, a sample taken from Escambia
Bay in August 1973 had a whole-body concentration of
only 0.1 mg/kg.

Future studies should include research on interaction of PCB and salinity on juvenile and postlarval
shrimp since chronic toxicity tests have shown that
these stages were more susceptible to the chemical
(Nimmo *et al.*, 1971a). In addition, studies by Dana
Beth Tyler-Schroeder, of the Gulf Breeze Laboratory,
have shown the susceptibility of larvae of grass
shrimp *(Palaemonetes pugio)* to the PCBs, Aroclors ®
1016 and 1242, decreases with age (personal communication). Also, the "in vivo" effect of PCBs on
ATPase activity in shrimp should be fully investigated.

ACKNOWLEDGMENT

This work was supported by Contribution No. 198 of the Gulf Breeze Environmental Research Laboratory.

LITERATURE CITED

Black, V. S. 1957. Excretion and osmoregulation. In: *Physiology of Fishes*, pp. 163-99, ed. by M. E. Brown. New York: Academic Press.

Cutkomp, L. K., H. H. Yap, D. Desaiah, and R. B. Koch. 1972. The sensitivity of fish ATPases to poly-chlorinated biphenyls. *Environ. Health Perspect. 1*:165-68.

Davis, P. W., J. M. Friedhoff, and G. A. Wedemeyer. 1972. Organochlorine insecticide, herbicide and polychlorinated biphenyl (PCB) inhibition of Na, K—ATPase in rainbow trout. *Bull. Environ. Contam. Toxicol. 8*:69-72.

_____ and G. A. Wedemeyer. 1971a. Inhibition by organochlorine pesticides of Na^+, K^+—activated adenosine triphosphatase activity in the brain of rainbow trout. *Proc. West. Pharmacol. Soc. 14*:47.

_____ and _____. 1971b. Na^+, K^+—activated ATPase inhibition in rainbow trout: a site for organochlorine pesticide toxicity. *Comp. Biochem. Physiol. 40B*:823-27.

Duke, T. W., J. I. Lowe, and A. J. Wilson, Jr. 1970. A polychlorinated biphenyl (Aroclor 1254 ®) in the water, sediment, and biota of Escambia Bay, Florida. *Bull. Environ. Contam. and Toxicol. 5*:171-80.

Eisler, R. and P. H. Edmunds. 1966. Effects of endrin on blood and tissue chemistry of a marine fish. *Trans. Am. Fish. Soc. 95*:153-59.

Grant, B. F. and P. M. Mehrle. 1970. Chronic endrin poisoning in goldfish, *Carassius auratus*. *J. Fish. Res. Bd. Canada 27*:2225-32.

Gunter, G., J. Y. Christmas, and R. Killebrew. 1964. Some relations of salinity to population distributions of motile estuarine organisms, with special reference to penaeid shrimp. *Ecology 45*: 181-85.

_____ and G. E. Hall. 1963. Biological investigations of the St. Lucie estuary (Florida) in connection with Lake Okeechobee discharges through the St. Lucie canal. *Gulf Res. Rep. 1*: 189-307.

Hayashi, M. and M. Matsumura. 1967. Insecticide mode of action: effect of dieldrin on ion movement in the nervous system of *Periplaneta americana* and *Blatella germanica* cockroaches. *J. Agric. Food Chem. 15*:622-27.

Keffler, L. R. 1972. A study of the influence of calcium on the effects of DDT on fishes. Ph.D. dissertation, The University of Mississippi, University, Mississippi.

Kinter, W. B., L. S. Merkens, R. H. Janicki, and A. M. Guarino. 1972. Studies on the mechanism of toxicity of DDT and polychlorinated biphenyls (PCBs): disruption of osmoregulation in marine fish. *Environ. Health Perspect. 1*:169-73.

Matsumura, F. and T. Narahashi. 1971. ATPase inhibition and electrophysiological change caused by DDT and related neuroactive agents in lobster nerve. *Biochem. Pharmacol. 20*:825-37.

McFarland, W. N. and B. D. Lee. 1963. Osmotic and ionic concentrations of penaeidean shrimps of the Texas coast. *Bull. Mar. Sci. Gulf Caribb. 13*:391-417.

Nimmo, D. R. and R. R. Blackman. 1972. Effects of DDT on cations in the hepatopancreas of penaeid shrimp. *Trans. Am. Fish. Soc. 101*:547-49.

_____, _____, A. J. Wilson, Jr., and J. Forester. 1971a. Toxicity and distribution of Aroclor ® 1254 in the pink shrimp, *Penaeus duorarum*. *Mar. Biol. 11*:191-97.

_____, P. D. Wilson, R. Blackman, and A. J. Wilson, Jr. 1971b. Polychlorinated biphenyl absorbed from sediments by fiddler crabs and pink shrimp. *Nature (London) 231*:50-52.

O'Hara, J. 1973. The influence of temperature and salinity on the toxicity of cadmium to the fiddler crab, *Uca pugilator*. *U. S. Fish. Wildl. Serv. Fish. Bull. 71*:149-53.

Pfeiler, E. and L. B. Kirschner. 1972. Studies on gill ATPase of rainbow trout *(Salmo gairdneri)*. *Biochim. Biophys. Acta 282*:301-10.

Silbergeld, E. K. 1973. Dieldrin: effects of chronic sublethal exposure on adaptation to thermal stress in freshwater fish. *Environ. Sci. Technol.* 7:846-49.

Simmons, E. G. 1957. An ecological survey of the upper Laguna Madre of Texas. *Publ. Inst. Mar. Sci., Univ. Tex.* 4:156-200.

Tanaka, R., T. Sakamoto, and Y. Sakamoto. 1971. Mechanism of lipid activation of Na, K, Mg— activated adenosine triphosphatase and K, Mg— activated phosphatase of bovine cerebral cortex. *J. Membrane. Biol.* 4:42-51.

Vernberg, W. B. and J. Vernberg. 1972. The synergestic effects of temperature, salinity and mercury on survival and metabolism of the adult fiddler crab, *Uca pugilator*. *U. S. Fish. Wildl. Serv. Fish. Bull.* 70:415-20.

Williams, A. B. 1960. The influence of temperature on osmotic regulation in two species of estuarine shrimps *(Penaeus)*. *Biol. Bull. (Woods Hole)* 117:560-71.

Yap, H. H., D. Desaiah, L. K. Cutkomp, and R. B. Koch. 1971. Sensitivity of fish ATPases to poly-chlorinated biphenyls. *Nature (London) 233*:61-62.

Zein-Eldrin, Z. P. and D. V. Aldrich. 1965. Growth and survival of postlarval *Penaeus aztecus* under controlled conditions of temperature and salinity. *Biol. Bull. (Woods Hole) 129*:199-216.

THE EFFECT OF COPPER AND CADMIUM ON THE DEVELOPMENT OF *TIGRIOPUS JAPONICUS*

ANTHONY D'AGOSTINO and COLIN FINNEY

New York Ocean Science Laboratory
Montauk, New York   11954

Tigriopus spp. are phagotrophic, herbivorous harpacticoid copepods widely distributed in the intertidal zone and tidal pools (Fraser, 1936; Belser, 1959). Most species can be reared successfully with little effort. *Tigriopus japonicus* was kept xenic, in crude culture, for over 15 years, feeding on *Tetraselmis maculata* and an unknown natural micro-flora (Provasoli *et al.*, 1970). The species has also been reared axenically, under very exacting bacteria-free conditions, feeding on known food organisms and/or in chemical media (Provasoli *et al.*, 1956; Shiraishi and Provasoli, 1959; Provasoli and D'Agostino, 1969). The availability of diverse rearing methods, the short duration of its life cycle, and a very consistent developmental pattern through successive generations make *Tigriopus* ideally suited for testing the deleterious effect of potentially harmful substances on development. Accordingly, the aims of this work were to document the developmental pattern under controlled conditions and the response of the species following sustained exposure to salts of the heavy metals, copper and cadmium.

445

MATERIALS AND METHODS

SOURCE OF EXPERIMENTAL ANIMALS

Stock cultures of *Tigriopus japonicus* were established with organisms kindly supplied by Dr. L. Provasoli, Haskins Laboratories, Yale University, New Haven, Connecticut. The stocks were cultured xenically; the technique differed only slightly from that employed by Provasoli *et al.* (1956). The Miquel's seawater-enriched medium (=MSWE) was replaced with SWES, a charcoal-treated seawater (Lewin and Lewin, 1960) enriched with micronutrients, essential metals, and vitamins (Provasoli, 1968). The food algae *Isochrysis galbana* and *Rhodomonas lens* were substituted for *Platymonas maculata*.

The cultures were renewed monthly—usually, a single serial subculture was started by transferring 15 to 30 ovigerous females into freshly prepared 150 ml of medium and algae contained in 300 ml capacity crystallizing dishes. The culture vessels which were covered with 155 mm watch glasses were kept in an environmental chamber set to provide a temperature regime of $20°$ to $22°C$, 200-foot candles of illumination, and a light:dark period of 16:8 hrs. Each female released 25 to 35 nauplii per brood, which, under the prevalent culture conditions, developed into ovigerous females within 21 to 26 days.

Each stock subculture provided approximately 200 to 400 females for experimental purposes. *Tigriopus* spp. produce progressively fewer eggs as they age (Comita and Comita, 1966); accordingly, although possible, it was not advisable to harvest from old cultures. Multiple subcultures were prepared whenever the need for a greater number of healthy ovigerous females was anticipated.

THE LIFE CYCLE OF THE TEST ORGANISM

The developmental stages of *Tigriopus japonicus* were studied by rearing many nauplii to adulthood,

446

each in a separate container. Twice daily the exuvia were collected and the characteristics of the individual stages were defined by contrasting the morphology of the exuvia and resulting intermolts with those of organisms at comparable developmental stages but derived from mass cultures. The temporal aspects of the life cycle are reconstructed in Figure 1; the duration of the intermolts differed only slightly from those reported by Koga (1970).

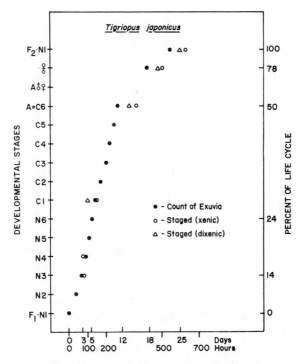

Fig. 1. *Reconstruction of the life cycle of* Tigriopus japonicus. *Stages and molting frequencies were established by collection of exuvia = ●; staged by examination of morphological features, the organisms were cultured xenically = ○; staged according to morphological features, the organisms were cultured dixenic = Δ.*

447

The adults and advanced copepodite stages were recognizable with the naked eye. But observation of all other stages required examination with a binocular dissecting microscope. The nauplii were approximately 0.11 mm; the metanaupliar, first through fifth copepodite, last copepodite, and subsequent adult stages were 0.13 to 0.23, 0.28 to 0.50, and > 0.70 mm, respectively.

TEST METHOD

Gravid females and the resulting progenies were exposed to concentrations of chemicals in replicate tubes for 48 days. The cultures were examined at prescribed intervals to follow the course of hatching development and adulthood, copulation, and finally reproduction. Under controlled conditions, the whole process was shown to take < 21 days.

Gravid females were washed serially several times with charcoal-treated seawater to minimize carry-over of contaminants. One gravid female was then inoculated into each test tube (screw-capped 20 × 125 mm Pyrex) containing 10 ml of SWES, 0.1 ml suspensions of each of the food algae *R. lens*, and *I. galbana*, and the concentrations of the materials to be tested. The food algae inoculum carried sufficiently large concentrations of cells (78 × 10^3 cells/ ml) in the final medium to provide *ad libitum* feeding for the duration of the experiment and, as expected from the work of Shiraishi and Provasoli (1959), these satisfied all trophic requirements of the animals. The salinity of the medium was 30 o/oo and ambient conditions were identical to those specified for the stock cultures.

A single dose response test was made up of 90 tubes. One set of 5(3) replicate control tubes received no additions and five sets of 5(3) replicate experimental tubes contained the five serial dilutions of the material being tested. The parent female was removed as soon as the F_1 nauplii had hatched. Three tubes from each set were harvested on days 2, 3, 6,

12, and 48. The organisms were fixed with 3% (v/v)
formaldehyde solution, and the animals, at different
stages of development, were sorted and counted. Once
the life history had been defined to facilitate enu-
meration of the stages, it was found convenient to
group closely related developmental events into a few
categories (developmental milestones) which adequately
summarized the course of normal development of
Tigriopus. These milestones are: hatching of the
first filial generation (= F_1); attainment of metanaupliar stages 2 and 3 (= N 1-3); metanaupliar stages
4 through 6 (= N 4-6); copepodite stages 1 through 5
(= C 1-5); copepodite stage 6 (= C 6 or A = Adult);
appearance of male and female (= ♂, ♀); gravid
females (= ♀); and hatching and development of the
second filial generation (F_2, N 1-3, N 4-6). These
easily discernible stages of the life history of
Tigriopus are reproduced in Figure 2.

  Each category comprised finite portions of the
life cycle of *Tigriopus* and was assigned a numerical
value corresponding to the fraction of the life cycle
that it represented. The fractional values assigned
to the selected developmental milestones were: 0.09 =
F_1 N 1-3, 0.20 = F_1 N 4-6, 0.37 = C 1-5, 0.50 = C 6,
which is equivalent to the first adult stage = A,
0.78 = ♀, 1.0 = F_2 N 1-3, and 1.09 = F_2 N 4-6. This
expedient permitted numerical scoring of population
ontogeny at any given time and allowed presentation
of the data in a relatively simple form. Thus, if a
tube contained 10 animals at stages F_1 N 1-3 and 20 at
stages F_1 N 4-6, the numerical value computed was:
(10 × 0.09) + (20 × 0.20) • (100)/30 = 16.33. This
was plotted as a single point on a "Developmental
Index" scale of 0-100 units (Fig. 4). Conversely, if
more detailed information on the ontogenetic structure
of the population was desired, a bar graph was constructed that showed the extent each stage was represented in a population as a percent of the whole at
any given time. For example, in the above case the
height of the individual bars was determined by com-
puting values for F_1 N 1-3 as (10/30) • 100 and for

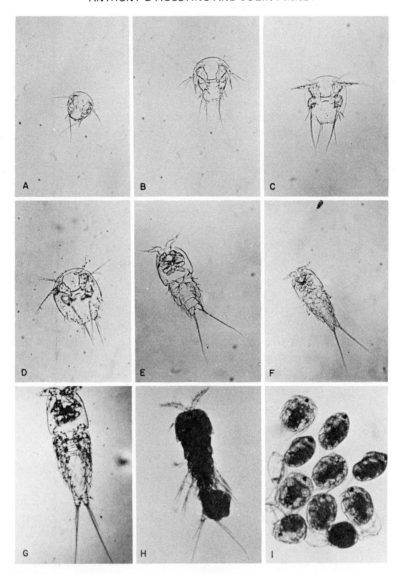

Fig. 2. *Easily discernible stages in the life
cycle of* Tigriopus japonicus. *Exuvia of:
A = nauplius; B = third stage; C = fourth stage;
D = sixth stage; E = first copepodite; F = third
copepodite; G = fifth copepodite; H = ovigerous
females; I = nauplii hatching from ovisac.
Magnified: A to E* $\underset{=}{X}$*120, F to I* $\underset{=}{X}$*60.*

F_1 N 4-6 as (20/30) • 100. The values were plotted on a vertical axis that was subdivided into seven segments, each of 0 to 100 units. Each segment represented a category of developmental events (Fig. 3). The former graphical method allowed immediate comparison of data and identification of the stressful conditions that would affect development. In contrast,

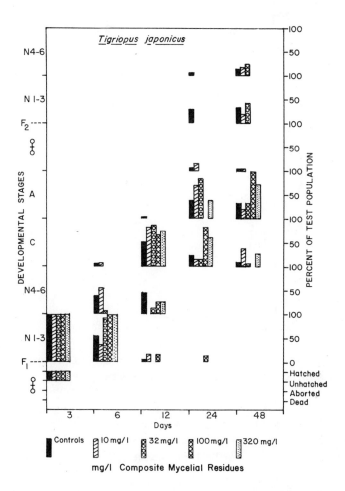

Fig. 3. Frequency of developmental stages of Tigriopus japonicus *when reared with and without mycelial residues in the media.*

451

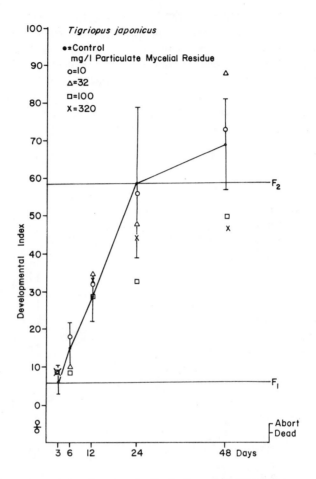

Fig. 4. *Developmental index of* Tigriopus
japonicus *exposed to concentrations of
mycelial residue.*

the latter, although cumbersome, gave a more detailed
representation of the population age structure and
identified the affected portion of the life cycle.

MATERIALS TESTED

Four substances were tested: a complex mixture
of residues recovered from industrial mycelial

452

fermentation processes (= mycelial residues), salts
of the heavy metals—copper and cadmium—and the
organometal phenylmercuric acetate. Copper and cadmium were chosen because they were known to be trace
contaminants of the mycelial residue. A measure of
the tolerance of *Tigriopus* to concentrations of $CdSO_4$
and CuCl and to combinations of these salts was expected to give some insight into the role that trace
quantities of heavy metals might contribute to the
toxic properties of complex mixtures. The sensitivity
of *Tigriopus* to phenylmercuric acetate was determined
because its tolerance to heavy metals has not been
documented in the literature.

The mixture of mycelial residues was a wet paste
(33.8% total solids) containing mycelial debris,
unidentifiable particulate matter, and dissolved
chemicals. Chemical analysis of the paste carried
out according to methods derived from Anon. (1971)
and Windom (1972) showed the mycelia to contain the
following heavy metals in mg/kg (wet weight): Cd,
0.03; Cr, 2.6; Cu, 2.8; Mn, 9.7; Hg, 0.02; Ni, <0.6;
Fe, 160; and Zn, 360 (Alexander, personal communication). The paste was homogenized thoroughly to
inhibit rapid sedimentation, and a stock suspension
was prepared in charcoal-treated seawater (= chSW).
Similarly, stock solutions of $3CdSO_4 \cdot 8H_2O$, CuCl, and
phenylmercuric acetate were prepared in chSW. The
salinity and pH of the stock suspensions and solutions were adjusted to correspond to that of SWES,
30 o/oo, and pH 7.5. Appropriate dilutions were
dispensed into the test medium volumetrically.

RESULTS

The rate of development and fecundity of
T. japonicus were not affected when reared with 10 to
32 mg/l of mycelial residue in the media. Development
was progressively and significantly retarded, however,
by 100 and 320 mg/l of residue. At these latter concentrations the hatching of the nauplii appeared

normal, and development of the copepodite stages proceeded at the same rate as the control animals, and stage C-5 molted to the adult stage C-6. However, the females failed to become ovigerous during the time course of the experiment—48 days. The mycelial residue was acutely toxic only at 1000 mg/l; the parent ovigerous females (= $P_1 \cdot$ ♀) died within 6 days. Generally, the brood sacs were aborted; occasionally the nauplii were free but these were dead. The concentration of Cd^{++} and Cu^{++} ions brought into the final medium by 100 mg/l of mycelial residue were calculated to be 0.000003 and 0.000280 mg/l, respectively. These were well below the concentration of Cd^{++} (as $3CdSO_4 \cdot 8H_2O$) and Cu^{++} (as CuCl), which, when added together, interfered with the normal development of *Tigriopus*.

Cadmium sulfate inhibited the development of ovigerous females and thence the production of the F_2 generation at concentrations > 0.1 mg/l, equivalent to a calculated Cd^{++} ion concentration of 0.0438 mg/l, assuming that there had been 100% ionization of the salt (Fig. 5). Similarly, cuprous chloride inhibited the production of the F_2 generation at concentrations $\geq$ 0.1 mg/l (or 0.0642 mg/l Cu^{++}); development failed to proceed past the F_1 N 1-3 stage when cuprous chloride was $\geq$ 1 mg/l (0.642 mg/l Cu^{++}) (Fig. 6). Cadmium sulfate and cuprous chloride were acutely toxic to the ovigerous females and nauplii at concentrations $\geq$ 10 mg/l in 72 and 48 hrs, respectively.

When *Tigriopus* was reared in media containing $3CdSO_4 \cdot 8H_2O$ and CuCl simultaneously, 0.01 mg/l of each salt was sufficient to inhibit development of ovigerous females (Fig. 7). Significantly, the concentrations of Cd^{++} (0.00438 mg/l) and Cu^{++} (0.0064 mg/l) ions brought into the medium by this combination of salts were 10 times less than the concentrations needed to cause an equivalent impairment when the salts had been tested individually.

Phenylmercuric acetate was, as expected, highly toxic. Concentrations of 10 to 0.01 mg/l (or 5.96 to 0.00596 mg/l Hg^{++}) killed the $P_1 \cdot$ ♀ in 4 to 72 hrs.

Development appeared unimpaired in concentrations ≦ 0.001 mg/l.

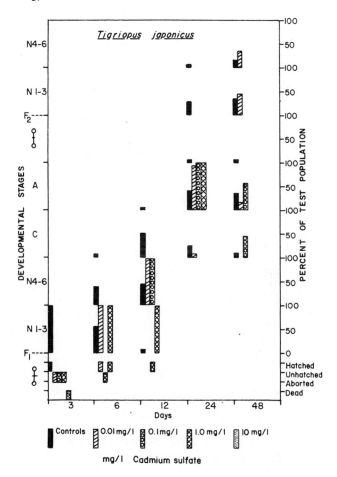

Fig. 5. *Frequency of developmental stages of* Tigriopus japonicus *exposed to CdSO₄.*

DISCUSSION

Successful production of the F_2 generation was the principal criterion employed to identify concentrations of potentially toxic chemicals that would

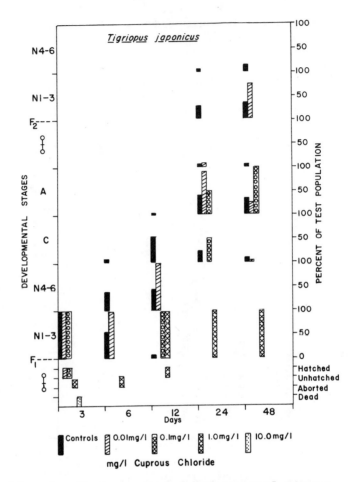

Fig. 6. Frequency of developmental stages of Tigriopus japonicus *exposed to CuCl.*

not be expected to interfere with the survival of
Tigriopus. Generation time of *Tigriopus* reared under
controlled conditions was shown to be ≅ 21 days. A
delay in excess of 24 days, or ≅ 2 X the generation
time, was considered indicative of inhibition of the
life cycle.

The results, if evaluated from this point of
view, indicated that exposure of *Tigriopus* to the
following conditions could be expected to have

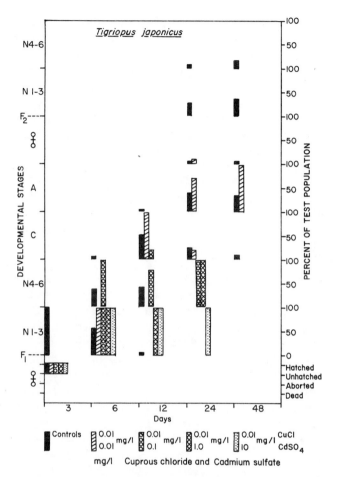

Fig. 7. Frequency of developmental stages of Tigriopus japonicus *exposed simultaneously to CdSO₄ and CuCl.*

deleterious effects on the survival of the species: 100 mg/l of mycelial residues contributing in mg/l, 0.000003 Cd^{++} and 0.00028 Cu^{++}, or 0.044 mg/l Cd^{++} (as $3CdSO_4 \cdot 8H_2$), or 0.064 mg/l Cu^{++} (as CuCl), or the combination 0.0044 + 0.0064 mg/l $Cd^{++} + Cu^{++}$.

The inhibition of development observed with 100 mg/l of mycelial residues could not be attributed solely to the combined action of Cd^{++} and Cu^{++} since similar concentrations, introduced into the medium as $CdSO_4$ and $CuCl$ separately or simultaneously, permitted normal development. Nevertheless, it is not excluded that synergistic action between Cd^{++}, Cu^{++}, and other ions of heavy metals, such as zinc, may have been responsible for the failure of *Tigriopus* to produce the F_2 generation. However, slower growth and reduced fecundity could also be attributed to other factors. *Tigriopus* is a particle feeder; the bulk of its nutritional requirements is met by ingestion of algal cells which have a diameter of between 3 and 7 microns. The size distribution of mycelial debris overlaps this range; the particles are mostly fragments of cell walls made up of largely inert and refractory organic substances.

Tigriopus, much like other phagotrophic invertebrates, is probably an indiscriminate feeder. Since it may select ingestible particles only by size, the inert matter may be as accepted. Robinson (1957) showed that the growth of *Daphnia* was impaired when suspensions of food algae were supplemented with particles of clay and silt. Similarly, nonnutrient particles were shown to compete effectively for available gut space of the filter-feeding *Artemia* (Reeve, 1963), and growth could be influenced by regulating the relative concentration of inert particulate matter in the chemically defined medium (Provasoli and D'Agostino, 1969). Therefore, reduced growth and fecundity of *Tigriopus* reared with 100 to 300 mg/l of mycelial residues in the medium may be attributed to an overabundance of inert particles, which, if ingested in lieu of algal cells, may result in impaired growth. This last hypothesis is reinforced by the evident lack of acute toxicity observed at these high concentrations of extraneous matter. Abortion of embryo sacs and subsequent death of the $P_1 \cdot \female$ was noted only with 1000 mg/l of mycelial residue. At this concentration the ions of Cd^{++}, Cu^{++} and Zn^{+} in the final medium should have been 0.00003, 0.0028 and 0.36 mg/L, respectively. The data suggest other equally probable but tentative interpretations, thus

highlighting the uncertainties that follow attempts
to identify the toxic elements of complex mixtures of
chemicals. In this instance the problem was com-
pounded by the nature of the residue, in part ideally
suited for ingestion.

Biological assays of the toxic properties of
soluble salts yield data less difficult to interpret.
Tigriopus was unaffected by 0.0044 mg/l Cd^{++}, but it
failed to produce the F_2 generation when reared with
0.044 m/l Cd^{++} in the media. Eisler (1971) summarized
the data on the acute toxicity of cadmium with respect
to nine marine species of invertebrates. The 96-hr
TL_{50} values reported for the most sensitive species,
Crangon septemspinosa, the sand shrimp, and the least
sensitive species, *Mytilus edulis*, the blue mussel,
were 0.32 and 25.0 mg/l Cd^{++}, respectively. Those
concentrations are 72 and 5681 times greater than
those identified as safe for *Tigriopus*. The disparity
between the values generated by the *Tigriopus* assay
serves to strengthen the notion that, in the context
of pollution control, it is unrealistic to search for
a standard method and organisms that presumably would
identify safe concentrations of chemicals for all
environments and species (Brown, 1973). Nevertheless,
a thorough study of the cadmium residues found in
mummichogs collected in nature and from fish exposed
to experimental concentrations in culture led Eisler
(1971) to conclude that a correction factor of 1/550
applied to the 96-hr TL_{50} of the shrimp would predict
a safe concentration for this species. The value
derived, 0.00058 mg/l Cd^{++}, is one order of magnitude
less than the safe concentration identified for
Tigriopus.

Tigriopus was more sensitive to copper than to
cadmium. The lowest concentrations of the two salts
that inhibited development of the F_2 generation were
of the same order of magnitude. However, the next
higher concentrations gave different responses.
Development stopped at C 6, the first adult stage, in
0.438 mg/l Cd^{++} and at the F_1 • N 1-3 stage in 0.642
mg/l Cu^{++}. According to Prytherch (1934), copper

459

plays a significant role in settling, metamorphosis, and distribution of the larvae of the oyster. He reported that 0.5 mg/l of total copper was not toxic and that normal settlement and metamorphosis occurred when the copper content of the estuarine water was 0.6 mg/l. Galtsoff (1932) gave the tolerance limits of oysters at 0.1 to 0.5 mg/l Cu^{++}. Exposed to these concentrations, *Tigriopus* would not survive. While the copepod and oyster larvae are both planktonic, their sensitivity to Cu^{++} are markedly different.

Copper and cadmium were listed by Doudoroff and Katz (1953) among the ions most highly toxic to fish. Their synergistic activity was also implied. Results obtained with the *Tigriopus* assay suggest that these two metal ions act synergistically. Maximum growth scored on the "developmental index scale" with *Tigriopus* reared in 0.0044 mg/l Cd^{++} and 0.0064 mg/l Cu^{++} was 50. Identical concentrations of the individual ions permitted scores of 80 (Fig. 8). However, the assay method and the procedure for sorting the results were standardized against development, so it is doubtful that the data can be treated quantitatively as is done with TL_{50} values.

SUMMARY

(1) Culture conditions were defined that allowed *Tigriopus japonicus* to be reared through two successive generations with and without heavy metals in the media.

(2) Copper and cadmium were found to inhibit growth and development of the F_2 generation at 0.064 mg/l and 0.044 mg/l, respectively.

(3) Development of *Tigriopus* exposed to copper and cadmium simultaneously indicated that 0.0044 mg/l Cd^{++} and 0.0064 mg/l Cu^{++} acted synergistic.

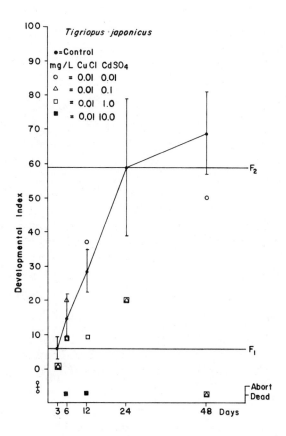

Fig. 8. Developmental index of Tigriopus japonicus *exposed to CdSO₄ and CuCl.*

ACKNOWLEDGMENT

This work was aided in part by Contract BR 73-05 with Affiliated Colleges and Universities, Inc., New York.

LITERATURE CITED

Anon. 1971. Methods for chemical analysis of water and wastes. Environmental Protection Agency, Water Quality Office. Cincinnati, Ohio.

Belser, B. L. 1959. The research frontier. Where is science taking us? *Saturday Rev. 42*:58.

Brown, V. M. 1973. Concepts and outlook in testing the toxicity of substances to fish. In: *Bioassay Techniques and Environmental Chemistry*, pp. 73-95, ed. by G. Glass. Ann Arbor, Mich.: Ann Arbor Science Publishers.

Comita, G. W. and J. J. Comita. 1966. Egg production in *Tigriopus brevicornis*. In: *Some Contemporary Studies in Marine Science*, pp. 171-85, ed. by H. Barnes. New York: Hafner.

Doudoroff, P. and M. Katz. 1953. Critical review of literature on the toxicity of industrial wastes and their components to fish. II. The metals as salts. *Sewage Indust. Wastes 25*:802-39.

Eisler, R. 1971. Cadmium poisoning in *Fundulus heteroclitus* (Pisces: Cyprinodonitidae) and other marine organisms. *J. Fish. Res. Bd. Canada 28*:1225-34.

Fraser, J. H. 1936. The occurrence, ecology and life history of *Tigriopus fulvus* (Fischer). *J. Mar. Biol. Ass. U. K. 20*:523-36.

Galtsoff, P. S. 1932. The life in the ocean from a biochemical point of view. *Wash. Acad. Sci. 22*:246-527.

Koga, F. 1970. On the life history of *Tigriopus japonicus* Mori (Copepoda). *J. Oceanogr. Soc. Jap.* (Nippon Kaiyu Gakkashi) *26*:11-21.

Lewin, J. C. and R. A. Lewin. 1960. Auxotrophy and heterotrophy in marine littoral diatoms. *Can. J. Microbiol. 6*:127-34.

Provasoli, L. 1968. Media and prospects for the cultivation of marine algae. In: *Cultures and Collections of Algae*, pp. 63-75, ed. by A. Watanabe and A. Hattori.

_____, D. E. Conklin, and A. D'Agostino. 1970. Factors inducing fertility in aseptic crustacea. *Hegloländer wiss. Meeresunters 20*:443-54.

_____ and A. D'Agostino. 1969. Development of artificial media for *Artemia salina*. *Biol. Bull. Mar. Biol. Lab., Woods Hole 136*:434-53.

_____, K. Shiraishi, and J. L. Lance. 1956. Nutritional idiosyncrasies of *Artemia* and *Tigriopus* in monoxenic culture. *Ann. N. Y. Acad. Sci.* 77:250-61.

Prytherch, H. 1934. The role of copper in the settling metamorphosis, and distribution of the American oyster, *Ostrea virginica*. *Ecological Monogr.* 4:47-107.

Reeve, M. R. 1963. The filter-feeding of *Artemia*. In suspensions of various particles. *J. Exp. Biol. 49*:207-14.

Robinson, M. 1957. The effects of suspended materials on the reproductive rate of *Daphnia magna*. *Publ. Inst. Mar. Sci. Univer. Tex.* 4: 265-77.

Shiraishi, K. and L. Provasoli. 1959. Growth factors as supplements to inadequate algal foods for *Tigriopus japonicus*. *Tohoku J. Agr. Res. 10*: 89-96.

Windom, H. L. 1972. Deliberation of a workshop held at Santa Catalina Marine Biological Laboratory of the University of Southern California. In: *Marine Pollution Monitoring: Strategies for a National Program*, pp. 131-33, ed. by Goldberg. Allen Hancock Foundation, NOAA.

SYNERGISTIC EFFECTS OF THREE HEAVY METALS ON GROWTH RATES OF A MARINE CILIATE PROTOZOAN

JOHN S. GRAY

University of Leeds
Wellcome Marine Laboratory
Robin Hood's Bay
Yorkshire, England

In the estuarine and coastal regions of the world, discharges of wastes never occur as single chemicals, but instead as mixed effluents. Yet the standard test for assessing toxicity of proposed discharges usually involves only single chemicals. In this test the concentration of chemical which kills 50% of the organisms under test in 48 or 96 hrs is estimated (Lethal Concentration LC_{50} or Lethal Dose LD_{50}). An arbitrary figure, such as one-tenth or one-hundredth of this LC_{50} concentration, is then suggested as being the upper limit of effluent concentration.

The standard test can be criticized on the following grounds: (a) it is too short when effluent discharge will be continuous; (b) an organism surviving the test concentrations may not survive for longer periods, may not grow and reproduce at that concentration or may in nature avoid that concentration; (c) on the assumption that stress increases with temperature, high summer temperatures are used, yet lower temperatures may induce higher stress; (d) the organisms used are frequently at high trophic levels and adults whereas organisms at low trophic levels and larvae or juveniles are known to be often more sensitive; (e) chemicals are tested singly, whereas in nature mixtures occur.

In an attempt to allay some of the above criticisms, study has been made of the effect of three heavy metal ions ($HgCl_2$, $Pb(NO_3)_2$ and $ZnSO_4$) on growth rates of a marine ciliate protozoan. Recent work on phytoplankton has shown that determining effects of chemicals on growth rates is a sensitive bioassay technique (Kayser, 1970, 1971). Reduction in population growth rate was used as the criterion of pollution effects in this study.

Much of the nutrient and mineral regeneration cycle in shallow seas occurs within sediments, and here many toxic chemicals are preferentially adsorbed from the water column (Gross, 1970). Should these chemicals interrupt the catabolic processes of the carbon, sulphur and phosphorous, etc., cycles, then gross pollution could result. It was decided to test effects of heavy metals on a bacterivorous sediment-living ciliate protozoan (*Cristigera* spp.), since such an organism is near the base of the food web within the sediment ecosystem.

The effect of interacting environmental variables on organisms used in toxicity tests is rarely attempted. Yet without such information it is difficult to assess the implications of acute toxicity tests. One assumption made is that toxicity increases with temperature, and frequently unrealistically high temperatures are used. In the present study, the effects of interactions of salinity and temperature on growth rates of *Cristigera* were investigated initially, prior to the testing of effects of heavy metals.

MATERIALS AND METHODS

Little is known of culture methods for marine ciliate protozoa; a simple and effective method was discovered. Sediment-living bacteria were isolated from intertidal sand at Robin Hood's Bay on ZoBell's 2216 E medium marine agar (Difco Laboratories). One organism, a pseudomonad, was cultured under sterile conditions in a medium of seawater and 0.03% peptone as nutrient. A sand sample was taken from the sediment where the bacterium was isolated and the microfauna extracted using the seawater ice method (Uhlig, 1964). A 250 ml flask of sterile seawater was inoculated with stock bacteria and incubated at $18^{\circ}C$ for 12 hrs. The mixed ciliate culture was added to the flask. One organism, subsequently identified as *Cristigera* spp., grew abundantly and was isolated as a single species culture. This species was maintained for over 2 years on the pseudomonad diet at $6^{\circ}C$ with routine subculturing.

Aged sterile seawater was used in the experiments. Water quality, indicated by the growth potential for *Cristigera*, varied greatly from batch to batch. It was found necessary to collect water from 12 to 14 miles offshore to ensure consistent quality. Presumably the inshore water is contaminated from the highly polluted North England coast rivers (I.C.E.S. Report, 1969). Four hundred ml beakers of aged seawater were inoculated with bacteria and peptone and cultured for 12 hrs in a water bath at the experimental temperature. One to two ml (2 to 300 individuals) of the *Cristigera* culture was added to the beaker. A lag phase of from 2 to 6 hrs (depending on temperature) then resulted prior to the exponential growth phase.

Rate of population growth of the ciliate was monitored with a Coulter Counter Model A set to count ciliates, but not bacteria. Experiments were conducted in duplicate.

When in exponential phase, the growth rate was monitored over 4 or 5 hrs with regular counts being made. Growth rates were measured as a constant (K), assessed as $K = (\log_2 n_t - \log_2 n_o)/t$ (where n_t = number of organisms at time t (in hours), n_o = number of organisms at start of experiment).

SALINITY-TEMPERATURE EFFECTS

A 3^2 orthogonal factorial design was used (Box, 1954) with temperatures at 11, 14 and 17°C and salinities at 18.4, 26.4 and 34.4 o/oo. Table 1 shows the growth rate constant K over the factorial design. An analysis of variance was conducted on these data. In this and the subsequent response surface analysis, for ease of calculation, the growth rate constant was multiplied by 1000. Table 2 shows the analysis of variance. Temperature and salinity both have significant quadratic effects and there are significant

TABLE 1

Growth Rates (K) of Cristigera *at Various Temperature-Salinity Concentrations*

Salinity o/oo	Temperature		
	11°C	14°C	17°C
18.4	0.356	0.433	0.348
	0.361	0.440	0.345
26.4	0.349	0.433	0.458
	0.347	0.444	0.441
34.4	0.251	0.345	0.437
	0.258	0.347	0.445

TABLE 2
Analysis of Variance of Data in Table 1
(K is here multiplied by 1000, L = Linear, Q = Quadratic)

Source of Variation	Degrees of Freedom	Sum of Squares	Mean Square	F Ratio
Temperature (T)	(2)	12353.444		
T_L	1		3201.333	89.201*
T_Q	1		9152.111	255.012*
Salinity (S)	(2)	32225.778		
S_L	1		25392.000	707.517*
S_Q	1		6833.777	190.415*
Interaction (TS)	(4)	24517.556		
$T_L S_L$	1		19701.125	548.948*
$T_L S_Q$	1		4976.042	130.292*
$T_Q S_L$	1		135.375	3.772
$T_Q S_Q$	1		5.014	0.139
Error	9	323.000	35.889	
Total	17	69419.778		

*Significant at p 0.05.

469

linear temperature × linear salinity and linear temperature × quadratic salinity interactions.

The regression equation representing the response $\hat{y} - 440.34 = 16.33x_1 + 46.00x_2 - 48.60x_1^2 - 41.90x_2^2 + 49.62x_1x_2$, where $\hat{y}$ is the estimated response (× 1000), x_1 is temperature and x_2 is salinity with -1, 0 and +1 substituted for the respective temperature and salinity values from lowest to highest and where the interaction term x_1x_2 is not split into its components as in the analysis of variance. Reduced to its canonical form the equation becomes $\hat{Y} - 453.851 = 20.129\ x_1^2 - 70.179\ x_2^2$, which indicates that the response surfaces are ellipses and decrease from the response center. The error variance of the regression model was significantly less than that from the analysis of variance, indicating that the equation represents a reasonable fit to the data. An important feature of response surface analysis is that the center can be estimated. The optimum response where K = 0.4538 occurred at 27.64 o/oo and 15.94°C. Figure 1 shows the response surface contours fitted over the experimental design.

Further experiments were conducted as a salinity of 34.4 ± 0.5 o/oo and 16 ± 1°C since it was simpler to conduct experiments under these conditions. The response (Fig. 1) can only be a maximum of 90% under these conditions.

EFFECTS OF HEAVY METALS ON GROWTH RATES

When the ciliate populations were growing exponentially, varying concentrations and mixtures of mercuric ions, lead ions and zinc ions were added to cultures and the resulting changes in growth rates recorded. Preliminary experiments gave the range of concentrations that caused reductions in growth rate without killing the organisms. Concentrations used were expressed as ppm added since it is difficult to measure or calculate actual ionic concentrations. Also, chemicals are added to the natural environment by

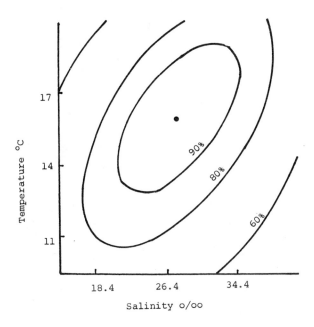

Fig. 1. Response surface contours of effects of salinity and temperature on growth rate of Cristigera.

weight or weight percentage. The effects of concentration in ppm is readily convertible to discharge levels. In each experiment a control was run under similar conditions except that the heavy metals were not added. While each experiment was ultimately conducted in duplicate, the concentrations and combinations of chemicals used were never the same in any one experiment. This precaution was necessary in case subcultures did not give similar results on different occasions. Replication was remarkably consistent. The effects of the heavy metals were assessed as reduction in the growth rate constant K in percentage. Table 3 shows the results of the complete 3^3 factorial design.

Concentrations of $HgCl_2$ as low as 0.0025 ppm added to offshore seawater reduced growth rate by 9.5%. At the highest tested concentration (0.005 ppm),

TABLE 3

Effects of Adding Varying Concentrations of Heavy Metals on Cristigera Growth Rates (3^3 Factorial Design with Replication). Percent Reduction in Growth Rate Constant K

| Metal | ZnSO$_4$ | | | | | | | | |
| | 0 ppm | | | 0.125 ppm Pb(NO$_3$)$_2$ | | | 0.25 ppm | | |
	0	0.15 ppm	0.3 ppm	0	0.15 ppm	0.3 ppm	0	0.15 ppm	0.3 ppm
0	0	6.09	11.22	10.05	13.34	17.39	15.76	19.88	23.82
	0.01	10.82	12.36	6.52	15.43	20.15	12.73	16.78	28.06
HgCl$_2$ 0.0025 ppm	8.47	9.81	13.79	13.37	15.72	26.65	16.69	26.04	47.33
	10.59	11.49	15.19	14.48	16.74	18.54	20.82	32.00	55.34
0.005 ppm	12.24	17.65	18.54	20.89	22.65	25.96	33.73	37.90	70.54
	11.88	19.66	25.01	16.83	19.91	20.49	37.17	35.19	65.02

mercury ($HgCl_2$) reduced the growth rate by 12% whereas lead ($Pb(NO_3)_2$) (at 0.3 ppm) reduced the growth rate by 11.7% and zinc ($ZnSO_4$) (at 0.25 ppm) 13.7%. On a purely additive basis, one would expect on combining these three chemicals at these concentrations would together reduce growth rate by 12 + 11.7 + 13.7 = 37.4%. In fact (Table 3, lower right corner), growth rate was reduced 67.2%.

Not all interactions gave supplemental synergistic effects; many were antagonistic. The antagonistic effect is shown well at the two factor level. $HgCl_2$ at 0.0025 ppm alone reduces growth rate by 9.51% and $Pb(NO_3)_2$ at 0.15 ppm on its own by 8.46%, yet mixed together growth rate was reduced by only 10.65%. Similar effects can be shown with Hg × Zn and Pb × Zn interactions.

Percentages of reduced growth were transformed to angles, using the formula $p = sin^2\Theta$ where p = % and Θ at 100% = 90° (Fisher and Yates, 1963), for the subsequent statistical analyses. The 3^3 orthogonal factorial design allows complete analysis not only of two factor interactions but also three factor interactions. Table 4 shows a complete analysis of variance listing linear and quadratic effects.

Since some of the interaction effects were antagonistic and some supplemental, regression equations were calculated from the transformed data. Table 5(a) shows the equations for all two factor interactions, where the mean is the mean response for each concentration of metal ion. In all cases the curvature is significant (Table 4) and negative for metal ions alone (-1), suggesting the first part of the typical dose response curve. As the concentration of metal ions increased, typically the curvature becomes positive, e.g., Hg against Zn. The intermediate (0) concentrations were usually antagonistic and the higher concentrations (+1) approach additive expectation.

However, when three factors were considered, the response is almost linear in Hg^{++} against Pb^{++} × Zn^{++}, whereas there is positive curvature in the other two

473

TABLE 4

Analysis of Variance of Data in Table 3 (Percent Reduction in Growth Rate Has Been Transformed to Angular Units, L = Linear, Q = Quadratic)

Source of Variation	df	Sum of Squares	F Ratio
Main Effects			
Mercury (Hg)			
Hg_L	1	1228.04	263.37*
Hg_Q	1	3.76	0.81
Lead (Pb)			
Pb_L	1	1001.35	227.67*
Pb_Q	1	5.29	1.14
Zinc (Zn)			
Zn_L	1	2216.22	475.28*
Zn_Q	1	60.39	12.96*
Two Factor Interactions			
Mercury × Lead			
$Hg_L \times Pb_L$	1	11.48	2.46
$Hg_Q \times Pb_L$	1	1.37	0.29
$Hg_L \times Pb_Q$	1	41.99	9.00*
$Hg_Q \times Pb_Q$	1	9.81	2.10
Mercury × Zinc			
$Hg_L \times Zn_L$	1	33.25	7.13*
$Hg_Q \times Zn_L$	1	2.36	0.65
$Hg_L \times Zn_Q$	1	156.26	33.51*
$Hg_Q \times Zn_Q$	1	0.71	0.20
Lead × Zinc			
$Pb_L \times Zn_L$	1	39.55	8.48*
$Pb_Q \times Zn_L$	1	88.73	19.00*
$Pb_L \times Zn_Q$	1	99.71	27.37*
$Pb_Q \times Zn_Q$	1	2.12	0.45
Three Factor Interactions			
Mercury × Lead × Zinc			
$Hg_L \times Pb_L \times Zn_L$	1	131.56	28.23*
$Hg_Q \times Pb_L \times Zn_L$	1	83.95	18.01*
$Hg_L \times Pb_Q \times Zn_L$	1	2.73	0.59
$Hg_Q \times Pb_Q \times Zn_L$	1	21.79	4.68
$Hg_L \times Pb_L \times Zn_Q$	1	7.60	1.63
$Hg_Q \times Pb_L \times Zn_Q$	1	1.32	0.28
$Hg_L \times Pb_Q \times Zn_Q$	1	13.52	2.90
$Hg_Q \times Pb_Q \times Zn_Q$	1	0.14	0.03
			Mean Square
Error	27	125.97	4.67

*Significant at p 0.01.

TABLE 5

Regression Equations of the Interaction Effects of Heavy Metal Ions, Based on the Data from Table 3

(a) Two Factor Interactions (Mean response is the mean for each concentration, transformed to angular units.)

Valence		Mean Response	Slope	Curvature
			Hg	
	-1	12.573	9.707	-6.265
Pb	0	19.678	2.232	1.107
	+1	24.320	3.367	-1.897
			Pb	
	-1	13.045	9.707	-7.292
Hg	0	20.441	4.417	2.137
	+1	23.163	3.495	1.190
			Hg	
	-1	13.230	10.620	-5.115
Zn	0	21.815	3.840	-0.145
	+1	27.515	8.107	2.727
			Zn	
	-1	13.045	9.707	-7.292
Hg	0	21.416	4.503	-0.740
	+1	28.102	7.195	3.715
			Pb	
	-1	13.230	10.620	-5.115
Zn	0	21.452	4.290	-1.235
	+1	25.442	5.270	-0.320
			Zn	
	-1	12.573	9.580	-6.265
Pb	0	21.523	4.507	-1.126
	+1	26.027	4.230	1.045
	S.E.	1.247	1.528	0.882

(b) Three Factor Interactions (Mean response is the mean of the total response for each concentration.)

			Pb × Zn	
Hg		26.011	2.899	-0.259
			Hg × Zn	
Pb		26.011	2.711	0.346
			Hg × Pb	
Zn		26.011	3.903	1.138
	S.E.	1.247	1.528	0.882

equations. Again intermediate concentrations (0) when
mixed were antagonistic but higher concentrations (+1)
were strongly supplemental. To the author's knowledge
no one has yet demonstrated antagonistic effects with
mixtures of one concentration and supplemental effects
on increasing the concentration.

The complete response surface equation representing the data in Table 3 is

$$\hat{y} = 28.253 + 5.841x_1 + 7.846x_2$$
$$+ 5.429x_3 - 0.599x_1{}^2 + 2.204x_2{}^2$$
$$+ 0.625x_3{}^2 + 1.77x_1x_2$$
$$- 0.692x_1x_3 + 1.284x_2x_3$$

$\hat{y}$ is expressed in angular units and again the interaction terms are summed and not split into linear and
quadratic components, and where x_1 = concentration of
mercury, x_2 = concentration of zinc, x_3 = concentration of lead with -1, 0, +1 substituted for the concentrations from lowest to highest (the orthogonal
design).

In the reduced canonical form $\hat{Y} - 25.804 =$
$2.497X_1{}^2 - 0.872X_2{}^2 + 0.605X_3{}^2$, where the signs indicate the three-dimensional surface is trumpet shaped
with the greatest change being on the x_1 (mercury)
axis. The response surface center itself is not of
interest here, since clearly we do not wish to opti-
mize the effects, but rather explore the surfaces.
The error from the regression model was less than
experimental error variance indicating that the model
was a reasonable fit to the data.

SUBLETHAL EFFECTS OF HEAVY METALS ON *CRISTIGERA* GROWTH

The optimum salinity and temperature combinations
for the growth of *Cristigera* were found to be at
27.64% and 15.94°C. The response surface contours
indicated that at 34.4 o/oo (normal East Coast seawater) and 16°C (conditions used in subsequent experiments), the growth rate was 90% of optimum (Fig. 1).

Most data on toxicity of chemicals to marine organisms are based on experiments conducted at a constant temperature and a constant salinity (Portmann, 1970; Calabrese *et al.*, 1972). It would seem essential to investigate the stress of the range of natural conditions before determining the effects of toxic chemicals on an organism. This was not done by Calabrese *et al.* (1972), who used artificial seawater and temperatures of 26°C in their experiments on the effects of heavy metals on oyster larvae. Had a range of temperatures been tested and comparisons made with natural seawater and using response surface analysis, the stress imposed by these conditions could have been ascertained. Without such knowledge one cannot tell whether or not the toxicity levels found are due to heavy metals or a combination of medium, temperature and heavy metals. Factorial designs are response surface analysis offer efficient means of investigating the effects of varying conditions, prior to toxicity testing.

Cristigera growth rate was found to be inhibited 12% by a concentration of 0.005 ppm of $HgCl_2$, 11.7% by a concentration of 0.3 ppm of $Pb(NO_3)_2$ and 13.7% by a concentration of 0.25 ppm $ZnSO_4$. These concentrations are much lower than those found to be toxic to the crustacean *Crangon crangon* (Hg 3.3-10, Zn 100-330 ppm), and the bivalve *Cerastoderma* (= *Cardium*) *edule* (Hg 3.3-10, 100-330 ppm) (Portmann, 1970). Portmann found that the shrimp *Pandalus montagui* was the most sensitive of eight organisms tested, the LC_{50} values being $HgCl_2$, 0.075 ppm; $Pb(NO_3)_2$, 375 ppm; $ZnSO_4$, 9.5 ppm. All of Portmann's data relates to LC_{50} tests which clearly must be less sensitive than growth rate studies. Using the flagellate *Dunaliella tertiolecta*, Portmann (1972) found significant reductions in growth rate at mercury concentrations of 1 ppm. Larvae of *Carcinus maenas*, *Crangon crangon* and *Homarus gammarus* were from 14 to 1000 times more susceptible than adults (Connor, 1972). The LC_{50} figures for *C. maenas* larvae were $HgCl_2$, 0.014 ppm and $ZnSO_4$, 1.0 ppm; *C. crangon* larvae $HgCl_2$, 0.01 ppm;

H. gammarus HgCl$_2$, 0.033-0.1 ppm. Oyster larvae,
Ostrea edulis, had LC$_{50}$ values of HgCl$_2$, 0.001-
0.0033 ppm. However, the concentrations of lead and
zinc found to inhibit growth in *Cristigera* are the
lowest recorded for marine organisms and the mercury
data are equivalent to that of *O. edulis* larvae
(Connor, 1972) and *Crassostrea virginica* larvae
(Calabrese *et al.*, 1972). Thus, the sensitivity of
the growth rate to some chemicals in this organism
suggests that it is a potentially highly useful test
animal.

The statistically significant interaction effects
produced by mixing the chemicals tested at both two
factor and three factor (Table 2, Table 4, Fig. 2,
Fig. 3) are the most important findings of this study.

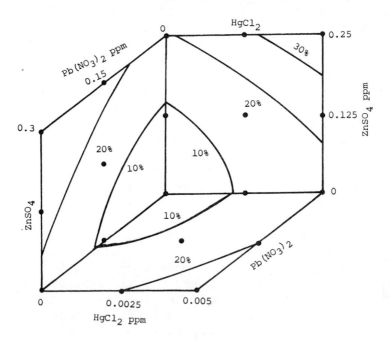

Fig. 2. *Effects of interactions of
mercury × zinc, mercury × lead,
lead × zinc on reduction in growth
rate (percent) of* Cristigera.

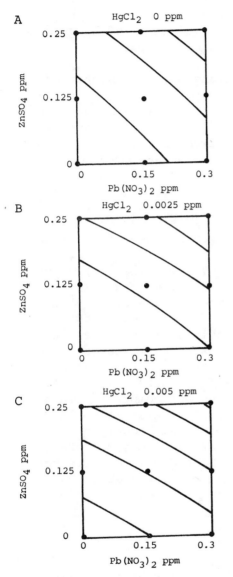

Fig. 3. *Effects of interaction of mercury × lead × zinc on reduction in growth rate (percent) of* Cristigera *taken as planes along mercury axis. A = mercury at 0 ppm. B = mercury at 0.0025 ppm. C = mercury at 0.005 ppm.*

The two and three factor interactions were antagonistic at low concentrations (shown by the negative curvature in regression equations in Table 5), whereas higher concentrations showed supplemental synergism by positive curvature. The mechanism of heavy metal ion effects on *Cristigera* growth needs to be investigated further.

Use of factorial designs enabled antagonistic and supplemental synergistic effects to be analyzed statistically. A dramatic effect was produced on *Cristigera* growth when the three highest concentrations of metal ions were mixed; a 67.2% reduction in growth rate occurred where 37.4% was predicted from purely additive properties.

Since most effluents are mixtures of chemicals rather than a single chemical, it is essential that supplemental synergistic effects, such as those recorded here, are investigated. Use of the single parameter LC_{50} test for toxicity cannot provide data on such effects. The discovery of mixtures of chemicals that act antagonistically, so that the mixture is less toxic than the component chemicals of the mixture, is an important and neglected aspect of pollution research. It would seem essential, therefore, that factorial designs using mixtures of chemicals should be adopted in toxicity testing programs.

The analysis of data from factorial designs has been greatly eased by the use of computer programs. Although the data in the present example could be adequately represented by a linear model, nonlinear models may well prove better fits to some data. Lindsey *et al.* (1971) and Alderdice and Forrester (1971a, 1971b) discuss such nonlinear models.

SELECTION OF ORGANISMS FOR USE IN TESTS

From the above results one might have expected that sediment samples taken from a highly polluted estuary (River Tees, England) would be devoid of *Cristigera*, since heavy metal concentrations were

480

greatly in excess of those used in the experiments
reported. However, *Cristigera* was fairly common in
sediments from the Tees. The apparent paradox can,
I believe, be explained by adaptation. Pollution
levels have been high in the Tees since the 1920s
and the organisms have had time to adapt. Microorganisms with their rapid generation times (just
over 2 hrs at optimum conditions for *Cristigera*) have
great inherent capacity to adapt physiologically, and
perhaps genetically, to pollution stress. Thus, use
of low trophic level microorganisms (protozoa, phytoplankton) in toxicity tests may prove successful from
the standpoint of sensitivity, but the rapid adaptation potential could render such organisms useless as
field indicators. The adaptation potential of
Cristigera and related ciliates is currently under
investigation in our laboratory by a colleague using
two-stage chemostat cultures and similar work is
being undertaken on estuarine oligochaetes. Bryan
and Hummerstone (1971) and Bryan (1974) have shown
that polychaetes in Cornish estuaries are adapted to
naturally occurring high levels of heavy metals.
Similarly, *Patella vulgata* in the Severn Estuary,
England, were living with up to 86.4 ppm of Cd in
their tissues (Peden *et al.*, 1972). Clearly the
polychaetes and limpets above would not be reasonable
organisms to use in toxicity tests.

A second problem with many organisms used in
toxicity tests is robustness. *Cristigera* was selected
because it was easy to culture. The commonly used
organisms are mostly robust and easy to culture and
may, in fact, be laboratory "weeds" that are excep-
tionally difficult to kill. In ecological terms such
organisms are r-selected (MacArthur and Wilson, 1967),
i.e., selected for their intrinsic rate of population
growth. Such species are opportunists which may be
very common at one time but ultimately are outcompeted
by K-selected species, species selected for their
competitive ability. From our experience it appears
that many K-selected species are difficult to culture
since by definition they appear later in successions

than r-selected species. Yet, these species may be the ones that are the most sensitive to pollution stress. Thus, more emphasis needs to be placed on the culturing of rare species.

The third problem concerns species with broad physiological tolerances such as estuarine species. Since estuarine organisms have a wide tolerance to salinity, temperature and oxygen levels, does this, tolerance preadapt the organisms to tolerate pollution stress? I believe that this is the case and, therefore, estuarine organisms may not be suitable test organisms. As to the ecological position of species, Dayton (1972) has shown the importance of what he terms "foundation species" in structuring communities. I contemporaneously called such species "primary species" (Gray, 1974). Clearly, the use of species important in structuring communities is sound ecologically. Such species may not, however, be sensitive and it is the sensitive species that need testing if we are to detect first stages in ecosystem degenerations. Therefore, I would suggest the use of sensitive species which are long-lived and have only slow adaptation capacity and occur in relatively stenotopic environments. Organisms such as subtidal bivalve molluscs, or polychaetes, may be suitable and can be used for backup field monitoring, since the organisms remain *in situ* and must tolerate the conditions or perish.

SUMMARY

(1) The standard LC_{50} toxicity test was criticized for its shortness, lack of sensitivity, assumptions of environmental stress interactions with pollutants, use of organisms of high trophic levels, and use of single chemicals whereas mixtures occur in nature.

(2) Using factorial experimental designs, the growth rate of a ciliate protozoan *Cristigera* was found to be optimum at 27.64 o/oo salinity and

482

15.94 C. Significant interaction occurred between salinity and temperature on growth rates.

(3) Mercury ions ($HgCl_2$) at an added concentration of 0.005 ppm reduced growth rate by 12%, lead ions ($Pb(NO_3)_2$) at 0.3 ppm by 11.7% and zinc ions ($ZnSO_4$) by 13.7%. On mixing these concentrations, growth rate was reduced by 67.2%, whereas on an additive basis 37.4% would be predicted.

(4) Regression equations and graphical response surfaces are shown for two and three factor interactions.

(5) The toxic effects are shown to be lower or as low as the most sensitive known organisms.

(6) Consideration of adaptation problems, robustness and ecological roles of organisms, is given in relation to the selection of species for use in toxicity testing programs.

ACKNOWLEDGMENTS

Mr. R. J. Ventilla carried out most of the experimental work in this investigation. I would like to thank Dr. F. J. Vernberg for inviting me to participate at this symposium. The research was financed by a grant from the British Natural Environment Research Council.

LITERATURE CITED

Alderdice, D. F. and C. R. Forrester. 1971. Effects of salinity and temperature on embryonic development of the Petrale Sole (Eopsetta jordani). J. Fish. Res. Bd. Canada 28:727-44.

Box, G. E. P. 1954. The exploration and exploitation of response surfaces: some general considerations and examples. Biometrics 10:16-60.

Bryan, G. W. 1974. Adaptation of an estuarine polychaete to sediments containing high concentrations of heavy metals. This volume.

_____ and L. G. Hummerstone. 1971. Adaptation of the polychaete *Nereis diversicolor* to estuarine sediments containing high concentrations of heavy metals. I. General observations and adaptations to copper. *J. Mar. Biol. Ass. U. K. 51*:845-66.

Calabrese, A., R. S. Collier, D. A. Nelson, and J. R. MacInnes. 1973. The toxicity of heavy metals to embryos of the American oyster *Crassostrea virginica. Mar. Biol. 18*:162-66.

Connor, P. M. 1972. Acute toxicity of heavy metals to some marine larvae. *Mar. Poll. Bull. 3*: 190-92.

Dayton, P. K. 1972. Toward an understanding of community resiliance and potential effects of enrichments to the benthos at McMurdo Sound, Antartica. In: *Proceedings of the Colloquium on Conservation Problems in Antartica*, ed. by B. C. Parker. Allen Press.

Fisher, R. A. and F. Yates. 1963. *Statistical Tables for Biological, Agricultural and Medical Research*. Sixth ed. Edinburgh and London: Oliver and Boyd.

Gray, J. S. 1974. Animal-sediment relationships— a review. *Oceanogr. Mar. Biol. Ann. Rev.* In press.

Gross, M. G. 1972. Waste removal and recycling by sedimentary processes. In: *Marine Pollution and Sea Life*, pp. 152-58, ed. by M. Ruivo. London: Fishing News Books, Ltd.

I.C.E.S. Report. 1969. Report of the I.C.E.S. working group on pollution of the N. Sea (I.C.E.S. Cooperative Research Report). Series A.

Kayser, H. 1970. Experimental-ecological investigations on *Phaeocystis poucheti* (Haptophyceae): cultivation and waste water test. *Helgoländer wiss. Meeresunters 20*:195-212.

_____. 1971. Pollution of the North Sea and rearing experiments on marine phytoflagellates as an indication of resultant toxicity. In: *Advances in Water Pollution Research*, Vol. 2, pp. 2-7, ed. by S. H. Jenkins. Oxford: Pergamon Press.

Lindsey, J. K., D. F. Alderdice, and L. V. Pienaar. 1970. The analysis of non-linear models—the non-linear response surface. *J. Fish. Res. Bd. Canada 27*:765-91.

MacArthur, R. H. and E. O. Wilson. 1967. *The Theory of Island Biogeography*. Princeton, N. J.: Princeton University Press.

Peden, J. D., J. H. Crothers, C. E. Waterfall, and J. Beasley. 1973. Heavy metals in Somerset marine organisms. *Mar. Poll. Bull. 4*:7-9.

Portmann, J. E. 1970. The toxicity of 120 substances to marine organisms. Shellfish Information Leaflet No. 19, p. 10.

_____. 1972. Results of acute toxicity tests with marine organisms, using a standard method. In: *Marine Pollution and Sea Life*, pp. 212-17, ed. by M. Ruivo. London: Fishing News Books, Ltd.

Uhlig, G. 1964. Eine einfache Methode zur Extraktion der vagilen, mesopsammalen Mikrofauna. *Helgoländer wiss. Meerseunters 11*:178-85.

SUBJECT INDEX

A

Abate, 226, 232, 234-235
Acartia clausi, 349
Acetylcholinesterase (AChE)
  inhibition by malathion, 154
Agardiella, 94
Agonus cataphractus, 21
Anguilla rostrata, 21, 106, 146, 198, 207
Anomolocera pattersoni, 349
Anoplopoma fimbria, 41, 44
Antagonistic effects of heavy metals, 473
Arsenic
  multiple factor effects, 134
  tolerance, 132-134
Artemia, 384, 458
Artemia salina, 259
ATPase
  assay, 59-60
  heavy metal effects on, 30, 115
  ionic effects, 438
  Na, K, Mg ATPase inhibition, 61, 113-117,
    198, 202-219, 437-440
  PCB effects, 436-437

B

Behavior
  assay measurements, 399, 404, 406
  heavy metal effects on, 152, 238-241, 383,
    398, 414-415, 418, 421
  hydrocarbon effects, shell closure, 362
  locomotor activity, 398, 405
  PCB—salinity on *Penaeus* sp., 433
  pesticide effects on, 152, 184, 238-241
  phototactic responses, 411-412, 420
  swimming rates, 404, 414-415, 433

Uca pugilator, 182-190, 192-195, 381-385,
  398, 404, 411-412, 414-415
Bioaccumulation
  adaptation, 481-483
  of metals, 2, 8, 22-27, 79-80, 382-413,
    414-421, 453
  of pesticides, PCBs, 154-158, 428
  of petroleum hydrocarbons, 294-297,
    304-308
Bioassay, *see also* Tissue uptake
  activity based, 417
  heavy metals, 15-18, 74-76, 111, 414-421,
    459
  pesticides, 144, 428
  spin-labeling, 370-378
Bioconcentration
  binding ratio, 1, 75
  DDT, 165-177, 181
  heavy metals, 421
  pesticides, 154
Brachidontes recurvus, 156
Brachionus plicatilus, 259
Brachydanio rerio, 37
Bryopsis, 94
Bucephalus, 230

C

Cadmium
  behavior, 414-421
  effects on
    Tigriopus, 453, 461
    Uca pugilator, 383, 392-395, 397-400,
      412-421
  fish embryos, 33-34
  hypertension, 26
  "itai-itai", 416

A 4
B 5
C 6
D 7
E 8
F 9
G 0
H 1
I 2
J 3